Common Solvents

Name [CAS Number]	Molecular Formula	Structure	Formula Weight	Point °C	g/mL
acetic acid [64-19-7]	$C_2H_4O_2$		60.05	116–118	1.05
acetic anhydride [108-24-7]	$C_4H_6O_3$		102.09	138–140	1.08
acetone [67-64-1]	C_3H_6O		58.08	55–56	0.79
acetonitrile [75-05-8]	C_2H_3N	CH_3CN	41.05	80–81	0.79
1-butanol [71-36-3]	$C_4H_{10}O$		74.12	117	0.81
carbon tetrachloride [56-23-5]	CCl_4	CCl_4	153.82	76–77	1.59
chloroform [67-66-3]	$CHCl_3$	$CHCl_3$	119.38	60–62	1.49
dichloromethane methylene chloride [75-09-2]	CH_2Cl_2	CH_2Cl_2	84.93	39–40	1.32
diethyl ether [60-29-7]	$C_4H_{10}O$		74.12	35	0.71
N,N-dimethylformamide [68-12-2]	C_3H_7NO		73.10	152–153	0.95
dimethyl sulfoxide [67-68-5]	C_2H_6OS		78.14	189	1.10
1,4-dioxane [123-91-1]	$C_4H_8O_2$		88.11	100–101	1.03
1,3-dioxolane [646-06-0]	$C_3H_6O_2$		74.08	74–75	1.06
ethanol [64-17-5]	C_2H_6O		46.07	78	0.79
ethyl acetate [141-78-6]	$C_4H_8O_2$		88.11	76–77	0.90
n-hexane[*] [110-54-3]	C_6H_{14}		86.18	68–69	0.66
methanol [67-56-1]	CH_4O	CH_3OH	32.04	64–65	0.79
n-pentane [109-66-0]	C_5H_{12}		72.15	35–36	0.63
1-propanol [71-23-8]	C_3H_8O		60.10	96–97	0.80
2-propanol [67-63-0]	C_3H_8O		60.10	80–83	0.78
tetrahydrofuran [109-99-9]	C_4H_8O		72.11	65–67	0.89
toluene [108-88-3]	C_7H_8		92.14	110-111	0.87
water [7732-18-5]	H_2O	H_2O	18.02	100	1.00

[*] The term **hexanes** refers to a mixture of hexane isomers that has a density of 0.67.

UNDERSTANDING THE PRINCIPLES OF ORGANIC CHEMISTRY

A LABORATORY COURSE

Steven F. Pedersen
University of California, Berkeley

Arlyn M. Myers
University of California, Berkeley

BROOKS/COLE
CENGAGE Learning

Australia • Brazil • Japan • Korea • Mexico • Singapore • Spain • United Kingdom • United States

BROOKS/COLE
CENGAGE Learning

Understanding the Principles of Organic Chemistry A Laboratory Course
Steven Pedersen and Arlyn Myers

Publisher: Mary Finch

Acquisitions Editor: Lisa Lockwood

Development Editor: Sandra Kisilica

Assistant Editor: Elizabeth Woods

Technology Project Manager: Lisa Weber

Marketing Manager: Amee Mosley

Marketing Assistant: Kevin Carroll

Marketing Communications Manager: Kinda Yip

Project Management, Editorial Production:
Pre-Press PMG

Art Director: to come

Print Buyer: Paula Vang

Image Permissions Editor: Amanda Groszko

Text Permissions Editor: Margaret
Chamberlain-Gaston

Production Service: Pre-Press PMG

Copy Editor: Christine Caputo

Cover Designer: Yvo Riezebos

Cover Images: Plant (coffee fruit)—© Doable/
amanaimages/Corbis, Commercial Spearmint
Plants—© Gary Holscher/AgStock Images/
Corbis, Close-up of Geckos Foot and Suckers
on Palm Leaf—Martin Harvey, Sugar Cane—
Raul Touzon

Compositor: Pre-Press PMG

For product information and technology assistance, contact us at
Cengage Learning Academic Resource Center, 1-800-423-0563
For permission to use material from this text or product,
submit all requests online at **www.cengage.com/permissions**.
Further permissions questions can be e-mailed to
permissionrequest@cengage.com.

Library of Congress Control Number: 2009939414

ISBN-13: 978-1-111-42816-7

ISBN-10: 1-111-42816-6

Brooks/Cole, Cengage Learning
20 Davis Drive
Belmont, CA 94002-3098
USA

Cengage Learning is a leading provider of customized learning solutions with office locations around the globe, including Singapore, the United Kingdom, Australia, Mexico, Brazil, and Japan. Locate your local office at:
international.cengage.com/region

Cengage Learning products are represented in Canada by Nelson Education, Ltd.

For your course and learning solutions, visit **academic.cengage.com**

Purchase any of our products at your local college store or at our preferred online store **www.ichapters.com**

Printed in Canada
2 3 4 5 6 7 8 13 12 11 10

To Julie, whose smile, support and presence have always worked wonders.

To Jesse and Erienne, I am indeed a very proud father!

To Dad, who taught me how to build things and catch fish, and to Mom, who always encouraged me to try new things, including chemistry.

—SP

To my husband, John, and to our children and their families. John has had amazing patience during this project. His support, positive attitude, generous nature, and encouragement over our many years together have made the journey fun and satisfying. Karlyne and her family, Shawn and his family, and Amanda bring me great joy and pride and enrich my life.

—AM

TABLE OF CONTENTS

v

PART II Experiments   **209**

PREFACE TO THE STUDENT

I hear and I forget.
I see and I remember.
I do and I understand.

—anonymous

The organic chemistry lab provides you the opportunity to do, to explore, and to discover. During this laboratory course you will acquire a basic understanding of chemical reactions on a molecular level. You will understand why certain conditions are required to get molecules to react with one another. To isolate and purify the compounds you prepare, you will learn how to separate the product from impurities that are present in the reaction mixture and why these techniques are successful.

This book is divided into two parts. The first part, *Theory and Techniques*, gives you all of the background knowledge necessary to perform the experiments. The second part, *Experiments*, gives you the background, procedures, and safety tips for the experiments you will be conducting.

SAFETY

Before you begin your adventure in the organic chemistry lab, it is important that you are aware of important safety rules for your well-being as well as that of other students. It is essential that you follow the safety guidelines listed in **Chapter 1,** *Safety in the Chemistry Lab,* and the safety notes included in each experiment.

PREPARATION FOR THE LABORATORY PERIOD

In order to be successful in the course, you will have to prepare before the laboratory period. Guidelines for getting ready to do an experiment are outlined in **Chapter 2,** *The Laboratory Notebook and the Laboratory Period.* Knowing the physical properties and hazards of the chemicals and the techniques used in the lab are critical to working safely in the laboratory. By reading about an experiment ahead of time, you will be able to plan your time wisely and work efficiently. The physical properties and safety considerations, as well as a plan of action for conducting the experiment, should be written in your laboratory notebook before the lab period.

DURING THE LABORATORY PERIOD

You will have to multitask in the laboratory. While you are performing the experiment, it is important to document what you do and what you observe in your laboratory notebook. Think about what you are doing, why you are doing it, what is happening and why, and what you can conclude from your observations. Keep a complete record of these thoughts and all data and observations in your notebook. These details should be included in your laboratory report to document and verify your conclusions.

LITERATURE RESOURCES

In preparing for the laboratory period and answering the questions in some of the experiments, you will be required to consult the literature. **Chapter 3,** *Where to Find It: Searching the Literature,* lists some important resources that are available in print and on the Internet.

BACKGROUND KNOWLEDGE

In **Chapter 4,** *Things You Need to Know Before You Begin,* there is a brief review of the material you have learned in previous courses that will be important in the organic chemistry laboratory. An understanding of acids and bases is necessary in choosing the correct reagents to isolate and purify many of the products. In some cases, you will be required to calculate the limiting reactant to determine the amount of other reactants to use. The limiting reactant is always calculated to determine the theoretical and percent yield.

UNDERSTANDING THE TECHNIQUES USED IN THE ORGANIC CHEMISTRY LAB

Before you attempt to make a compound in the organic chemistry lab, it is necessary to understand the conditions required to get molecules into solution so that they can react. It is also important to know how to perform the techniques used to isolate and purify a compound from a reaction mixture.

In order to synthesize a compound and to isolate and purify it, the intermolecular forces between molecules are manipulated. **Chapter 5,** *Properties of Organic Molecules,* is devoted to a discussion of these forces that are important in getting molecules into solution to enable reactions to occur, in isolating the product from the reaction mixture, and in separating the desired product from impurities. It is important to understand these intermolecular forces so that you understand what you are doing in the laboratory, why you are doing it, what is happening during the experiment, and why it is happening.

Chapter 6, *Characteristic Physical Properties of Pure Compounds,* describes the physical properties of molecules and how the presence of other compounds affects these physical properties.

Chapter 7, *Isolation and Purification of Compounds,* explains the techniques we use to isolate organic molecules from a reaction mixture and to purify them.

For many of the experiments, you will have to identify the product of a reaction from spectral data. **Chapter 8,** *Structure Determination,* gives a detailed explanation about how to determine the structure of a compound from nuclear magnetic resonance (NMR) and mass spectral data.

After you have mastered the techniques used in the organic chemistry lab and performed several of the synthesis experiments described in this text, you may want to develop an experimental procedure to synthesize a known compound or a compound that has not yet been described in the literature. **Chapter 9**, *Running a Synthetic Reaction*, will introduce you to how to read a published synthesis and also the steps you will need to take to develop an entirely new synthesis for a compound.

THE EXPERIMENTS

Because of the importance of mastering the laboratory techniques, the first eleven experiments are designed to give you the opportunity to practice these methods. In many of these experiments you will cooperate with other students to collect data. In this way, you will obtain the necessary data in one laboratory period to enable the team to solve a puzzle. You will find that teamwork and cooperation are important in the lab, in all fields of science, and also in other aspects of your life.

The remaining experiments in the text are synthesis experiments where you will make compounds using traditional and also recently developed methods. You will be introduced to some of the more environmentally-friendly reagents chemists have developed recently. You will also investigate the mechanism of a reaction.

IN CONCLUSION

Experimenting in the lab is fun and will increase your understanding of organic chemistry. The laboratory skills you develop will help you in future laboratory courses in chemistry and in other fields of science. The cleanup sections at the end of every experiment will acquaint you with the proper way to dispose of the chemicals you use in the laboratory and elsewhere. It is essential that everyone cooperate in disposing of materials correctly in order to protect our environment. Careful observation and documentation of data, teamwork and collaboration with others, and locating information about chemicals and chemical reactions in the literature are important skills that you will use in many aspects of life.

PREFACE TO THE INSTRUCTOR

The underlining concept of this laboratory text is our belief that students need a basic understanding of acid-base properties and the intermolecular forces between molecules in order to understand and appreciate what they are doing in the organic chemistry laboratory and why they are doing it. The techniques used in the organic chemistry laboratory manipulate the forces between molecules to enable them to react and to isolate and purify the desired products from the reactants and side-products. Understanding acid-base properties and intermolecular forces enables the students to answer these important questions: "What am I doing? Why am I doing it? What is happening? Why is it happening? and What can I conclude from my observations?" The experiments were developed to give the students practice in the basic techniques used in the organic chemistry lab and to introduce them to traditional as well as recently developed synthetic methods. Several experiments use reagents and techniques from the recent literature that offer more environmentally-friendly methods for accomplishing reactions. These include an osmium-catalyzed asymmetric-dihydroxylation reaction, a Suzuki reaction, a solvent-free oxidation using Magtrieve™, and transfer-hydrogenation reactions using palladium on carbon. Ion-exchange resins are used as catalysts in several reactions and many of the experiments use natural products as starting materials.

ORGANIZATION

Part I: Theory and Techniques consists of nine chapters.

Critical information about safety in the chemistry laboratory is described in **Chapter 1,** *Safety in the Chemistry Lab.*

Chapter 2, *The Laboratory Notebook and the Laboratory Period,* emphasizes the importance of preparing for the laboratory period by obtaining information from the literature about the materials used in the experiment. The discussion of the format for the laboratory notebook and the laboratory report is generalized to give the instructor flexibility.

It is important for students to be introduced to finding information in the literature, not only to prepare for the experiment by locating the physical properties of the compounds they will be using in the experiment, but also to acquaint them with the primary literature. **Chapter 3,** *Where to Find It: Searching the Literature,* introduces students to obtaining information from print resources and the Internet.

Chapter 4, *Things You Need to Know Before You Begin,* includes basic material that students are already familiar with from high school or general chemistry, but emphasizes its importance in the organic chemistry laboratory by using organic chemical examples. Reviews of acid-base chemistry, the determination of the limiting reactant, and the calculation of theoretical yields are included.

Chapter 5, *Properties of Organic Molecules,* is devoted to an explanation of the types of intermolecular interactions between molecules in both pure form and in solution. These interactions are referred to throughout the laboratory course. They form the basis for the discussion of the physical properties of molecules in **Chapter 6**, *Characteristic Physical Properties of Pure Compounds,* and the methods used to isolate and purify compounds in **Chapter 7**, *Isolation and Purification of Compounds.* An understanding of these forces and how they affect the physical properties and the techniques helps students understand, for example, how we choose solvents for extraction, how we isolate a product from a reaction mixture, and how we purify the product from the impurities present. With this knowledge, students are capable of not only rescuing an experiment that has gone awry, but they also have the background to devise new experimental procedures.

Chapter 8, *Structure Determination,* gives students the background knowledge necessary to analyze NMR, IR, and mass spectral data.

Chapter 9, *Running a Synthetic Reaction,* offers suggestions to students on how to read a synthetic procedure from the literature in order to repeat a published experiment. In addition, it includes hints on the factors that must be considered in order to synthesize a compound if the procedure is not available in the literature.

In *Part II: Experiments,* ten of the experiments are designed to explore the measurement of the physical properties of molecules and to practice the techniques used in the organic chemistry laboratory. These experiments reinforce students' understanding of the material introduced in Part I. The majority of these technique-oriented experiments are exploratory. In many cases, groups of students collaborate to collect data as a team. Students are encouraged to discuss the implications of their observations and the trends in the data with their peers. This collaboration is intended to promote teamwork in the laboratory and a sense of cooperation among students.

The remainder of the experiments in Part II include synthesis- and mechanism-based projects to provide students practice in running reactions and using standard techniques to purify and isolate products. In many cases, the students work on these projects by themselves to allow them to perfect their individual laboratory skills. The majority of the experiments require that students solve problems both in and out of the lab. NMR spectroscopy and mass spectrometry are used extensively to validate a result or to provide the students with important information to determine the structure of their product. Spectral data that are relevant to each experiment are available on this book's companion website, **www.cengage.com/chemistry/pedersen_myers_1e**. In the case of the NMR spectra, the FIDs and the processed spectra are provided. Consequently, instructors have the flexibility of requiring the student to acquire the spectrum, process the FID, or interpret the final result.

FEATURES

- Safety Notes are included throughout each experiment to warn students about potential hazards.
- Chemical Safety Notes emphasize the proper handling of reagents.
- Technique Tips give students guidelines for handling equipment and using proper procedures.

- To reinforce the importance of what they are learning in the lecture course and how it relates to the laboratory, the experiments include references to topics in undergraduate text books or the primary literature.

- Introductory essays are included in the experiments to expose students to the history and practical applications of laboratory science. Within many experiments there is also a "Did you know?" box that provides a brief, interesting fact about one of the chemicals used in the experiment.

- Each experiment is set up in a consistent fashion including Background Reading, Prelab Checklist, Experimental Procedure, and Cleanup.

- The discussion sections at the end of every experiment provide the students with guidelines for information that should be included with their data and observations in their laboratory report. There are also lab report questions that expand on the items included in the discussion sections to get students to think about other aspects of the experiment.

THE EQUIPMENT AND ENVIRONMENTALLY-FRIENDLY (GREEN) CHEMISTRY

The laboratory equipment needed for the experiments includes a combination of standard laboratory glassware, for example, beakers, test tubes, and Erlenmeyer flasks and a few select pieces of microscale glassware. In many cases, the choice of equipment is flexible to give students the opportunity to improvise. The majority of the experiments are microscale (1 mmol or less) and most make use of test tubes as the reaction vessel. There is one larger-scale experiment to give students practice in the techniques and equipment used when the amount of material is increased. In these cases, students are challenged with recovering the solvent to minimize waste generation.

Developing the skills necessary to generate less hazardous waste in the laboratory and learning the proper disposal of all end products of a reaction is fundamentally important in all fields of science from both an economic and moral standpoint. For example, students are urged to share chromatography solvent chambers and a cleanup section in each experiment exposes students to their obligation to protect the environment by the proper disposal or recycling of all reagents and end products. In addition to using microscale for most of the experiments, waste minimization is accomplished by recycling the products from many of the experiments. This concept is not an extension of multi-step syntheses, but rather a broader-based approach to synthetic chemistry. Products can be recycled within the same course or used in different courses. Students take greater pride and care in their work if they know that what they have just prepared will not be discarded as waste.

Many of the reaction-based experiments make use of reagents that are more efficient and environmentally acceptable compared with traditional counterparts and students are exposed to both homogenous and heterogeneous catalysis. The experiments that are considered environmentally friendly are based on current research trends in academic and industrial laboratories in order to acquaint students with modern reagents and techniques.

INSTRUCTOR RESOURCE MATERIALS

Ancillary material is available on the book's companion website, **www.cengage.com/chemistry/pedersen_myers_1e**, to be used at the discretion of the instructor.

- Each experiment is accompanied by a set of instructor's notes that discuss the goal of the experiment and provide tips on how to optimize the students' experience.

- Detailed prelab meeting notes are included to serve as a guide for discussions before the laboratory period.

- The average time of the experiment is based on several semesters' observations.

- Stockroom notes include the amounts of all chemicals needed together with instructions for preparing any stock solutions. Specialized equipment and supplies are also listed.

- The answers to the questions at the end of every experiment are provided.

- The ^{1}H and ^{13}C [^{1}H] NMR spectra and, in many cases, the mass spectra for starting materials and products are available on the Internet site. The NMR data are available as FIDs or as processed spectra to give the instructor options.

The goal of the text and ancillary material is to provide the instructor flexibility in choosing interesting, educational experiments and to give the students an understanding of and experience in organic chemistry techniques while affording them the excitement of discovery in the organic chemistry laboratory.

ACKNOWLEDGMENTS

Many individuals helped with the refinement of this text. First and foremost, we are grateful to the thousands of UC Berkeley students who performed the experiments and provided feedback during office hours, by e-mail, and from evaluations. Their constant curiosity and desire to learn made the development of the text a very satisfying and worthwhile undertaking.

Many of the experiments in this text were developed with the assistance of several undergraduate researchers. Their patience with the scientific process and their vigilance with respect to detailed observations helped to transform many of our ideas into reliable experiments for our laboratory courses.

Hundreds of teaching assistants shared their observations from the day-to-day interactions with the students in their laboratory sections. They told us what worked and what didn't. They kept track of melting points, yields, reaction times, and other data that need to be monitored when a new experiment is introduced into a large course. Some of our teaching assistants went above and beyond in providing feedback and suggestions and we are sincerely grateful for their input.

Robert Steiner of the organic chemistry stockroom at UC Berkeley and his team of undergraduate assistants never said "we can't do that" when approached with a new experiment, or even an idea for a new experiment.

Bob and his team always worked with us to make things happen in an orderly and timely fashion. We could not have been as creative as we have tried to be if it were not for this group.

Lucia Briggs and Cheri Hadley have provided their support for this project by dealing with many of the day-to-day administrative issues involved with large laboratory courses. We are very thankful to have had the opportunity to work with two consummate professionals.

Many people read or used various versions of this text and provided valuable feedback. They include Peter Alaimo, Seattle University; Jon Ellman, UC Berkeley; Rainer Glaser, University of Missouri; Sarah Goh, Williams College;. Jennifer Krumper, Idaho State University; MaryAnn Robak, UC Berkeley; John Tannaci, California Lutheran University; and Carl Wagner, Arizona State University.

Finally, we are indebted to the individuals who made the wheels turn in the publishing arena. The late John Vondeling had confidence in our program at Berkeley and wanted us to share our approach with other educators. Sandra Kiselica, our editor, made important suggestions for the text. We appreciate her support and encouragement during this lengthy process. Lisa Lockwood has been instrumental in her behind the scenes efforts at helping us obtain what we had envisioned for this text, from the front cover to the website.

Steven F. Pedersen
Arlyn M. Myers

THE FRONT COVER:
NATURE'S ORGANIC CHEMISTRY LABORATORY

LEFT PANEL:

Sugar Cane

Sugar cane is a tropical perennial grass that is the main source of sucrose (table sugar). The sucrose is extracted (Section 7.6.2) from the crushed sugar cane with hot water. Lime (calcium hydroxide) is added to the slightly acidic solution obtained (pH 5 – 5.5) to adjust the pH to about 7 (Section 4.1) in order to prevent the acid hydrolysis of the sucrose to glucose and fructose. The solution is then filtered (Section 7.2) to remove the insoluble material and heated to boiling to remove some of the water (Section 7.3.1.a). The water obtained is condensed and recycled. Because sugar cane consists of 50% water, the sugar industry is a producer, not a consumer of water. The fibrous material removed in the filtration step is burned to provide fuel for the sugar mill and to generate electricity for the local power grid. Some of the fibrous residue is used for animal feed and paper manufacture, making the production of sugar an environmentally-friendly industry.

To crystallize the impure sugar (Section 7.1), the supersaturated solution is seeded with pure crystalline sucrose (Section 7.1.6). The brownish impure sucrose is then purified by dissolving it in water and the solution is decolorized by filtering it through a bed of activated charcoal (Section 7.1.4 and Experiment 6). The solution is concentrated by vacuum distillation (Section 7.3.3) and the white crystalline material we know as sugar crystallizes out of this solution.

RIGHT PANEL TOP TO BOTTOM:

Spearmint

Over 50% of the essential oil obtained from the steam distillation (Section 7.3.4) or extraction (Section 7.6.2) of the spearmint plant consists of R-(-)-carvone (Experiment 17). Small amounts of myrcene (Experiment 26), α-pinene (Experiment 16), and other monoterpenes are also present.

Coffee Berries

Coffee berries are the fruit of the coffee plant and each berry contains two seeds or beans. The coffee beans contain caffeine which acts as a natural pesticide by paralyzing and killing insects that feed on the plant. The beans are removed from the ripe, red berries and roasted. The roasted beans are ground to be used in the popular drink made by extracting the flavors from the grounds with hot water (Section 7.6.2). Depending on the method of preparation, an 8 oz cup of coffee can contain from 80 to 150 mg of caffeine in addition to the flavor ingredients. The beans can be washed (Section 7.6.3) to remove the caffeine to produce decaffeinated coffee, which contains approximately 2 to 14 mg of caffeine in an 8 oz cup. Caffeine can be isolated from coffee beans by extraction (Section 7.6.2).

The leaves of the tea plant, the kola nut whose extract is used to make carbonated soft drinks, and the bean in the fruit of the cocoa tree, which is used to make chocolate, all contain caffeine. It is also an ingredient in over-the-counter pain medications (Experiment 7).

Gecko

The intermolecular forces (Chapter 5), called van der Waals forces, between the tiny hairs on the foot of the gecko and smooth surfaces, such as glass, allow some species of geckos to walk up a glass wall and across the ceiling (Section 5.4). These same forces are responsible for the physical properties of compounds (Chapter 6). The intermolecular forces between identical and different molecules allow them to be mixed or separated (Chapters 6 and 7) and are important in the techniques used in the chemistry laboratory. Mixing is important to bring the reactants together and separation is essential for the isolation and purification of compounds (Chapter 7).

Theory and Techniques

CHAPTER 1

Safety in the Chemistry Laboratory

Safety must be everyone's primary concern in the chemistry lab. Understanding and following all safety rules in the organic chemistry lab is critical to your well-being as well as that of others in the lab room and to the protection of the environment.

1.1 PERSONAL SAFETY

- Dress appropriately for the lab period. Bare feet, sandals, shorts, short skirts, and halter or tank tops are not allowed in the lab room. Long hair and loose clothing should be tied back. Synthetic fingernails pose a personal fire hazard and should not be worn.

- Personal items, including, but not limited to, book bags and books, should be put in the area of the lab room designated for that purpose. Do not put personal items on the lab bench.

- Do not work alone in the lab.

- Safety glasses or goggles must be worn at all times when in the laboratory. The first thing you should do when you enter the lab room is to put on your safety glasses or goggles. The last thing you should do before leaving the lab room is to take them off.

- Inform your instructor about any health conditions that might affect your performance in the lab. This includes identifying any allergies to specific chemicals.

- Never smoke, eat, drink, or apply cosmetics in the lab.

- If you spill a chemical on your skin, wash the area well with soap and water for at least five minutes and report the incident to your instructor immediately.

- Make sure that you wash your hands before leaving the lab room.

- Never taste any substance in the lab and do not purposely smell chemicals directly unless instructed to do so.

- Familiarize yourself with the environment of the lab room. Know where safety equipment can be found and make sure you know how to operate the eyewash fountain and safety shower properly. In some cases, it may be important that you know how to operate a fire extinguisher.

- Be aware of the activities of others in the lab room. For example, do not use flames when someone else in the lab is using flammable materials.

3

- Do not perform an unscheduled or unauthorized experiment.
- While in the lab, concentrate on what you are doing and why you are doing it.
- Personal electronic equipment, including cell phones, should not be used during the laboratory period.

1.2 FIRE/EARTHQUAKE/EMERGENCY EVACUATION

- In the event of any emergency situation in the lab, remain calm and follow the established procedure for the situation as well as any directions from your instructor.
- Familiarize yourself with the location of any emergency exits in your lab room on the first day of class.
- Familiarize yourself with the emergency evacuation plan for your building on the first day of class.
- Immediately notify your instructor of any fire or other emergency in the lab.
- Before exiting the lab in an emergency, turn off all heat sources (sand baths, melting-point instruments, etc.) if possible.

1.3 EQUIPMENT SAFETY

- Use only digital thermometers to measure the temperature of a sand bath. *Never use a mercury thermometer for this purpose.*
- Only round-bottomed flasks or thick-walled filter flasks should be placed under vacuum. Erlenmeyer flasks should never be placed under vacuum. Check that glassware does not have any cracks before placing it under vacuum.

- Never use a Bunsen or microburner before checking to ensure that there are no flammable substances in your vicinity.
- Be aware that liquids with low flash points can ignite if they come in contact with a hot surface, such as a hot plate. In addition, a spark from a hot plate thermostat can ignite flammable vapors in the vicinity. For example, diethyl ether has a vapor density greater than air and can creep along a lab bench to pose a hazard far from the point of use.
- Never pipette by mouth.
- Before heating a liquid, add a *new* boiling stick or chip, or stir with a magnetic stirring bar.
- Never heat a closed system.
- Never distill a system to dryness.
- Keep your face away from the opening of a vessel when mixing reactants or when applying heat to a reaction mixture.
- Never vent a separatory funnel or other system that is under pressure toward yourself or another person.
- Assemble equipment and apparatus properly using clamps when needed. Use a suitable heating source and protect the reaction from

water or air when necessary. If noxious fumes are given off during a reaction, use an appropriate trap or work in a fume hood.

1.4 CHEMICAL SAFETY

- Familiarize yourself with the properties of the chemicals you will be working with and make a table of important physical properties and hazards. The most complete information is given in the **Material Safety Data Sheets (MSDS),** which include the information listed below. These are available on the Internet, for example at http://hazard.com/msds/.

Section 1.	Identifying Information
Section 2.	Composition/Information on Ingredients
Section 3.	Hazards Identification
Section 4.	First Aid Measures
Section 5.	Fire Fighting Measures
	(This section includes the flash point and autoignition temperature.)
Section 6.	Accidental Release Measures
Section 7.	Handling and Storage
Section 8.	Exposure Controls/Personal Protection
Section 9.	Physical and Chemical Properties
Section 10.	Stability and Reactivity
Section 11.	Toxicology Information
Section 12.	Ecological Information
Section 13.	Disposal Considerations
Section 14.	Transport Information
Section 15.	Regulatory Information
Section 16.	Other Information
	(This section includes the **National Fire Protection Association (NFPA) Ratings**.)

- Before coming to the lab, read the experiment carefully and enter a **prelab** (Section 2.1) in your laboratory notebook. The prelab should include a flowchart of the procedure you will be following.

- Record all safety notes for the experiment in the prelab write-up in your laboratory notebook.

- Carefully read all labels and make sure that you use the correct solvents and reactants.

- Be careful to use the correct amount of each reactant. Weigh materials in a suitable container and use an appropriate vessel to dispense liquid reactants.

- If a chemical is spilled, notify your instructor immediately. Follow the proper method for cleaning up the chemical. Chemical spills must be cleaned up immediately.

- Control the rate of addition of reactants if necessary. For example, the instructions for the experiment will indicate if reactants should be added all at once, in a dilute solution, or gradually.

- The temperature of the reaction mixture should be controlled by heating or cooling to obtain a reasonable reaction rate.
- When instructed, the reaction should be stirred to maintain homogeneity. Check to make sure that mixing is efficient.

NFPA Ratings are found on chemical containers, on trucks transporting chemicals, and in establishments where chemicals are used, such as in gas stations. It is important to know the meaning of the colored areas of the NFPA diamond and the level of the hazard indicated by the number in each area.

HEALTH (BLUE)

4   *Danger*   May be fatal on short exposure. Specialized protective equipment is required.

3   *Warning*   Corrosive or toxic. Avoid skin contact or inhalation.

2   *Warning*   May be harmful if inhaled or absorbed.

1   *Caution*   May be irritating.

0   No unusual hazard

FLAMMABILITY (RED)

4   *Danger*   Flammable gas or extremely flammable liquid

3   *Warning*   Flammable liquid, flash point below 100° F

2   *Caution*   Combustible liquid, flash point of 100° to 200° F

1   *Caution*   Combustible if heated

0   Not combustible

REACTIVITY (YELLOW)

4   *Danger*   Explosive material at room temperature

3   *Danger*   May be explosive if shocked, heated under confinement, or mixed with water.

2   *Warning*   Unstable or may react violently if mixed with water.

1   *Caution*   May react if heated or mixed with water, but not violently.

0   *Stable*   Not reactive when mixed with water

SPECIAL NOTICE KEY (WHITE)

W Water Reactive

OX Oxidizing Agent

1.5 CLEANUP AND WASTE DISPOSAL

- Observe all guidelines for the proper disposal of by-products and solvents according to laboratory rules and state law. General rules for cleanup are given at the end of each experiment. These should be included in the prelab write-up in your laboratory notebook.
- Minimize waste by using only the amounts of materials necessary to conduct the experiment. By sharing materials with others, such as chromatography developing chambers, the waste generated can be reduced.

- Before storing any compounds, put them in an appropriate container that is labeled with your name, the name of the material or the number of the experiment, and the date it was generated.

- Dispose of all chemicals and chemically contaminated materials in the appropriately labeled containers. Needles must be disposed of in the special containers designed for handling syringe needles.

- Return all chemicals to their proper place. Leave the lab bench and all common areas clean.

CHAPTER 2

The Laboratory Notebook and the Laboratory Period

A laboratory notebook is the equivalent of an explorer's journal. An explorer might record latitude and longitude whereas a chemist might record weight and volume. An explorer might comment on the clarity of a stream noting the date, time, and weather conditions. A chemist might enter the observation that a reaction became cloudy at 50°C. Just as it is important for the explorer to document findings while in the field, it is important for a chemist to record observations in the laboratory at the time they are made.

The type of laboratory notebook required will depend on the preference of your instructor. It should, however, have consecutively-numbered pages. The first one or two pages should be reserved for a **Table of Contents** that will be completed during the course by listing the name and notebook page number of each experiment.

2.1 BEFORE THE LABORATORY PERIOD

Just as an explorer plans for a trip by selecting suitable clothing and tools for the conditions to be encountered, you must prepare for the experiment you will be conducting. To facilitate completion of the experiment, assure your success, and optimize laboratory safety, read the assignment before the laboratory period to familiarize yourself with the techniques, experimental procedure, and theory of the experiment.

Complete a **prelab** in your laboratory notebook.

- Write the title of the experiment.

- Enter the title of the experiment and the notebook page number in the Table of Contents at the front of your laboratory notebook.

- Write a brief introduction that states the purpose of the experiment.

- Include a balanced chemical equation if you are performing a reaction in the experiment. In addition, write equations for all expected side-reactions.

- Make a table summarizing all reactants, reagents, and products. Include structures and appropriate physical properties of these compounds and note any particular hazards/precautions necessary for handling specific materials. You should, for example, know the melting points for the starting materials and products. It is not necessary to record the melting point of a substance such as sodium sulfate, which you might use as a drying agent. See Chapter 3 for a list of appropriate resources.

9

- In your table of starting materials, reagents, and products, enter the molecular weight for all starting materials and products. Record the weight or volume of all starting materials and reagents that you will use in the experiment. Calculate the number of millimoles you will use for any starting material included in the equation for the reaction and determine the limiting reactant. Calculate the number of millimoles and weight of the product expected and the theoretical yield. Although these amounts will have to be recalculated during the laboratory period to reflect the exact amounts of the limiting reactant used in the actual experiment, it is important to have a "ball-park" figure as you work in the lab.

- A flowchart or brief description of all of the steps of the procedure should be included for reference during the laboratory period. It is important to have a complete, but brief, summary of the experimental procedure to use while you are conducting the experiment. This is sometimes most easily accomplished with a flowchart. By formulating a plan, you can organize your time efficiently. If the procedure calls for a 30-minute reflux, plan what you can do during that time to prepare for subsequent steps of the procedure. After the reaction is completed, the product must be separated from the reaction mixture and all side-products must be removed. This is called the **workup**. For each step of the workup, indicate where the product is and the by-products and impurities removed during the procedure. The flowchart represents *"What you are supposed to do"*. Some students prefer dividing the page into two columns so that *"What you actually do"* can be entered in the notebook next to the summary of the procedure. Make sure that you include abbreviated versions of the Safety Notes, the Experimental Notes, the Technique Tips, and the proper method of disposing of all chemical wastes from the cleanup section in your prelab write-up.

2.2 DURING THE LABORATORY PERIOD

THINK ABOUT IT!

As you work in the lab, it is important that you understand what you are doing and why you are doing it. A laboratory synthesis is not like making a stew in which the exact amounts of ingredients and the order in which they are added is not critical to the outcome. In a laboratory experiment, each ingredient is important, the amount used is significant, and the order of each step is critical. As you are completing your prelab write-up and conducting the experiment, think about each step and why it is important to the success of the experiment. Should problems arise, you will be able to recognize more quickly that something is wrong and you will also have insights into how you can hopefully salvage the experiment.

As you work in the lab, ***Think: What and Why.***

What am I doing?

Why am I doing it?

What is happening?

Why is it happening?

What can I conclude from my observations?

- Put on your safety glasses as soon as you enter the lab room.

- Enter the date in your laboratory notebook. If data is collected during more than one laboratory period, include the exact dates of collection.

- Actual experimental procedures (exactly what you did), data (weights, volumes, physical constants, and yield), and observations are to be recorded directly in the laboratory notebook in permanent ink as you work in the lab. If you make a mistake, cross out the error with a single line. Simply placing a check mark next to the *"What you are supposed to do"* column, if you use this format, is not considered an observation. You must state exactly what you did and what you observed. The entries should be sufficiently detailed so that you or someone else could repeat the experiment exactly as it was conducted. Be sure to include notes on the physical characteristics of the reaction mixture and your products. For example, you might note that you observed a white precipitate, yellow oil, or needle-shaped crystals. In cases where there are problems, detailed notes will help you and your instructor figure out what occurred. Thin layer chromatography plates should be drawn in the laboratory notebook to the exact size with the origin, solvent front, and all spots marked. Note the method of visualization (UV, I_2, etc.) on the plate and record the solvent system used next to the plate. Recalculate the theoretical yield based on the actual amounts of materials used. Compare your yield with the calculated theoretical yield. If the yield is over 100%, it may indicate that your product was not dry when you recorded the weight.

- In cases where you are working with a team of peers, you must include all of the members' names in your laboratory notebook. Any data obtained from your team members should be recorded during, not after, the laboratory period and credited to the member collecting the data.

- Make sure that you use the correct reactants and reagents in the designated amounts.

- Follow all guidelines for the proper storage and disposal of chemicals and chemically contaminated materials.

  - Products must be put in appropriate containers as directed by your instructor. In the case of individual containers, each container should be labeled with your name, the date, and the name of the compound or the title of the experiment. Do not store unlabeled chemicals.

  - Solvents and reaction solutions must be disposed of according to the guidelines issued by your instructor.

  - Pasteur pipettes, filter paper, and other materials that are not reusable should be disposed of in the containers for chemically contaminated materials.

  - Syringe needles must be disposed of in the special container for syringe needles.

2.3 BEFORE LEAVING THE LAB

- Make sure you clean up your lab bench and return all equipment and reagents to their proper places.
- Make sure electronic thermometers, heating devices, and all other pieces of equipment are turned off.
- Make sure that all water is turned off.
- Your instructor may require that your laboratory notebook be initialed before you leave the laboratory.

2.4 AFTER THE LABORATORY PERIOD

The laboratory report is an important part of an experiment. It includes organizing the original data, documenting your observations, summarizing the findings, analyzing the results, and reporting the conclusions that can be inferred from the results. In synthetic experiments, you should explain any differences between the physical properties of the product that you measured and the literature values. Also, identify any problems that might account for a low yield. The format of the report required will depend on the preference of your instructor.

2.5 YOUR RESPONSIBILITY AT ALL TIMES

As a scientist and as a human being, you have the responsibility to practice the highest standards of ethics at all times. The prelab write-up must be your own work. You should conduct experiments by yourself unless specifically instructed to work in pairs or groups. For experiments conducted in pairs or groups, it is the ethical responsibility of each member of the team to contribute equally to the collection of data and all participants must be cited in the laboratory notebook. Your laboratory notebook must be a record of exactly what *you* did in the lab and what *you* observed during the experiment. The laboratory report must be based on these observations and the data collected during the laboratory period, clearly indicating data collected by other members of your team. For example, if you obtained a low melting point or yield, include a possible explanation, but do not report hypothetical results or falsify any information. The original source of any ideas that are not your own should be clearly referenced.

Your obligations as a good citizen include, but are certainly not limited to, the proper disposal of all chemicals and compliance with all safety rules and precautions. Cooperation with, support of, and respect for other students and the staff are important parts of being a responsible human being.

Should you have any questions with regard to an ethical judgment or action related to the course, you should feel free to discuss the issue with the instructor in charge of the course.

CHAPTER 3

Where to Find It: Searching the Literature

It is important to know the physical and chemical properties of the materials you will be working with in the lab. Become familiar with the many resources available to obtain this information. Some are outlined below, but it is equally important for you to get to know your library and to investigate Internet resources at your institution with the help of your instructor or the librarian.

In order to complete the prelab write-up for an experiment, you will need information on the starting materials, reagents, and products involved. The *CRC Handbook of Chemistry and Physics*, the *Merck Index*, *Lange's Handbook of Chemistry*, and the *Aldrich Catalog Handbook of Fine Chemicals* are excellent resources. Because the names of compounds are listed differently in these references, it is sometimes necessary to use their molecular formula indices to locate a compound.

The *CRC Handbook of Chemistry and Physics* and *Lange's Handbook of Chemistry* listings include the structural formula, the molecular weight, the crystalline form, the specific rotation, the melting and boiling points, the density, and rough estimates of the solubility.

Although the *Merck Index* does not include as many compounds as the *CRC* and *Lange's*, it also lists toxicity and pharmacological information. The solubility data provided is also more specific.

The *Aldrich Catalog Handbook of Fine Chemicals* lists compounds sold by the Sigma-Aldrich Chemical Company. Again, the compounds are sometimes listed under different names than in other references, so the molecular formula index is helpful. In addition to information included in the previous sources, some structural formulas are provided along with references to the *Aldrich Library of NMR Spectra* and prices for the chemicals.

There are hard copies of these references. However, if an Internet connection is available, it is sometimes more convenient to obtain information online. The Aldrich catalog is accessible at http://www.sigmaaldrich.com/. Material Safety Data Sheets (MSDS) are available at http://hazard.com/msds/. Other public access sites with information include the National Institute of Standards and Technology Chemistry Web Book at http://webbook.nist.gov/chemistry/ and The United States National Library of Medicine's data at http://chem.sis.nlm.nih.gov/chemidplus/chemidheavy.jsp. Your institution may also have online access to the *CRC Handbook of Chemistry and Physics*, *Lange's Handbook of Chemistry*, and the *Merck Index*. Consult your instructor or librarian.

If you do not have an exact procedure for the preparation of a compound that you need, it will be necessary to conduct a literature search to determine if the compound or one that is similar has been made previously. In addition to the textbooks, reference books, and journals available in the library, there are many resources available online for finding published procedures. For example, *Organic Syntheses* is available at http://www.orgsyn.org/. The primary literature references for many named reactions are included in *Organic Reactions*, which might be available in the library or at http://www.chempensoftware.com/organicreactions.htm. Your college or university may subscribe to SciFinder Scholar, which is a valuable online database that includes *Chemical Abstracts*.

Sometimes it can be fun and also useful to simply enter the name of a compound into an Internet search engine. This technique may uncover a wealth of information from many different sources.

CHAPTER 4

Things You Need to Know Before You Begin

Before starting to work in the lab, it is helpful to review some basic concepts you have learned previously that are important in organic chemistry.

4.1 ACID-BASE REACTIONS

Make sure you understand acid-base reactions and have a general knowledge of the relative strengths of acids. Two definitions of acids and bases are used in organic chemistry. The **Brønsted-Lowry definition** is limited to protic acids and reactions that involve the transfer of a proton ($H^\oplus$). The **Lewis definition** is more general and describes acid-base reactions in terms of electron donors and acceptors.

4.1.1 The Brønsted-Lowry Definition

An **acid** *is a compound that can give up a proton to a proton acceptor. The proton acceptor is called a* **base**. The reactions involving protic acids therefore have the form shown in Equation 4.1.

$$\text{acid} + \text{base} \rightleftharpoons \text{conjugate base of the acid } + \text{conjugate acid of the base}$$

(4.1)

The acid, which is the proton donor, is transformed into its conjugate base in the reaction. The base accepts the proton to become its conjugate acid (Equation 4.2).

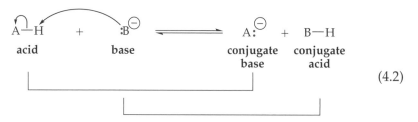

(4.2)

Charges are conserved in the reaction. For example, if the base is neutral, its conjugate acid carries a positive charge. If the protic acid is positively charged, its conjugate base is neutral.

We define the strengths of acids in dilute aqueous solution where water is the proton acceptor, or base, and becomes the hydronium ion (Equation 4.3).

$$A\!-\!H \ + \ H_2\ddot{O}: \quad \rightleftharpoons \quad \overset{\ominus}{A}: \ + \ H_3\overset{\oplus}{O}:$$

acid   base   conjugate   conjugate
                                      base   acid   (4.3)

We can define an equilibrium constant for this equation (Equation 4.4).

$$K_{eq} = \frac{\left[H_3O^{\oplus}\right]\left[A^{\ominus}\right]}{\left[AH\right]\left[H_2O\right]} \tag{4.4}$$

The concentration of water is a constant in a dilute aqueous solution. Its value of 55.5 mol/L (1000 g/L divided by the molecular weight) is incorporated into a quantity defined as the acidity constant, K_a (Equation 4.5).

$$K_a = K_{eq}\left[H_2O\right] = \frac{\left[H_3O^{\oplus}\right]\left[A^{\ominus}\right]}{\left[AH\right]} \tag{4.5}$$

The K_a values of some acids are shown in Table 4.1. They vary over many powers of 10 so it is convenient to use a logarithmic scale to represent acidities. We define the pK_a as the negative log to the base 10 of the acidity constant (Equation 4.6).

$$pK_a = -\log K_a \tag{4.6}$$

The pK_a values for some common acids are included in Table 4.1 where acids are listed in order of decreasing acid strength. *Remember that stronger acids have smaller pK_a values than weaker acids.* The strengths of the conjugate bases increase going down the table. In other words, a stronger acid produces a weaker conjugate base when deprotonated. That should make sense because if a molecule is eager to give up a proton (a strong acid), its conjugate base is not a good proton acceptor (a weak base).

Looking at the strengths of the acids across a row in the periodic table from methane (CH_4) (pK_a 50) to ammonia (pK_a 35) to water (pK_a 15.7) to HF (pK_a 3.2), we see that the acidity increases as the electronegativity of the atom to which the proton is attached increases. This is reasonable because a more electronegative atom can accommodate a negative charge better than a less electronegative atom, therefore resulting in an increased stability of the conjugate base anion, which makes it a weaker base. Differences in electronegativity, therefore, explain the increased acidity going from left to right across a row of the periodic table. Keep in mind that the stronger the acid is, the weaker its conjugate base is.

The size of the atom and its polarizability are important when considering acidity trends down a column of the periodic table. For example, consider HF (pK_a 3.2), HCl (pK_a −2.2), HBr (pK_a −4.7), and HI (pK_a −5.2) for

Table 4.1 Acid Dissociation Constants and pK$_a$ Values*

Acid	K$_a$	pK$_a$	Conjugate Base
HI	1.6×10^5	−5.2	I$^{\ominus}$
H$_2$SO$_4$	1.0×10^5	−5.0	HSO$_4$$^{\ominus}$
HBr	5.0×10^4	−4.7	Br$^{\ominus}$
(CH$_3$CH$_2$)$_2$OH$^{\oplus}$	4.0×10^3	−3.6	(CH$_3$CH$_2$)$_2$O
HCl	1.6×10^2	−2.2	Cl$^{\ominus}$
CH$_3$OH$_2$$^{\oplus}$	1.6×10^2	−2.2	CH$_3$OH
H$_3$O$^{\oplus}$	55.5	−1.7	H$_2$O
CH$_3$SO$_3$H	16	−1.2	CH$_3$SO$_3$$^{\ominus}$
CF$_3$C(O)OH	0.6	0.23	CF$_3$C(O)O$^{\ominus}$
HF	6.3×10^{-4}	3.2	F$^{\ominus}$
CH$_3$C(O)OH	1.7×10^{-5}	4.76	CH$_3$C(O)O$^{\ominus}$
H$_2$CO$_3$	4.3×10^{-7}	6.37	HCO$_3$$^{\ominus}$
HCN	6.3×10^{-10}	9.2	CN$^{\ominus}$
NH$_4$$^{\oplus}$	1.6×10^{-10}	9.8	NH$_3$
CH$_3$SH	1.0×10^{-10}	10.0	CH$_3$S$^{\ominus}$
HCO$_3$$^{\ominus}$	5.6×10^{-11}	10.3	CO$_3$$^{-2}$
CH$_3$OH	3.2×10^{-16}	15.5	CH$_3$O$^{\ominus}$
H$_2$O	2.0×10^{-16}	15.7	HO$^{\ominus}$
RC≡CH	1.0×10^{-25}	25	RC≡C$^{\ominus}$
NH$_3$	1.0×10^{-35}	35	NH$_2$$^{\ominus}$
RCH=CH$_2$	1×10^{-40}	40	RCH=CH$^{\ominus}$
CH$_4$	$\sim 1 \times 10^{-50}$	50	CH$_3$$^{\ominus}$

Left axis: Acid strength increases

Right axis: Base strength increases

*Note that pK$_a$ values that are less than the pK$_a$ of the hydronium ion or greater than the pK$_a$ of water cannot be measured in aqueous solution and the values depend on the solvent used for the measurement.

which the acidity increases down the periodic table. Relative acid strength in the hydrohalic acids can be understood if one considers that the iodide ion is large and polarizable. The negative charge is spread out over a larger area, therefore stablizing the anion when compared with the smaller bromide, chloride, or fluoride anions. The more stable the anion is, the weaker it is as a base and, therefore, the stronger its conjugate acid is. The fluoride ion is the smallest and the least polarizable. It is therefore the strongest base, which is equivalent to saying that hydrofluoric acid is the weakest acid in the series.

Resonance results in delocalization of charge and increases stability. If the conjugate base is stablized by resonance, the acid is stronger than would be predicted from comparable compounds where there is no resonance in the conjugate base. Methanol (CH_3OH) and acetic acid ($CH_3C(O)OH$) are good examples of this.

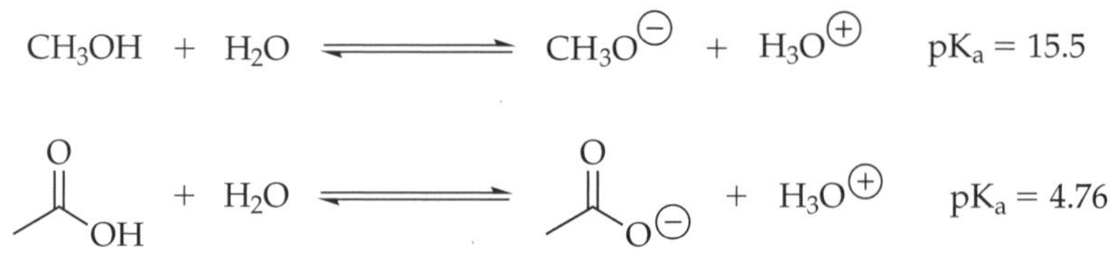

$$CH_3OH + H_2O \rightleftharpoons CH_3O^{\ominus} + H_3O^{\oplus} \qquad pK_a = 15.5$$

$$pK_a = 4.76$$

The acetate anion is stablized by resonance and is therefore a weaker base than the methoxide anion. This means that acetic acid is a stronger acid than expected when compared with methanol.

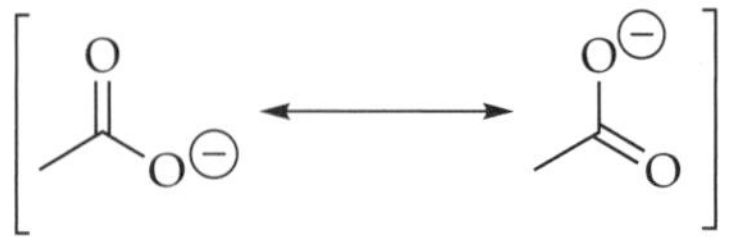

If the acid form is stabilized by resonance, the acid is weaker than comparable compounds. This explains why guanidine is a stronger neutral base when compared with ammonia. The guanidinium ion is stabilized by resonance and is a weaker acid than the ammonium ion where there is no resonance stabilization.

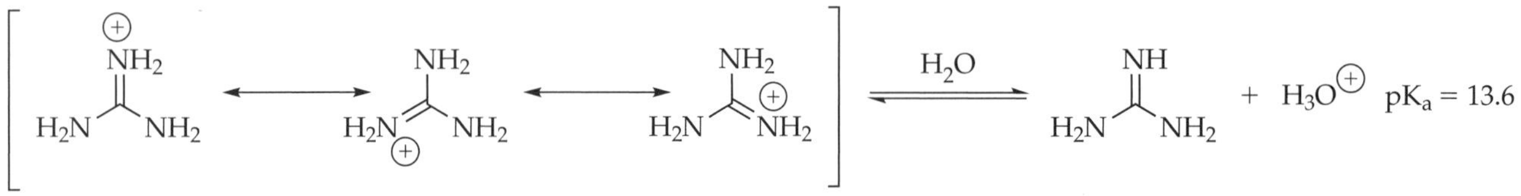

$$pK_a = 13.6$$

$$NH_4^{\oplus} \xrightarrow{H_2O} NH_3 + H_3O^{\oplus} \qquad pK_a = 9.8$$

4.1.1.a Calculating the Concentrations of the Acid and Base Forms Present

By converting Equation 4.5 to a logarithmic expression and multiplying by -1, we obtain Equation 4.7 that shows the relationship between pH, pK_a, and the relative concentrations of the acid form and the conjugate base form of the acid.

$$pK_a = pH - \log \frac{[A^{\ominus}]}{[HA]} \quad \text{or} \quad pH = pK_a + \log \frac{[A^{\ominus}]}{[HA]} \qquad (4.7)$$

The more general equation considers neutral as well as positively charged acids (Equation 4.8).

$$pH = pK_a + \log \frac{[\text{conjugate base form}]}{[\text{acid form}]} \qquad (4.8)$$

We can therefore readily calculate the pH of an aqueous solution if we know the ratio of the concentrations of the acid and base forms and the pK_a. Equation 4.8, which is sometimes called the **Henderson-Hasselbalch**

equation, is useful in calculating the concentrations needed to achieve a solution of a desired pH.

Example

We can calculate the ratio of the concentrations of acetic acid and sodium acetate needed to obtain a solution of pH 4.0. According to Table 4.1, the pKₐ of acetic acid is 4.76.

From Equation 4.8:

$$4.00 = 4.76 + \log \frac{\left[CH_3C(O)O^{\ominus}\right]}{\left[CH_3C(O)OH\right]}$$

Therefore, a pH of 4.0 results if the ratio of sodium acetate to acetic acid is 0.17.

$$\frac{\left[CH_3C(O)O^{\ominus}\right]}{\left[CH_3C(O)OH\right]} = 0.17$$

Note that when the concentrations of the acid and its conjugate base are equal in Equation 4.8 because the acid is 50% dissociated, the pH is equal to the pKₐ.

4.1.1.b Calculating the Equilibrium Constants for Acid-Base Reactions

The previous discussion focused on acids in aqueous solution where water serves as the proton acceptor. Let's consider the reactions of proton donors with other proton acceptors in dilute aqueous solution. Once you have written the equation for the acid-base reaction that occurs, it is necessary to pick out the proton donor (the acid) and the proton acceptor (the base). Remember that the same molecule, for example sodium bicarbonate, can be an acid in one reaction and a base in a different reaction.

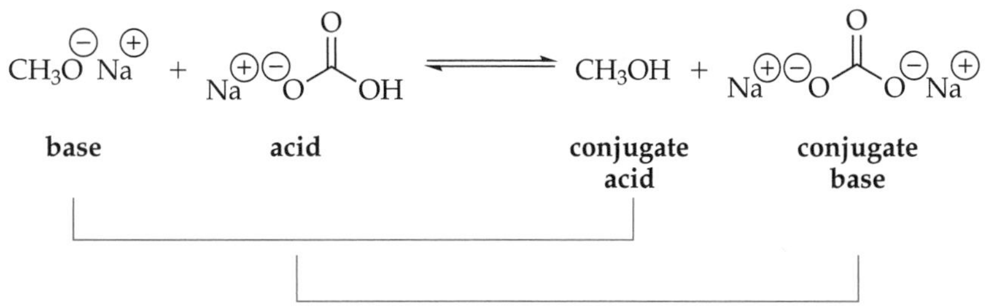

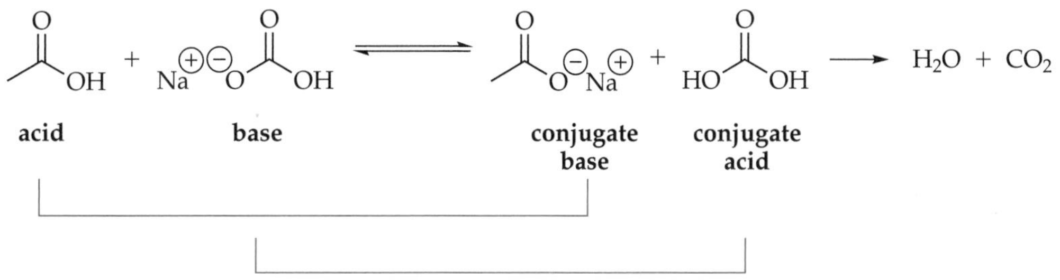

We can calculate equilibrium constants for acid-base reactions from the acid-dissociation constants or from the pK_a values. Let's look at Equation 4.2 again:

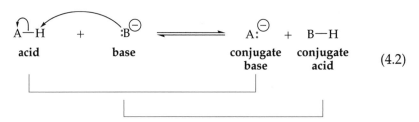

$$A-H \quad + \quad :B^{\ominus} \quad \rightleftharpoons \quad A:^{\ominus} \quad + \quad B-H$$

acid   **base**   **conjugate**   **conjugate**
                                           base   **acid**   (4.2)

Equation 4.9 shows the expression for the equilibrium constant for this reaction. If we multiply the numerator and denominator by the concentration of the hydronium ion, we see that the equilibrium constant can be expressed as the ratio of the acidity constant of HA to the acidity constant of HB.

$$K_{eq} = \frac{\left[A^{\ominus}\right]\left[BH\right]}{\left[AH\right]\left[B^{\ominus}\right]} = \frac{\left[H_3O^{\oplus}\right]\left[A^{\ominus}\right]\left[BH\right]}{\left[AH\right]\left[H_3O^{\oplus}\right]\left[B^{\ominus}\right]} = \frac{K_{a\,HA}}{K_{a\,HB}} \quad (4.9)$$

If we know the K_a values or pK_a values, Equations 4.9 or 4.10 can be used to obtain the equilibrium constant. Remember that pK_a is the negative log of the acidity constant.

$$\log K_{eq} = \log K_{a\,HA} - \log K_{a\,HB} = pK_{a\,HB} - pK_{a\,HA}$$
$$K_{eq} = 10^{(pK_{a\,HB} - pK_{a\,HA})} \quad (4.10)$$

We can, therefore, predict the direction of equilibrium for any proton transfer reaction if we know the pK_a values of the acids involved.

Example

Can sodium bicarbonate be used to protonate sodium acetate?

Begin by writing the equation and identifying the acid and base involved.

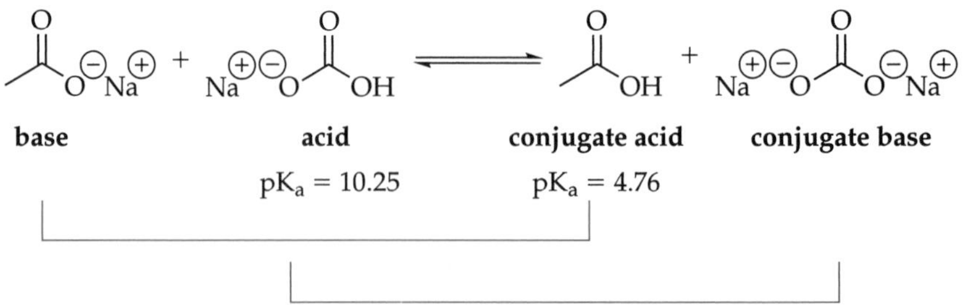

base   **acid**   **conjugate acid**   **conjugate base**
                  $pK_a = 10.25$ $pK_a = 4.76$

The bicarbonate anion is the reacting acid (HA) and acetic acid is the product acid (HB). Make sure you use the correct value from Table 4.1 for the bicarbonate anion acting as an *acid*. Substituting the pK_a values into Equation 4.10 gives:

$$K_{eq} = 10^{(4.76 - 10.25)} = 10^{-5.49}.$$

The equilibrium lies to the left and bicarbonate would not protonate the acetate anion. We know from this analysis, however, that sodium carbonate can be used to deprotonate acetic acid to give the acetate anion.

It is very easy to predict the direction of the equilibrium by just looking at the relative strength of the acids on each side of the equation. The equilibrium always shifts to the side of the weaker acid and weaker base. In other words, the stronger acid is more willing to give up a proton and the stronger base accepts the proton (Equation 4.11).

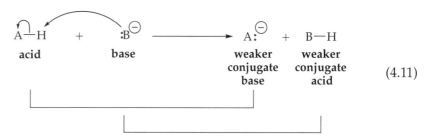

(4.11)

Using Table 4.1, the equilibrium shifts to the right if the reacting acid is *above* (stronger than) the product acid (the conjugate acid of the reacting base).

Example

If we want to find a base that will deprotonate an alkyne (RC ≡ CH), we can refer to Table 4.1. We need the conjugate base of an acid that is weaker than the alkyne, which means it has a pK_a value that is greater than 25 (is below the alkyne in Table 4.1). We can use NaNH$_2$, the conjugate base of ammonia.

Would sodium methoxide work? Methanol is a stronger acid than an alkyne. Methanol is above the alkyne in Table 4.1. Therefore, the methoxide anion will not remove a proton from an alkyne. Write out these equations to convince yourself.

Example

What if a student picks up a bottle of sodium bicarbonate to neutralize an ammonia spill. Will the bicarbonate work?

Look at the reaction:

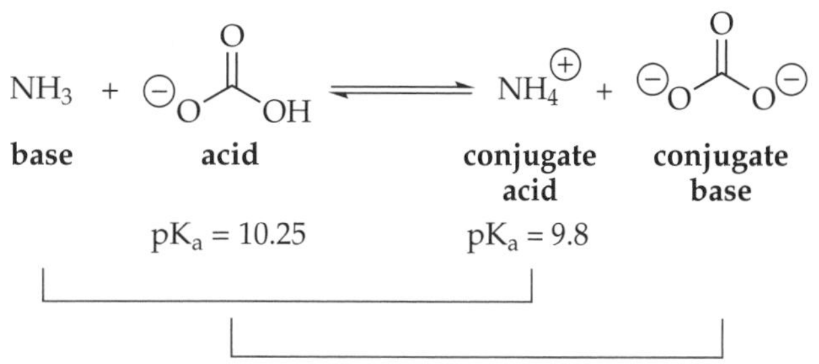

NH$_3$ +   [acid structure]   ⇌   NH$_4^+$ +   [conjugate base structure]

    base   acid   conjugate   conjugate

                                                 acid   base

                      pK_a = 10.25   pK_a = 9.8

Remember to pick out the right acids from Table 4.1. Looking at the pK_a values, we see that the stronger acid (the ammonium ion) is on the right. So the equilibrium will lie to the left, the side of the weaker acid and base. We could have located the ammonium ion and the bicarbonate ion in Table 4.1. Because the ammonium ion is above the bicarbonate ion, it is the stronger acid. The direction of the reaction is therefore toward bicarbonate, which is the weaker acid, so bicarbonate will not neutralize ammonia.

4.1.2 The Lewis Definition

The **Lewis definition** of acid-base reactions is more general and includes the **Brønsted-Lowry definition**. *A molecule that can accept a pair of electrons is a* ***Lewis acid***.

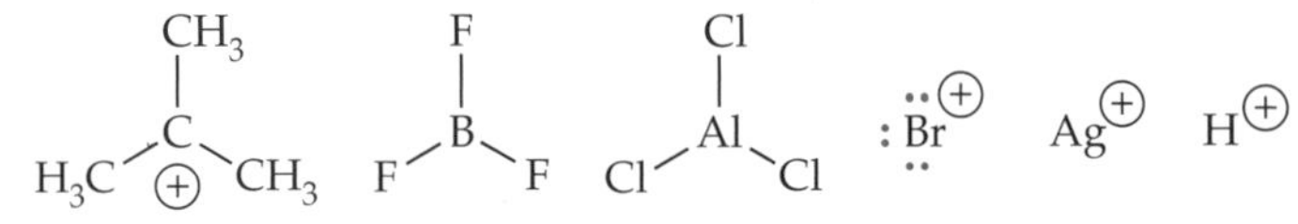

A molecule that can donate a pair of electrons is a ***Lewis base***.

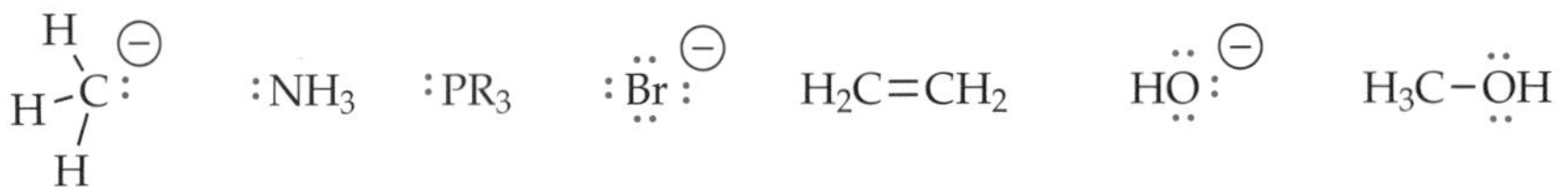

The following reactions are therefore acid-base reactions under the Lewis definition. In writing these reactions, we use a curved arrow notation to show the movement of a pair of electrons to an electron deficient site. The sharing of the electrons results in the formation of a bond between the Lewis acid and the Lewis base.

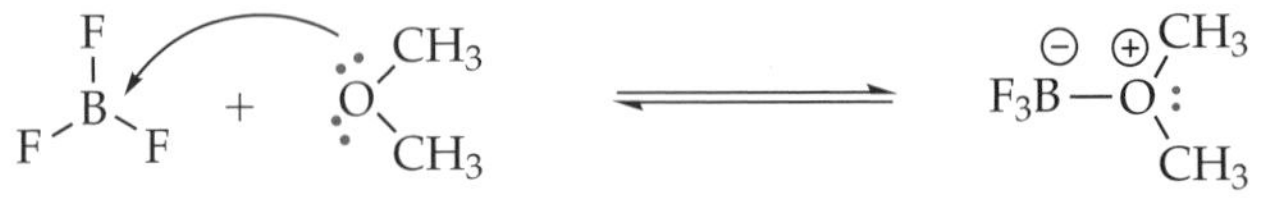

Lewis acid **Lewis base**

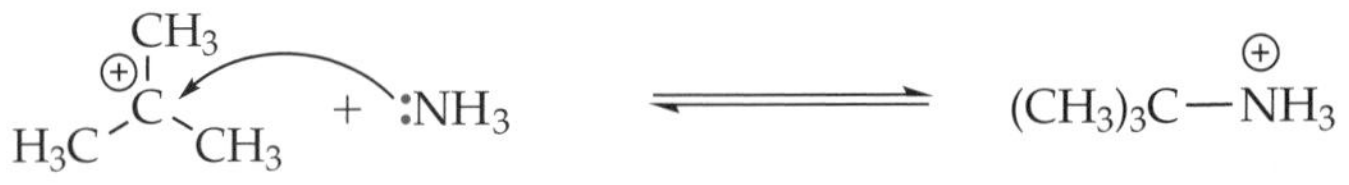

Lewis acid **Lewis base**

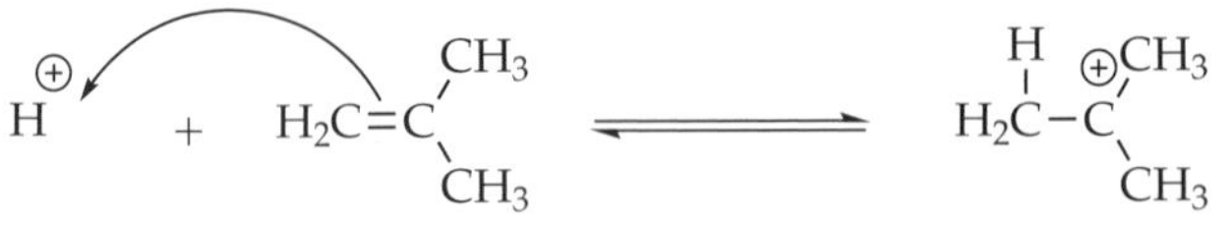

Lewis acid **Lewis base**

4.1.3 Problems

1. Will water and diethyl ether ($(CH_3CH_2)_2O$) react with one another in an acid-base reaction?
2. Calculate the exact pK_a of the hydronium ion ($H_3O^{\oplus}$). Show your work. Hint: Write the equation for the reaction of the hydronium ion with water

and the expression for the equilibrium constant of the reaction. Remember that $K_a = K_{eq} [H_2O]$ and there are 1000 g of water in 1 L.

3. Does benzoic acid (pK_a 4.20) undergo a favorable acid-base reaction with water? With diethyl ether? Write equations for the reactions and show the direction of the equilibria.

4. Calculate the equilibrium constant for the reaction of acetic acid ($CH_3C(O)OH$) with diethylamine (($CH_3CH_2)_2NH$). The pK_a of protonated diethylamine (diethylammonium ion) is 10.94. What are the major forms present in a dilute solution that is equimolar in acetic acid and diethylamine?

5. Write equations for two possible acid-base reactions of diethylamine with water. Indicate the direction of the equilibrium for both reactions. The pK_a for the diethylammonium ion is 10.94 and the pK_a for diethylamine can be extrapolated from that for ammonia found in Table 4.1.

6. You need to generate ammonia from ammonium chloride to use in a reaction. Should you use sodium bicarbonate or sodium carbonate? Explain using equations for the reactions showing the direction of the equilibrium in each case.

7. Which of the hydrohalic acids could you use to generate a significant amount of protonated methanol? Which acid would result in an equilibrium constant of 1? Which of the hydrohalic acids would not be suitable for protonating methanol?

8. Explain the relative pK_a values of methanol (CH_3OH) and methanethiol (CH_3SH). Which would be a stronger acid, PH_3 or SiH_4? Explain.

9. What are the relative percentages of hydrogen cyanide (HCN) and cyanide ion in a dilute aqueous solution at pH = 9.2? At pH = 10.2? Your answers should be rounded to the closest 1%.

4.2 LIMITING REACTANTS, THEORETICAL YIELD, AND PERCENT YIELD

In a chemical reaction, we usually do not obtain the maximum amount of material possible based on the amounts of the starting reactants used. Sometimes the reaction does not go to completion and side reactions often compete with the desired reaction. There are also losses during purification steps, such as extraction, distillation, and recrystallization, in addition to physical losses on the surfaces of the equipment. We measure the success of a chemical synthesis by the amount of product obtained compared with the maximum amount possible from the reaction. The **percent yield** for a reaction is calculated by dividing the amount of product obtained by the maximum amount possible, known as the **theoretical yield**, and multiplying by 100 (Equation 4.12).

$$\text{Percent yield} = \frac{\text{amount of product obtained}}{\text{amount of product possible}} \times 100 \qquad (4.12)$$

We must, therefore, determine the theoretical yield of the product in the same units as we measure the final amount of the product. This is usually the weight.

4.2.1 Calculating the Theoretical Yield

1. Write a balanced equation for the reaction.

2. Calculate the number of moles of each of the reactants used in the reaction.

   - If the reactants are expressed in terms of weight, divide by the molecular weight to find the number of moles used.

   - If the amounts of the reactants are measured by volume, multiply the volume by the density (g/mL) to determine the weight, and then calculate the number of moles.

3. Find the **limiting reactant**. Usually one reactant determines how much product can be obtained from the reaction and the other reactants are present in excess. In theory, the limiting reactant is completely consumed if the reaction goes to completion. Therefore, this reactant limits the amount of material one can obtain from the reaction.

   - Divide the number of moles of each reactant by its coefficient in the balanced equation to give the **number of equivalents** of each reactant available.

   - The reactant present in the least number of equivalents is the limiting reactant.

4. Calculate the **theoretical yield** assuming that the reaction goes to completion and that the limiting reactant is completely consumed.

   - Multiply the number of equivalents of the limiting reactant by the coefficient of the product shown in the balanced equation to calculate the number of moles of product possible.

   - Multiply the number of moles of product by the molecular weight of the product to calculate the theoretical yield in grams.

Example

To illustrate, let's look at a non-chemical example. One car consists of 1 frame, 2 windshield wipers, 2 headlights, 1 horn, and 4 wheels. Although many other materials are needed, we will limit the discussion of our car to these items. What if we have 6 frames, 10 wipers, 12 headlights, 4 horns, and 8 wheels. How many cars can we make?

First, we write the "balanced equation." Then, we divide the number of each item available by its coefficient in the equation to obtain the equivalents of each item. We find we can get 6 cars from the frames, 5 from the wipers, 6 from the headlights, 4 from the horns, and 2 from the wheels.

1 frame + 2 windshield wipers + 2 headlights + 1 horn + 4 wheels = 1 car.

given:	6	10	12	4	8
equivalents:	6	5	6	4	2

The number of wheels limits the yield of the cars, so the maximum number of cars we can make from the above combination of items is two. Wheels are the limiting item.

Example

The same process is used for chemical reactions. In the equation below, the amount of sodium cyanide (NaCN) limits the amount of the product that can be obtained. Theoretically, if the reaction goes to completion, the sodium cyanide will be completely consumed and there will be some bromomethane (CH_3Br) left. The maximum amount of acetonitrile (CH_3CN) possible is 1 mole.

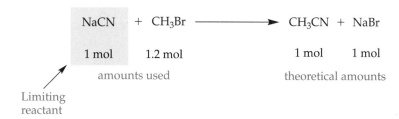

The theoretical yield of acetonitrile is equal to 1 mole times the molecular weight of acetonitrile.

$$\text{Theoretical yield} = 1 \text{ mol} \times 41 \text{ g/mol (MW of acetonitrile)}$$
$$= 41 \text{ g of acetonitrile}$$

It is common to use an excess of one or more reactants to increase the rate and/or the yield of a reaction.

Example

In the preparation of 1,6-heptadiene, two moles of sodium *tert*-butoxide react with one mole of 1,7-dibromoheptane. Therefore, we have to divide the number of moles of sodium *tert*-butoxide by two to find the equivalents of this reactant that are available for the reaction. We find that the sodium *tert*-butoxide is the limiting reactant in the example and the maximum amount of the 1,6-heptadiene that we can obtain is 0.0625 mol or 6.00 g.

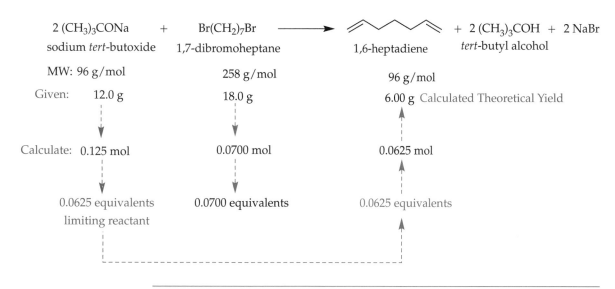

For redox reactions, we have to use the half-reactions to balance the equation, adding $H^{\oplus}$, $OH^{\ominus}$, H_2O, and electrons as required to balance the atoms and the charge. Remember the mnemonics, *"LEO goes GER,"* *Loss of Electrons is Oxidation and Gain of Electrons is Reduction*, or *"OILRIG,"* *Oxidation Is Loss, Reduction Is Gain.*

Example

For the oxidation of benzyl alcohol with CrO_2, we write one equation for the oxidation half-reaction and one equation for the reduction half-reaction. The unbalanced overall reaction is:

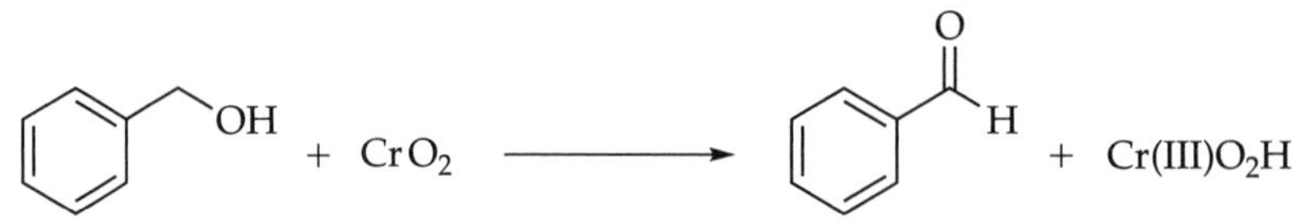

The oxidation half-reaction (LEO or OIL) is:

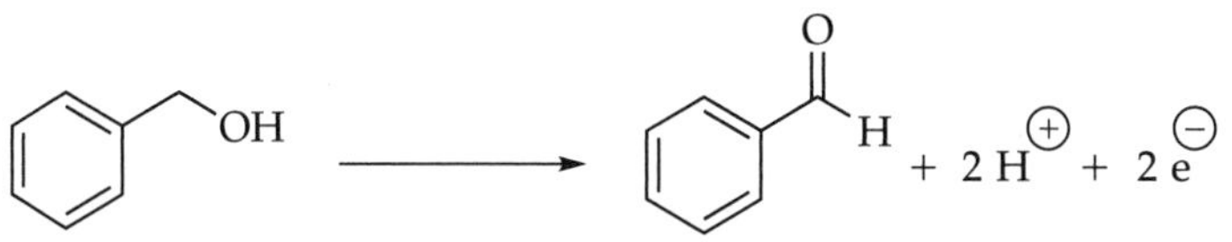

The reduction half-reaction (GER or RIG) is:

$$2\left[H^{\oplus} + e^{\ominus} + CrO_2 \longrightarrow CrO_2H \right]$$

We have to multiply the reduction half-reaction by two so that when we add the two half reactions together, the protons and the electrons cancel. The balanced overall equation is:

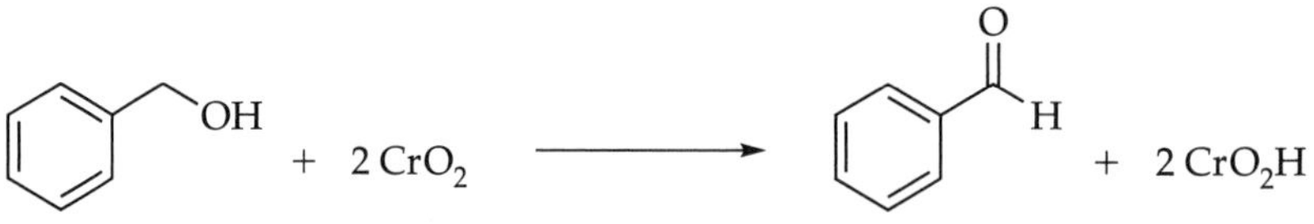

Now we can calculate the minimum amount of CrO_2 we would need to oxidize 0.50 mL of benzyl alcohol.

MW: 108 g/mol 84 g/mol

density: 1.04 g/mL

Given: 0.50 mL

Calculate: 520 mg 810 mg

Calculate: 4.8 mmol 9.6 mmol

4.8 equivalents 4.8 equivalents

Remember to consider only the reactants present in the balanced equation when you determine the limiting reactant. For example, a catalyst may be present in the smallest molar quantity, but because it is not incorporated into the product, it is not limiting the amount of product being formed.

Example

In the dehydration reaction shown below, the sulfuric acid is used as a catalyst so that the limiting reactant is the starting alcohol.

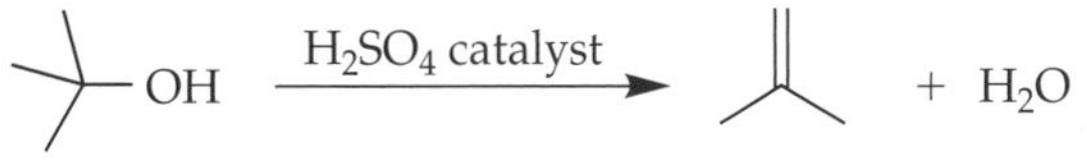

1 mole 0.10 mole 1 mole 1 mole

amounts used theoretical yield

4.2.2 Problems

In the following problems, find the limiting reactant and calculate the percent yield of the desired product. Be sure to show your work.

1. A mixture of 5.0 mL of *t*-butyl chloride and 100 mL of methanol was refluxed for 2 hours. After neutralizing any remaining hydrogen chloride, 3.5 grams of *t*-butyl methyl ether was isolated. Calculate the percent yield of the *t*-butyl methyl ether.

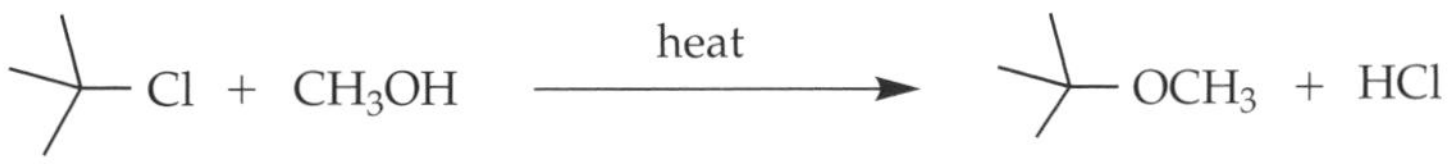

MW: 93 g/mol 32 g/mol 88 g/mol

density: 0.85 g/mL 0.79 g/mL

2. When 10 mL of 1-heptene was reacted with 5.1 grams of anhydrous hydrogen bromide, 8.7 grams of 2-bromoheptane was obtained after workup. Calculate the percent yield of 2-bromoheptane.

	MW:	98 g/mol	81 g/mol	179 g/mol
	density:	0.67 g/mL		

3. To prepare a dithioether, 794 mg of sodium methylthiolate was added to a solution of 0.50 mL of 1,3-dibromopropane dissolved in 50 mL of tetrahydrofuran. The reaction was stirred for 1 hour and then the tetrahydrofuran was removed by distillation. The resulting oil was further purified to give 450 mg of the product. Calculate the percent yield of the dithioether.

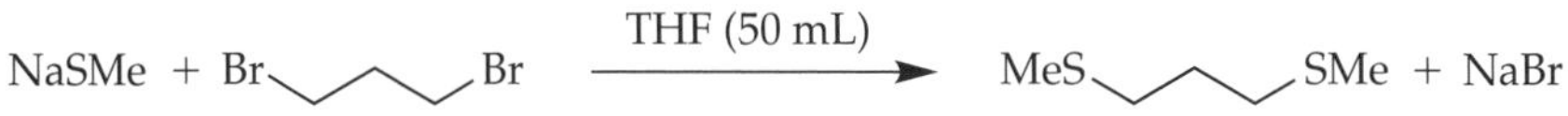

	NaSMe +	Br⌒⌒Br		MeS⌒⌒SMe + NaBr
MW:	70 g/mol	202 g/mol		136 g/mol
density:		1.99 g/mL		

4. A mixture of 3.0 g of 1,1-diphenylethylene and 100 mg of sulfuric acid was heated for 2 hours. Upon cooling, 2.2 grams of crystalline 1,1,3,3-tetraphenyl-1-butene precipitated from the mixture. What is the percent yield of the crystalline product?

	MW: 180 g/mol	98 g/mol	360 g/mol

5. Acetophenone (200 mg) was dissolved in 10 mL of 2-propanol. The mixture was cooled in an ice-water bath and 20 mg of sodium borohydride was added slowly with stirring. After workup, 158 mg of 1-phenylethanol was obtained. Calculate the percent yield of the product.

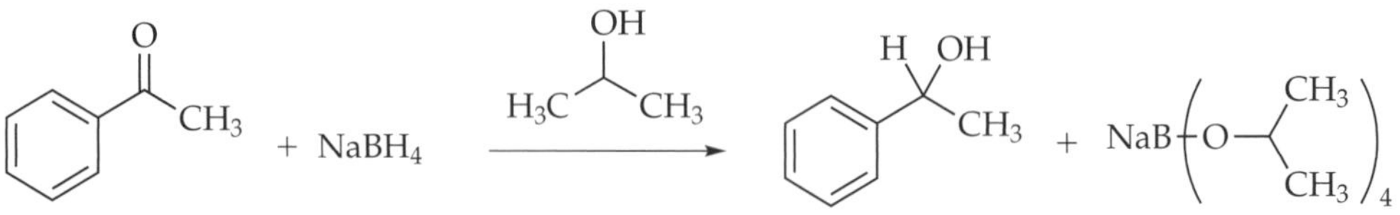

	MW:	120 g/mol	38 g/mol	60 g/mol	122 g/mol
	density:	1.0 g/mL		0.78 g/mL	

6. What was the percent yield of 4,4'-di-*t*-butylbiphenyl if 450 mg was isolated from the reaction of 462 mg of biphenyl with 0.50 mL of *t*-butyl chloride in the presence of 100 mg of Al?

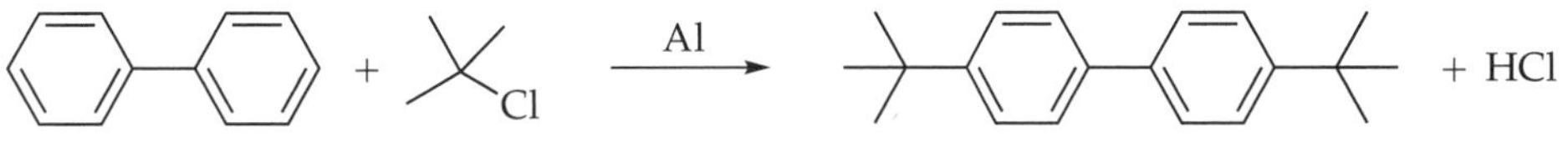

MW: 154 g/mol   93 g/mol   27 g/mol   266 g/mol
density: 0.85 g/mL

7. Vitamins are chemicals that are required by humans in small amounts, but that are not made in our bodies. Such chemicals must therefore be acquired from our diet. With this definition in mind, vitamin D_3 is not truly a vitamin because it is, in fact, produced in our skin by the interaction of sunlight in the ultraviolet region with a derivative of cholesterol. Because many people do not have adequate exposure to sunlight, vitamin D_3 is commonly added to foods, such as milk, to avoid the onset of serious diseases, such as rickets.

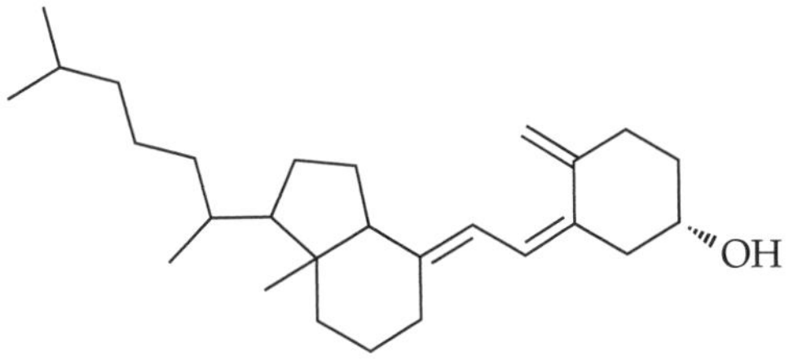

vitamin D_3

   a. What is the molecular formula of vitamin D_3?

   b. Calculate the molecular weight of vitamin D_3 (to hundredths of a g/mol).

   c. Circle four Lewis basic sites in vitamin D_3.

   d. If one looks at the nutrition label of a multivitamin supplement, the term IU (international unit) is often used to express a quantity. One IU of a specific vitamin is the amount necessary to elicit a specific type of biological activity.

   The recommended daily allowance (RDA) of vitamin D_3 for children and most adults is 200 IU/day, which translates to 5 micrograms/day (5 μg/day). Based on this, calculate how many milligrams are in 1 IU of vitamin D_3.

   e. It is known that humans can generate up to 12,000 IU's of vitamin D_3 upon exposure to 30 minutes of summer sun. How many millimoles of vitamin D_3 does this represent?

f. Alkenes can be hydrogenated to alkanes using a palladium catalyst.

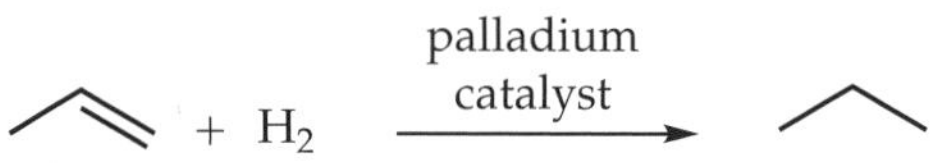

Starting with 2.5 grams of vitamin D_3 and 10 mg of palladium catalyst, how many millimoles of hydrogen (H_2) are necessary to hydrogenate all of the carbon-carbon double bonds in vitamin D_3?

g. Assume vitamin D_3 is the limiting reactant. What is the percent yield of the product from the reaction if one starts with 2.5 grams of vitamin D_3 and obtains 1.8 grams of product?

CHAPTER 5

Properties of Organic Molecules

The synthetic methods and the mechanisms of reactions that are described in textbooks and presented in lectures are the result of the research of organic chemists in the laboratory. You learn about reactions that synthetic organic chemists have developed to transform molecules and the mechanisms that the physical organic chemists have devised that are consistent with all current data in explaining how these reactions occur. With this knowledge, you can plan the synthesis of complex molecules on paper, but now you are going to be a chemist instead of a chemistry student. You will be conducting experiments and applying the principles of organic chemistry to gain experience in the laboratory techniques; to test, demonstrate, or establish a theory; to examine the validity of a hypothesis; or to determine the possibility of something as yet untried. You will find that, as an experimental chemist, you will have to consider many factors in order to carry out reactions and syntheses in the lab.

In order to get two materials to react to form a third, the reactants and reaction conditions must be chosen so that the molecules can interact. The activation energy of the reaction must be considered and the temperature adjusted to obtain a reasonable reaction rate. Once a successful transformation is achieved, the desired product must be separated from any remaining starting materials and any side-products of the reaction by manipulating the interactions between molecules. The product must then be characterized and its purity determined. An understanding of the properties of molecules and the intermolecular forces between molecules provides a "picture" at the molecular level of what is happening when we perform various procedures in the laboratory. In this chapter, we will examine these forces to develop an understanding that is essential to your success as an experimental chemist.

5.1 BONDING BETWEEN ATOMS TO FORM MOLECULES

We say that atoms bond to one another to form molecules, but we further distinguish a continuum of types of bonds depending on how equally the atoms share the bonding electrons. Linus Pauling, who won Nobel Prizes in Chemistry (1954) and the Peace Prize (1962)[1], defined a scale of electronegativity of elements to somewhat quantify how tightly the nucleus of an atom holds on to the electrons. In a row of the periodic table, the

---

[1] See http://www.nobelprizes.com and http://www.nobel.se/.

nuclear charge of the atoms increases from left to right. The greater the nuclear charge, the greater is the attraction of the positive nucleus for the surrounding electrons. The basis of this attraction is **Coulomb's law**, an experimental relationship discovered in 1784 by Charles Augustin de Coulomb. It states, *"The attractive force (F) between two oppositely charged objects is directly proportional to the product of the charge on each particle (q_1q_2) and inversely proportional to the square of the distance (r) between them."*

$$F = k\,\frac{q_1q_2}{r^2}$$

Across a row of the periodic table, the elements to the right have a greater nuclear charge and are said to be electronegative relative to those on the left. Therefore, the halogens hold electrons more strongly than elements to the left of them in the same row. One might think that the same increase in nuclear charge would determine the electronegativity trend as one goes down a column in the periodic table. However, the square of the distance from the nucleus is also important in determining the strength of the force. We find, therefore, that electronegativity decreases down a column of the periodic table because distances between the negative electrons and the positive nucleus increase. As a result, *fluorine is the most electronegative element whereas the elements at the bottom left part of the periodic table are the least electronegative.* Selected electronegativities are shown in Table 5.1.

When two *identical* atoms combine to form a molecule, the atoms share the electrons equally. This is the case, for example, in H_2 where the molecule is **non-polar** and the bond is called a **covalent bond**.

If two atoms of dissimilar electronegativities form a bond, there is unequal sharing of the electrons. For example, in the molecule HCl, the

Charles Augustin de Coulomb 1736–1806

Linus Pauling 1901–1994

Table 5.1 Electronegativities of Selected Elements[2]

Electronegativity Increases as

Nuclear Charge Increases →

H 2.18						
Li 0.98	**Be** 1.57	**B** 2.04	**C** 2.55	**N** 3.04	**O** 3.44	**F** 3.98
Na 0.93	**Mg** 1.31	**Al** 1.61	**Si** 1.90	**P** 2.19	**S** 2.58	**Cl** 3.16
K 0.82	**Ca** 1.00			**As** 2.18	**Se** 2.55	**Br** 2.96
						I 2.66

Electronegativity Increases as the Distance from the Nucleus Decreases ↑

[2] The electronegativities were defined by Linus Pauling. The values were revised by Allred, A.L. *J. Inorg. Nucl. Chem.* **1961**, *17*, 215.

Cl (3.16) is more electronegative than H (2.18). The Cl has a greater attraction for the bonding electrons and the molecule is **polar** with the Cl bearing a slight negative charge. The symbol $\delta+$ signifies a partial positive charge and $\delta-$ denotes a partial negative charge.

$$\overset{\delta+}{H}\!\!-\!\!\overset{\delta-}{Cl}$$

Hydrogen chloride has a **polar covalent bond**. This is illustrated using a plus symbol (+) to show the positive end connected to an arrow pointing toward the negative end.

$$\overset{+\longrightarrow}{H-Cl}$$

At the extreme, where the electronegativities of the atoms are very different, the electropositive element transfers its electrons to the electronegative element resulting in an **ionic** or **electrostatic bond**.

$$Na\cdot \; + \; \cdot\ddot{\underset{..}{C}l}\!: \; \longrightarrow \; Na^{\oplus} \; + \; :\ddot{\underset{..}{C}l}\!:^{\ominus}$$

In crystalline sodium chloride, the positively charged sodium ions are surrounded by six negatively charged chloride ions and the chloride ions are surrounded by six sodium ions. The electrostatic attraction between the ions of opposite charge is the basis for **ionic bonding**.

Therefore, bonds between atoms range from non-polar covalent to polar covalent to ionic depending on the electronegativity differences between the atoms involved in bonding.

5.1.1 Bond Dipole Moments

If the atoms forming a bond have the same or nearly equal electronegativities, the bond is non-polar. A bond is polar and has a dipole if the atoms forming the bond have different electronegativities. This difference in electronegativities gives rise to a **bond dipole moment**, μ, which is a measure of the size of the charge and the distance between the charges. Dipole moments are vectors, which means they have magnitude and direction. They are expressed in units called Debyes (D). The polarity of the bonds in a molecule determines the types and strengths of the intermolecular forces between molecules. These forces, which determine the physical properties of molecules and their interactions with different molecules, will be discussed in Section 5.3. *In assessing intermolecular forces between molecules, it is the polarity of the bonds that is important rather than the overall polarity of the molecule.*

5.2 THE EFFECT OF STRUCTURE ON THE POLARITY OF MOLECULES

The polarity of the bonds and the three-dimensional structure of the molecule determine the polarity of the molecule.

5.2.1 Molecular Dipole Moments[3]

Molecules that contain only atoms of similar electronegativity do not have polar bonds and are called **non-polar**. As shown in Figure 5.1, molecules with polar bonds can be polar or non-polar depending on the three-dimensional structure of the molecule. For example, carbon tetrachloride (CCl_4) has four

---

[3] Moulton, W.N. *J. Chem. Ed.* **1961**, *38*, 522–523.

Figure 5.1   Molecular vs. Bond Polarity

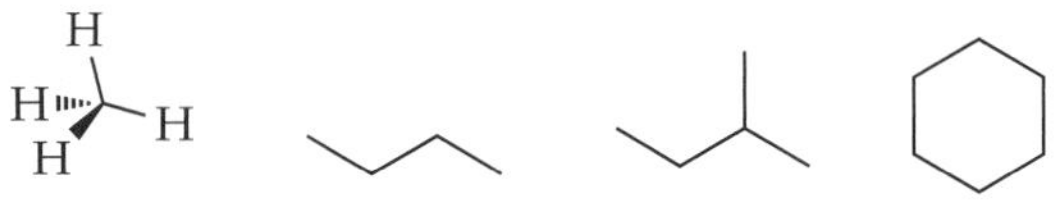

Non-polar molecules
with no polar bonds

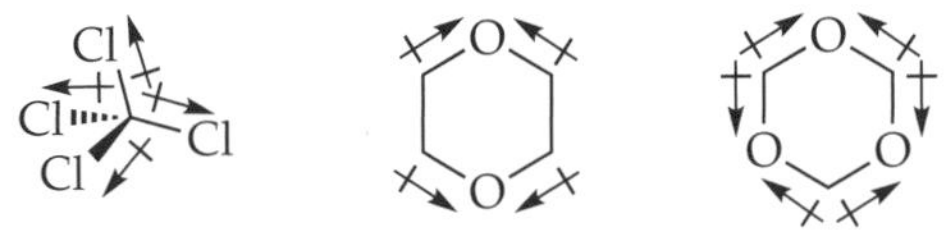

Non-polar molecules
with polar bonds

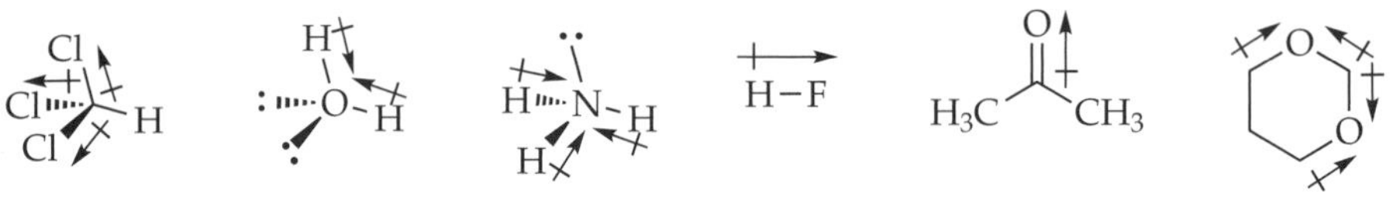

Polar molecules with polar bonds

polar carbon-chlorine bonds. The chlorines are located at the corners of a tetrahedron with the carbon at the center of the tetrahedron. The **dipole moments** of the individual carbon-chlorine bonds, which are vectors, are equal and point in opposite directions so that they cancel one another. Consequently, the molecule as a whole does not have a **molecular dipole moment**. On the other hand, in molecules such as trichloromethane ($CHCl_3$), the dipole moments of the individual bonds are not equal and are not applied in opposite directions. They do not cancel one another. These molecules have a molecular dipole moment and are called **polar** (Figure 5.1).

The stereochemistry of the isomers of 1,2-dichloroethene also explains why the *trans* isomer has no molecular dipole moment. The bond dipoles are of equal magnitude and point in opposite directions so that they cancel one another. The *cis* isomer has a molecular dipole because the bond dipoles do not point in opposite directions and therefore do not cancel.

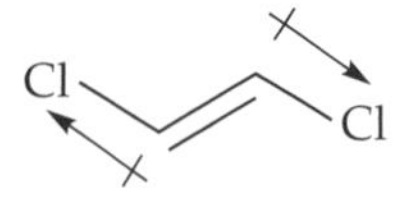

***trans*-1,2-dichloroethene**
molecular dipole moment = 0

***cis*-1,2-dichloroethene**
molecular dipole moment ≠ 0

Some situations require the examination of the molecular dipole moments of all contributing conformations. The net molecular dipole moment of a molecule is the weighted average of the molecular dipole moments of all contributing conformations. In other words, if any contributing conformation of a molecule has a molecular dipole moment, the molecule will have a net molecular dipole moment. For example, when 1,2-dichloroethane is drawn in its most stable anti-conformation, the molecular dipole moment is zero. However, there are two gauche conformations that contribute to the structure of this molecule, both of which have a molecular dipole moment. Therefore, 1,2-dichloroethane has a net molecular dipole moment. The eclipsed conformations are not included in this analysis because they are transition states leading to the more stable staggered conformations.

anti	gauche	gauche
molecular dipole	molecular dipole	molecular dipole
moment $= 0$	moment $\neq 0$	moment $\neq 0$

1,2-dichloroethane
net molecular dipole moment $\neq 0$

5.2.1.a Group Dipole Moments

For substituents that have more than one polar bond, it is convenient to define a **group dipole moment**. This term can be used to refer to either a simple **bond dipole moment** or a collection of **bond dipoles** associated with a given substituent. The bond dipole moments are added together to give the group dipole moment to allow the determination of the presence or absence of a **molecular dipole moment**.

For example, the group dipole moments for the nitro groups are equal and point in opposite directions in 1,4-dinitrobenzene. The molecule does not have a net molecular dipole moment.

resonance structures   group dipole moment

1,4-dinitrobenzene   **1,3-dinitrobenzene**
net molecular dipole moment $= 0$ net molecular dipole moment $\neq 0$

In 1,3-dinitrobenzene, the group dipole moments are equal, but they do not point in opposite directions and do not cancel one another. The net molecular dipole moment of 1,3-dinitrobenzene is not zero.

What about terephthalaldehyde? The two carbonyl functions are equivalent and opposite one another. However, unlike the nitro group, the functional group is not symmetrical. Because of the contributions from the different conformations, terephthalaldehyde has an overall net molecular dipole moment.

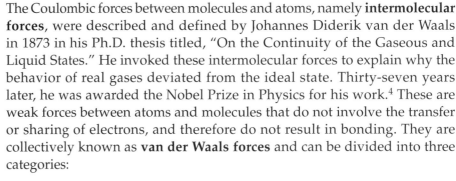

molecular dipole
moment = 0

molecular dipole
moment ≠ 0

**terephthalaldehyde
(4-formylbenzaldehyde)**
net molecular dipole moment ≠ 0

Because conformations must be included in the assessment of a molecule's net molecular dipole moment, we can conclude that some alkanes, such as butane, should have a molecular dipole moment. This is, in fact, correct. However, the electronegativity difference between carbon and hydrogen is small so that the group dipole moments of the alkyl substituents are small. Consequently, the molecular dipole moments of alkanes are small and the molecules are considered to be non-polar.

5.3 THE TYPES OF INTERMOLECULAR FORCES

*Johannes Diderik
van der Waals
1837–1923*

The Coulombic forces between molecules and atoms, namely **intermolecular forces**, were described and defined by Johannes Diderik van der Waals in 1873 in his Ph.D. thesis titled, "On the Continuity of the Gaseous and Liquid States." He invoked these intermolecular forces to explain why the behavior of real gases deviated from the ideal state. Thirty-seven years later, he was awarded the Nobel Prize in Physics for his work.[4] These are weak forces between atoms and molecules that do not involve the transfer or sharing of electrons, and therefore do not result in bonding. They are collectively known as **van der Waals forces** and can be divided into three categories:

- dipole-dipole interactions
- dipole-induced dipole interactions
- induced dipole-induced dipole interactions or London dispersion forces

Each of these interactions is weaker than interactions between fully charged ions (Section 5.3.4), but they play an important role in determining the physical properties of molecules and their interactions with one another. In most cases, a combination of forces acts together to contribute to the overall properties of the system so it is helpful to examine each of them to understand their relative importance and the factors that affect their strength. In the case of molecules with no polar bonds, the only attractive forces possible between the molecules are induced dipole-induced dipole

[4] See http://www.nobel.se/physics/laureates/1910/waals-bio.html.

Fritz London 1900–1954

interactions (London dispersion forces). For molecules with polar bonds, all three of the van der Waals forces contribute to the overall attractive forces: induced dipole-induced dipole, dipole-induced dipole, and dipole-dipole. We will see that the latter depend on the correct orientation of the dipoles and also on temperature. We must further distinguish molecules where a hydrogen is involved in a polar bond (polar protic) from those that have no partially positively charged hydrogen (polar aprotic). This distinction is important because of the strength of hydrogen bonding in polar protic compounds.

The importance of each of the individual forces varies with the structure of the molecules involved and may depend on one or more of the following:

- the charges present (q)
- the dipole moments of the bonds in the molecule (µ)
- the distance between the interacting parts of the molecules (r)
- the polarizability of the molecule (α)
- the ionization potential of the molecule (IP)
- the temperature (T)

Because these are electrostatic forces, they are *not* dependent on the masses of the interacting substances.

These non-bonding interactions also influence how molecules interact with those of a different type and help us to understand, for example, the miscibility or non-miscibility of liquids and the solubility of substances in liquids.

5.3.1 Induced Dipole-Induced Dipole Interactions (London Dispersion Forces) (ID-ID)

$$E_{ID-ID} = -\frac{3\alpha^2}{4r^6} IP$$

When instantaneous charge separation or dipoles are established in atoms or molecules, the effect is transmitted to neighboring atoms or molecules. These **induced dipoles** result in weak forces between the atoms or molecules. They are named after Fritz London[5] and are known as **London dispersion forces** or **induced dipole-induced dipole interactions (ID-ID)**. *These forces are proportional to the square of the polarizability (α) and to the ionization potential (IP). Because they are inversely proportional to the sixth power of the distance (r), London dispersion forces are important only for the areas of interacting molecules that are in close proximity with one another.*

5.3.1.a Induced Dipole-Induced Dipole Interactions in the Noble Gases

To understand these interactions, picture a helium atom that has two electrons in the spherical 1s shell (Figure 5.2). On average, the two electrons are evenly distributed about the nucleus. However, at any point in time the electrons can be in close proximity to one another to establish an instantaneous dipole where one side of the atom has a greater negative charge than the other. If a second helium atom comes in contact with this "polarized" atom, the slight dipole in the first atom induces a similar instantaneous dipole moment in the second causing a weak, momentary attraction between the two atoms.

The nuclei and electrons of the atoms repel one another at very small distances, but there is an optimum distance known as the **van der Waals radius** for which the attractive force between the atoms is greatest.

---

[5] See http://onsager.bd.psu.edu/~jircitano/London.html.

Figure 5.2   Induced Dipole Interactions in Helium

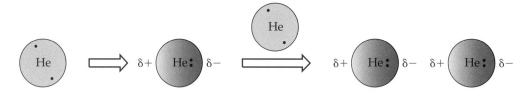

These interactions between helium atoms are weak because the orientation of the electrons is fleeting, but they are important in understanding the physical properties of helium. If there were no attractive forces between the helium atoms, it would not be possible to obtain liquid helium. The London dispersion forces are not strong enough to hold helium atoms together to form a liquid phase at room temperature and atomspheric pressure. However, the fact that helium can be liquified at high pressure and low temperature is evidence that weak induced dipole interactions do exist between the atoms.

The data in Table 5.2 shows that the melting and boiling points of the noble gases increase as a function of atomic weight. In the case of the boiling points, it is tempting to conclude that the heavier an element is, the harder it is to "lift it up" into the gas phase, but there is no "lifting" process involved in boiling. It is an "escape" phenomenon. *The **boiling point** of a substance is the temperature at which its vapor pressure is equal to the external pressure.* It is, therefore, a measure of the ability of the molecules to "escape" from the surface of the liquid into the gas phase. Nothing is needed to "lift" them up. The vapor pressure of a substance is related to the mass of the particles because heavier molecules require more energy to achieve the velocities that are required for them to escape from the surface. The vapor pressure is also inversely proportional to the intermolecular forces between the molecules. These forces are electrostatic in nature and depend upon partial charges rather than on the mass of the substance. The greater these electrostatic forces are, the greater is the energy required for the molecules to break away from one another and leave the surface of the liquid. The existence of these interactions explains why there are many examples of compounds that have the same or similar molecular weights, but radically different boiling points.

In the case of the noble gases, the differences in the intermolecular electrostatic interactions, the London dispersion forces, contribute to the differences in the boiling points. As the number of electron shells of an atom

Table 5.2 The Physical Properties of the Noble Gases

	MP°C	BP°C	IP Kj/mol [eV]	$\alpha \times 10^{-24}cm^3$
He		−268.91	2372 [24.59]	0.20
Ne	−248.59	−246.08	2081 [21.57]	0.40
Ar	−189.35	−185.87	1521 [15.76]	1.64
Kr	−157.37	−153.23	1351 [14.00]	2.48
Xe	−111.76	−108.0	1170 [12.13]	4.04
Rn	−71	−61.7	1037 [10.75]	5.30

increases down a column in the periodic table, the electrons in the outermost shell become easier to distort or polarize because they are further removed from the positive nucleus. This increased distortion or polarization of the electron cloud results in an increased magnitude of the induced dipole-induced dipole interactions. Consequently, these non-bonding interactions are stronger for the larger atoms, which contributes to a higher boiling point.

SUMMING IT UP!

The vapor pressure of a liquid depends on the kinetic energy of the molecules of the liquid and also on the intermolecular forces between the molecules.

5.3.1.b Induced Dipole-Induced Dipole Interactions Between Molecules with No Polar Bonds

As is the case for the noble gases, there are intermolecular attractions between molecules with no polar bonds. These are caused when temporary and fleeting dipoles are established in non-polar bonds that induce dipoles in the bonds of neighboring molecules resulting in an attractive force between the molecules.

Did you know?

Nitrogen is a colorless, tasteless, odorless, inert diatomic gas at room temperature and pressure. Because of intermolecular forces, nitrogen can be condensed to a liquid with a boiling point of $-195.79°C$. It solidifies to a snow-white solid with a melting point of $-210.01°C$. Liquid nitrogen is used in laboratories and industry as a cooling agent.

©Michael Krinke/ iStockphoto

Induced Dipole-Induced Dipole Interactions from Non-Bonding Electrons Let's examine the diatomic halogens where two identical atoms share electrons to form a bond. The halogens have non-bonding or lone pairs of electrons which are polarizable. Temporary, fleeting dipoles result when the electron cloud is not evenly distributed across both atoms. These induced dipoles in one molecule can cause a temporary dipole in an adjacent molecule, producing an intermolecular attraction between the molecules as illustrated in Figure 5.3.

The effect of these induced dipole-induced dipole interactions in the halogens is illustrated by the physical properties of these molecules (Table 5.3). Although the ionization potential (IP) decreases down the column of the periodic table, the polarizability (α) of the molecules increases with the size

Figure 5.3   Induced Dipole Interactions Between Bromine Molecules

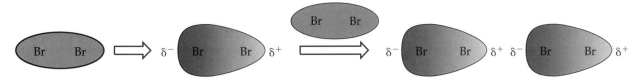

Table 5.3 Physical Properties of the Halogens

	MW g/mol	MP°C	BP°C
fluorine	38	−219.6	−188.1
chlorine	70.9	−101	−34
bromine	159.8	−7.2	58.8
iodine	253.8	113	184

of the atom. The effect of the polarizability outweighs that of the ionization potential, so the melting points and boiling points increase as the size of the halogen increases.

These induced dipole interactions in the halogens are stronger than those in the noble gases. In bromine, for example, the London dispersion forces are strong enough to keep it in the liquid state at room temperature and atmospheric pressure. The London dispersion forces between iodine molecules are strong enough that it exists in the solid state at room temperature and atmospheric pressure.

Induced Dipole-Induced Dipole Interactions from Bonding Electrons
London dispersion forces arise not only from non-bonding electrons, but also from those shared between atoms, the bonding electrons. Both of the electrons of hydrogen are involved in bonding, but there are also intermolecular forces between hydrogen molecules as evidenced by the existence of liquid hydrogen (bp −252.8°C).

The same is true in the alkanes where the electronegativity of carbon (2.55) and hydrogen (2.18) are fairly similar (Table 5.1), so that no significant permanent dipoles exist in the carbon-hydrogen bonds of molecules and there are no non-bonding electrons. However, the bonding electrons contribute to London dispersion forces. When the molecules are in close proximity, the bonding electrons from one molecule repel those from a neighboring molecule, thereby producing a temporary dipole. This effect spreads like a wave across the bonds of the molecule as the electron motion is coordinated between adjacent bonds. The strength of each of these interactions is small. For example, in alkanes it is 1–1.5 kcal/mol (4.2–6.3 kJ/mol) per CH_2 group. Because the number of induced dipoles produced in molecules increases as the size of the alkyl group increases, the overall effect becomes significant (Figure 5.4).

The external energy required to overcome dispersion interactions between methane molecules is minimal and the molecule exists in the gas phase under ambient conditions. However, as the number of points where London dispersion interactions can take place are increased with the addition of methylene groups (CH_2), the intermolecular forces increase. As shown in Table 5.4, the melting points and boiling points for linear alkanes increase with molecular weight. The first four alkanes are gases, but with the addition of one more CH_2 group, the intermolecular forces are strong enough so that pentane is a liquid at room temperature and atmospheric pressure. When the chain is extended to seventeen or eighteen carbons, the London dispersion forces are effective in making the molecule a solid at room temperature and atmospheric pressure.

The fact that this increase in boiling point is not completely explained by the increase in the molecular weight is verified by the physical

Figure 5.4   London Dispersion Forces in Pentane

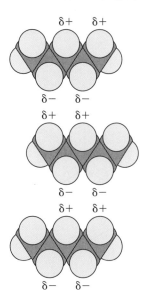

Table 5.4 Physical Properties of the Straight-Chain Alkanes

Alkane	MW g/mol	MP°C	BP°C
CH_4	16	−182.5	−191.7
CH_3CH_3	30	−183.3	−88.6
$CH_3CH_2CH_3$	44	−187.7	−42.1
$CH_3CH_2CH_2CH_3$	58	−138.3	−0.5
$CH_3(CH_2)_3CH_3$*	72	−129.8	36.1
$CH_3(CH_2)_4CH_3$	86	−95.3	68.7
$CH_3(CH_2)_5CH_3$	100	−90.6	98.4
$CH_3(CH_2)_6CH_3$	114	−56.8	125.7
$CH_3(CH_2)_7CH_3$	128	−53.5	150.8
$CH_3(CH_2)_8CH_3$	142	−29.7	174.0
$CH_3(CH_2)_9CH_3$	156	−25.6	195.8
$CH_3(CH_2)_{10}CH_3$	170	−9.6	216.3
$CH_3(CH_2)_{11}CH_3$	184	−5.5	235.4
$CH_3(CH_2)_{12}CH_3$	198	5.9	253.7
$CH_3(CH_2)_{13}CH_3$	212	10	270.6
$CH_3(CH_2)_{14}CH_3$	226	18.2	287
$CH_3(CH_2)_{15}CH_3$	240	22	301.8
$CH_3(CH_2)_{16}CH_3$*	254	28.2	316.1

*Pentane (C_5H_{12}) is a liquid at room temperature and atmospheric pressure and octadecane ($C_{18}H_{38}$) is a solid under these conditions.

Table 5.5 The Physical Properties of the Isomeric Pentanes

	Shape	MW g/mol	BP°C
		72	36
		72	30
		72	10

properties of the isomeric pentanes (Table 5.5). The dependence of the vapor pressure on the kinetic energies of the molecules would be the same. It is the molecular shape of these molecules that affects the vapor pressure. The London dispersion forces are dependent on surface area, which is maximized in the straight-chain five-carbon compound. Increased branching reduces the number of London dispersion interactions possible between neighboring molecules and results in a decrease in the boiling point.

5.3.2 Dipole-Dipole Interactions (D-D)

$$E_{D-D} = -\frac{2\mu^4}{3kTr^6}$$

We saw that when atoms with different electronegativities form a bond, such as H and Cl, a polar covalent bond results and the bond has a permanent dipole. *The Coulombic attraction of two permanent dipoles for one another, a **dipole-dipole interaction (D-D)**, is proportional to the fourth power of the dipole moment of the bond (μ) and inversely proportional to the temperature (T) and the sixth power of the distance between them (r).* The strength of this interaction is approximately 1.2–6.0 kcal/mol (5–25 kJ/mol). Dipole-dipole interactions are the only one of the van der Waals forces that are dependent on temperature. A specific orientation of the bonds is required for the dipoles to interact so that the positive end of the dipole of one bond is lined up with the negative end of the dipole of the bond of another molecule. As the temperature increases, molecular motion increases and the orientation of the dipoles becomes more random, therefore interfering with the required direction of alignment.

Because carbon and oxygen have different electronegativities (Table 5.1), bonds between these two atoms are polar. The interaction of the carbonyl (C=O) dipoles of aldehydes and ketones illustrates these intermolecular forces and the rigid alignment that is necessary (Figure 5.5).

The importance of the D-D interactions can be seen by comparing the boiling points of the alkanes with aldehydes and ketones, which both contain the polar carbonyl group, C=O, as shown in Table 5.6. Because the compounds with polar carbonyl bonds have dipole-dipole and dipole-induced dipole interactions (Section 5.3.3) in addition to the London dispersion forces analogous to those found in alkanes, the boiling points for the aldehydes and ketones are higher than those for alkanes of the same molecular

Figure 5.5 Dipole-Dipole Interactions in Aldehydes and Ketones

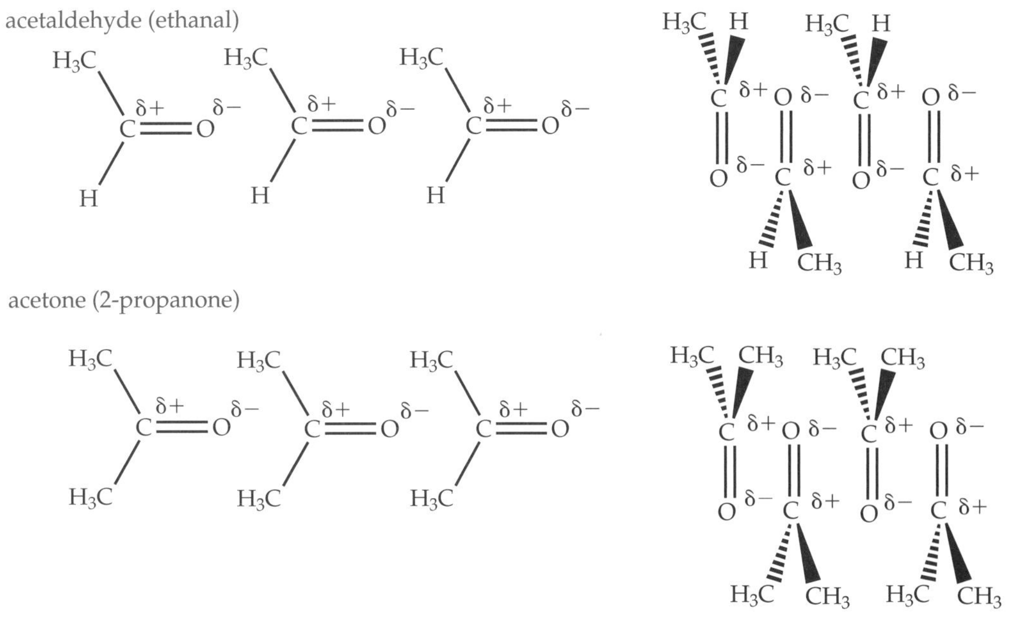

acetaldehyde (ethanal)

acetone (2-propanone)

weight. The difference in the boiling points decreases as the molecular weight increases so the importance of D-D interactions must be more significant in the small molecules and must decrease as the size of the molecule increases. This is reasonable because, as the size of the aldehyde or ketone molecules increases, there are fewer dipoles per unit volume. The probability of two dipoles finding one another and lining up with the positive end adjacent to the negative end therefore decreases and the contribution from the D-D interactions to the overall attractive force decreases. In contrast, the increased size of the alkane portion results in a greater contribution of ID-ID interactions relative to the D-D interactions.

5.3.2.a Hydrogen Bonding (a Special Dipole-Dipole Interaction)

Covalent bonds between hydrogen and nitrogen, oxygen, and fluorine are very polar because of the differences in the electronegativities between these atoms and hydrogen (Table 5.1). In addition, the atom to which hydrogen is attached has lone pairs of electrons available for sharing. These

Table 5.6 Comparison of the Boiling Points of Alkanes with Aldehydes and Ketones

							Difference in BP from Alkane	
MW g/mol	Alkane	BP°C	Aldehyde	BP°C	Ketone	BP°C	Aldehyde	Ketone
44	propane	−42	ethanal	20			62	
58	butane	−0.5	propanal	49	acetone	56	49.5	57
72	pentane	35	butanal	76	2-butanone	80	41	45
86	hexane	69	pentanal	103	2-pentanone	105	34	36
100	heptane	98	hexanal	131	2-hexanone	127	33	29
114	octane	126	heptanal	153	2-heptanone	150	27	24
128	nonane	151	octanal	171	2-octanone	173	20	22

Figure 5.6 A Partial Two-Dimensional Representation of Hydrogen Bonding in Water

| ——— | O—H bond |
| - - - - - | Hydrogen bond |

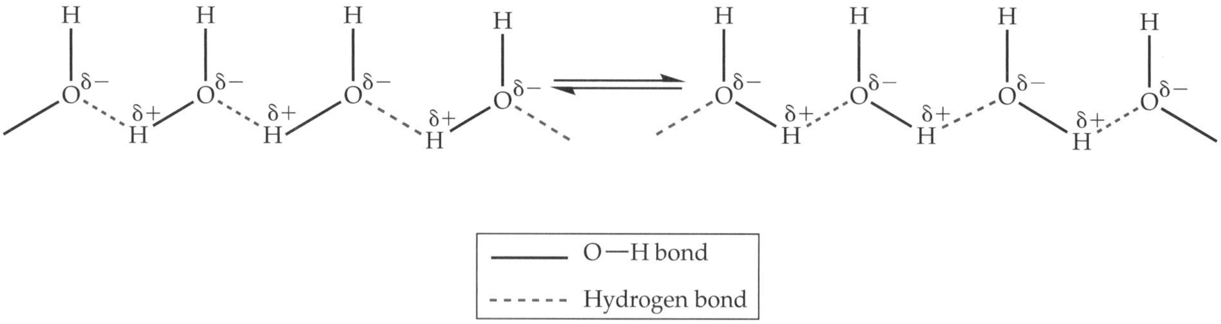

conditions lead to **hydrogen bonding** that involves strong dipole-dipole interactions between the electronegative atom that bears a partial negative charge and the hydrogen that has a partial positive charge. Compounds that contain hydrogen attached to an electronegative atom are called **polar protic molecules**. They include many common solvents and reagents such as water (HOH), alcohols (ROH), amines (RNH_2), and the hydrohalic acids (HX). The strongest hydrogen bonds are between small, very electronegative atoms where the size of the orbitals involved in the bonding are more similar to that of hydrogen, allowing a stronger overlap (N,O,F). Hydrogen bonding results in a complex three-dimensional lattice where the bonds between the hydrogen and the heteroatom are continually interchanging. A partial two-dimensional representation of hydrogen bonding in water is shown in Figure 5.6. This figure illustrates that the hydrogen atoms are able to move easily from one oxygen to another within the lattice. This phenomenon, which is called **proton transfer**, occurs very rapidly.

The strength of each hydrogen bond interaction in water is only about 5–6 kcal/mol (21–35 kJ/mol), but because of their great number, the effect of hydrogen bonding on the properties of water is significant. For example, the boiling point of water is 100°C while the boiling point of methane is −161.7°C. Because carbon and hydrogen have similar electronegativities, the bonds in methane are essentially nonpolar when compared with those in water. In addition, the carbon does not have unshared electrons available to partially bond with hydrogen as are found in the water molecule so hydrogen bonding is not possible in methane.

The importance of hydrogen bonding in nitrogen compounds is illustrated by considering amines that are constitutional isomers and have a molecular weight of 59 g/mol. Because the C–H bonds are not polar, the tertiary amine (trimethylamine) has no hydrogen donor even though it has a lone pair of electrons on the nitrogen (polar aprotic). Its low boiling point reflects the absence of hydrogen bonding in the molecule. In the case of the ethylmethylamine, the nitrogen-hydrogen bond is polar (polar protic). The

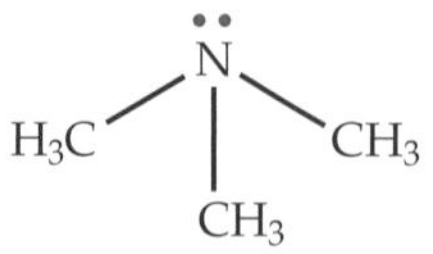

bp 2.9°C

trimethylamine
a tertiary amine
polar aprotic amine

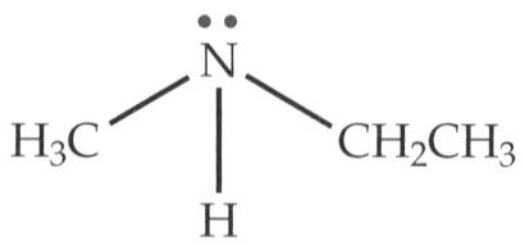

bp 37°C

ethylmethylamine
a secondary amine
polar protic amine

Figure 5.7   Dipole-Dipole Interactions in Propanal Compared with 1-Propanol

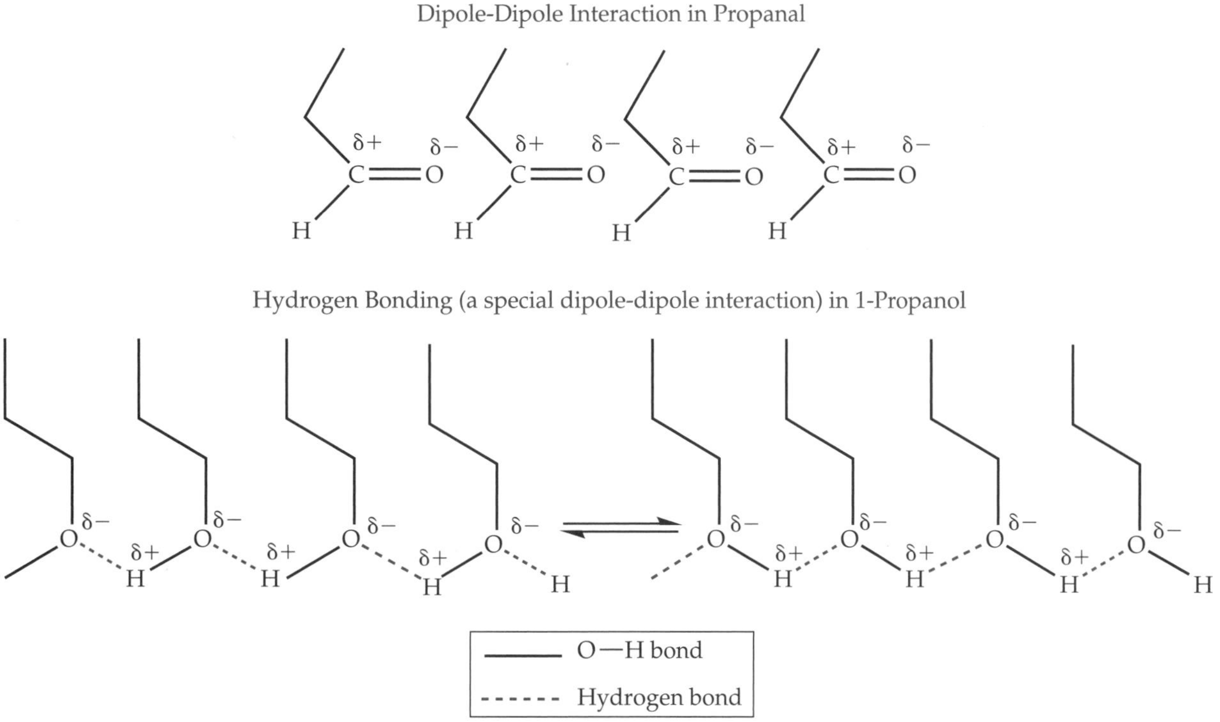

Dipole-Dipole Interaction in Propanal

Hydrogen Bonding (a special dipole-dipole interaction) in 1-Propanol

	O—H bond
- - - - -	Hydrogen bond

presence of the hydrogen donor in the molecule and the lone pair of electrons on nitrogen result in hydrogen-bonding interactions and a dramatic increase in the boiling point.

In propanal (MW 58 g/mol, bp 49°C), the C=O bond is polar and the oxygen has lone pairs. However, the C–H bonds are not polar and consequently there is no hydrogen donor and no hydrogen bonding in propanal. On the other hand, 1-propanol (MW 60 g/mol, bp 97°C) has a polar O–H bond and lone pairs of electrons on the oxygen. Their interaction results in hydrogen bonding in the alcohol that affects the physical properties of the molecule (Figure 5.7.)

The data in Table 5.7 indicate the importance of the strength of hydrogen bonding in polar protic compounds when compared with the dipole-dipole interactions in polar aprotic compounds. Again, the effect of the hydrogen bonding decreases with an increase in the number of CH_2 groups because there

Table 5.7 Comparison of the Boiling Points of Aldehydes and Alcohols

Aldehyde	MW g/mol	BP°C	Alcohol	MW g/mol	BP°C	Difference
ethanal	44	20	ethanol	46	78	57
propanal	58	49	1-propanol	60	97	48
butanal	72	76	1-butanol	74	118	42
pentanal	86	103	1-pentanol	88	138	35
hexanal	100	131	1-hexanol	102	156	25
heptanal	114	153	1-heptanol	116	176	23
octanal	128	171	1-octanol	130	195	24

are less dipoles per unit volume and the induced dipole-induced dipole interactions in the molecules become more significant in contributing to the overall intermolecular forces. Consequently, the boiling point difference between the aldehydes and alcohols decreases with increasing molecular weight.

Because of hydrogen bonding, the boiling points of carboxylic acids are also higher than, for example, aldehydes. Compare ethanoic acid (acetic acid) (MW 60 g/mol and bp 118°C) and propanal (MW 58 g/mol and bp 49°C).

ethanoic acid
MW 60 g/mol
bp 118°C

propanal
MW 58 g/mol
bp 49°C

5.3.3 Dipole-Induced Dipole Interactions (D-ID)

$$E_{D-ID} = -2\,\frac{\alpha\mu^2}{r^6}$$

We have seen that London dispersion forces arising from temporary dipole moments in an atom induce a similar effect in neighboring atoms (Figure 5.2). We saw how these forces explain the attractive forces between atoms of the noble gases. It seems reasonable that a permanent bond dipole in one molecule could induce a temporary dipole moment in a bond of another molecule resulting in a **dipole-induced dipole interaction (D-ID)**. For example, the dipole of the carbonyl group in one molecule of 2-hexanone can interact with a C–H bond of another molecule of 2-hexanone to induce a temporary dipole moment in the C–H bond. The importance of these interactions becomes significant as one considers that both ends of the dipole can interact with the C–H bonds of other molecules in solution and a specific alignment is not required. *The energy associated with such a force depends on the square of the strength of the dipole moment of the bond (μ) and the polarizability (α). It is inversely proportional to the sixth power of the distance (r).* The strength of each interaction is approximately 0.5–2.4 kcal/mol (2–10 kJ/mol).

Unlike D-D interactions, the D-ID interactions do not rely on one dipole finding another, but the number of dipoles relative to the non-polar portion of the molecules decreases as the size of the alkane portion increases. Consequently, the contributions from dipole-induced dipole interactions to the overall intermolecular forces are also reduced as the size of the alkane portion increases. In the aldehydes with higher molecular weights and ketones with larger alkyl groups, the induced dipole-induced dipole interactions that take place in all regions of the molecule become more important in contributing to the overall attractive force and the significance of the other interactions is decreased.

In addition to the three van der Waals forces, we must also consider the effect of ions that result in three additional contributions to intermolecular forces.

5.3.4 Ion-Ion Interactions (I-I)

$$E_{I-I} = k\,\frac{q_1 q_2}{r^2}$$

In the solid state, sodium chloride consists of sodium cations and chloride anions arranged in a crystal lattice in which each of the ions is surrounded by six of the oppositely charged ions. Coulomb's law (Section 5.1) describes the electrostatic force between a positive and a negative ion. Because this force is relatively strong, significant energy is needed to disrupt it. For example, about 188 kcal/mol (788 kJ/mol) is required to pull apart the sodium cation and the chloride anion in the gas phase.[6]

$$NaCl_{(gas)} \longrightarrow Na^{\oplus}_{(gas)} + Cl^{\ominus}_{(gas)} \quad \Delta H^\circ = 188 \text{ kcal n}$$

The strength of the **ion-ion interactions (I-I)** accounts for the physical properties of ionic materials. For example, the melting point of sodium chloride is 801°C and the boiling point is 1413°C. The same is true for the salts of carboxylic acids. Straight-chain saturated acids up to nonanoic are liquids at room temperature and pressure. In contrast, the salts of all of the carboxylic acids are solids at room temperature and atmospheric pressure and their melting points are higher than those of the acids.

acetic acid
mp 16°C

sodium acetate
mp 58°C

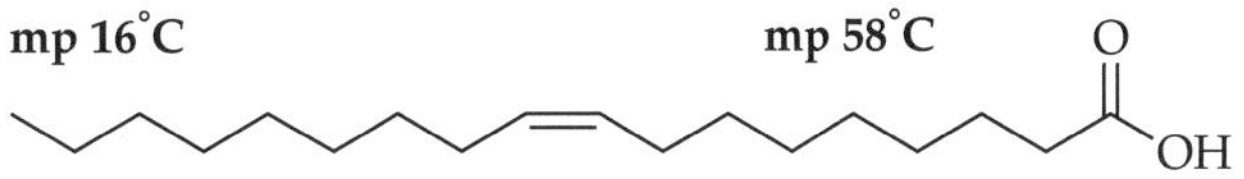

oleic acid
(Z)-9-octadecenoic acid
mp 13.4°C

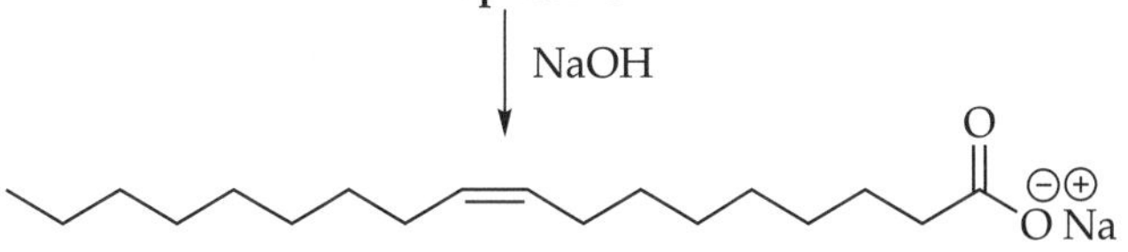

sodium oleate
sodium (Z)-9-octadecenoate
mp 232-235°C

Why are the carboxylic acid salts and sodium chloride so soluble in water? Compare dodecanoic acid with a solubility of 0.0055 g/100mL with sodium dodecanoate with a solubility of 1.2 g/100mL. As described in Section 5.4, there are other intermolecular forces between these salts and the water molecules that must be considered.

5.3.5 Ion-Dipole Interactions (I-D)

$$E_{I-D} = -\frac{q_1 \mu}{r^2}$$

*The intermolecular force between an **ion and a dipole (I-D)** is proportional to the charge on the ion (q_1) and the dipole moment of the bond (μ), and inversely proportional to the square of the distance between them (r^2).* Because the dipole moment of a polar bond is less than a full charge on an ion, each of these interactions is weaker than that between two ions. The strength of one of these interactions is approximately 9.6–143 kcal/mol (40–600 kJ/mol).

5.3.6 Ion-Induced Dipole Interactions (I-ID)

$$E_{I-ID} = -\frac{q^2 \alpha}{2r^4}$$

We would expect the **ion-induced dipole interaction (I-ID)** would be weaker than that between two ions or an ion and a dipole. Note that *the I-ID force is proportional to the polarizability and the square of the charges, but*

---

[6] Recall that the strength of a typical covalent bond in an organic molecule is approximately 100 kcal/mol (418 kJ/mol).

is inversely proportional to the distance between the interacting parts of the molecules to the fourth power, causing the influence to decrease more rapidly with distance than in the case of the ion-ion or ion-dipole force.

5.4 INTERMOLECULAR FORCES BETWEEN DISSIMILAR MOLECULES

So far we have considered pure compounds in which the interacting molecules are identical. It is important to extend this analysis to mixtures of compounds in order to understand the implications of these intermolecular forces on the manipulations we commonly carry out in the lab. In other words, what conditions are required to get different types of molecules to interact with one another or to separate them from one another? When two substances are mixed, the forces between identical molecules are replaced by those between dissimilar molecules. The change in the enthalpy ($\Delta H_{mixture}$) depends on the difference in the strength of the intermolecular forces between identical molecules when compared with those between dissimilar molecules (Equation 5.1).

$$\Delta H_{mixture} = \Delta H_{separation\ A:A} + \Delta H_{separation B:B} + \Delta H_{association\ A:B} \quad (5.1)$$

It takes energy to separate the identical molecules of each compound so $\Delta H_{separation\ A:A}$ and $\Delta H_{separation\ B:B}$ are positive (endothermic). On the other hand, energy is released if there are attractive forces between the dissimilar molecules, that is, $\Delta H_{association\ A:B}$ is negative (exothermic).

If the intermolecular forces between the identical molecules are less than those between the dissimilar molecules, the overall $\Delta H_{mixture}$ is negative.

$\Delta H_{mixture}$ is negative if $\Delta H_{separation\ A:A} + \Delta H_{separation B:B} < \Delta H_{association\ A:B}$

The overall $\Delta H_{mixture}$ is zero if the intermolecular forces between the identical and the dissimilar molecules are the same.

$\Delta H_{mixture} = 0$ if $\Delta H_{separation\ A:A} + \Delta H_{separation B:B} = \Delta H_{association\ A:B}$

The $\Delta H_{mixture}$ is positive (endothermic) if the attractive forces between the two different types of molecules in the mixture, $\Delta H_{association\ A:B'}$ are less than those for the pure substances, $\Delta H_{separation\ A:A} + \Delta H_{separation B:B}$.

$\Delta H_{mixture}$ is positive if $\Delta H_{separation\ A:A} + \Delta H_{separation B:B} > \Delta H_{association\ A:B}$

However, we must also take into consideration the contribution of the entropy to the overall free energy of the system (Equation 5.2).

$$\Delta G_{mixture} = \Delta H_{mixture} - T\Delta S_{mixture} \quad (5.2)$$

Because the entropy (disorder) of the system increases when we mix two dissimilar molecules, the $T\Delta S_{mixture}$ term in the free energy equation is positive. In order for the free energy of the mixture ($\Delta G_{mixture}$) to be negative, the entropy term must be larger than any positive contribution from the change in the enthalpy ($\Delta H_{mixture}$). These principles will become important as we consider solubility, recrystallization, extraction, distillation, and chromatography in subsequent chapters.

Did you know?

A gecko in a glass box can easily scale the smooth sides of the glass and even clamber along the ceiling leaving no trace of sticky residue. It is presently believed that this ability is dependent on the structure of the gecko's foot and van der Waals forces. The foot has about 5000 tiny hairs per square millimeter. The intermolecular forces between these hairs and the surface have been shown to account for the adhesion necessary to keep the gecko attached.

Ref: Autumn, K., et al. *Nature* **2000**, *405*, 681–685, Autumn, K., et al. *Proc. Natl. Acad. Sci. USA* **2002**, *19*, 12252–12256. See http://www.lclark.edu/~autumn/dept/geckostory.html.

5.5 THE DISTINCTION BETWEEN SOLUBILITY AND REACTIVITY

It is important to keep in mind that the intermolecular forces between molecules are weaker than the bonding forces that hold atoms together to form molecules. In a chemical reaction, bonds between atoms are broken and new bonds between different atoms are formed. In the following example, the bond between a carbon and chlorine is broken. The bond between the carbon and iodine is formed. The 2-chloro-2-methylpropane is transformed into 2-iodo-2-methylpropane.

2-chloro-2-methylpropane   2-iodo-2-methylpropane
t-butyl chloride   *t*-butyl iodide

In the acid-base reaction below, a bond between oxygen and hydrogen in acetic acid is broken and a bond between oxygen and hydrogen is formed to make water.

acetic acid   sodium acetate

Many techniques that you use in the lab manipulate only the intermolecular forces between the molecules present to separate the desired material from other components or to get substances to mix. No making or breaking of covalent bonds are involved. No chemical reactions are occurring. For example, no chemical reaction occurs if you mix 2-chloro-2-methylpropane and diethyl ether. There is not a change in the chemical composition of the individual components. You are exchanging the intermolecular forces between the pure components for those present in the mixture.

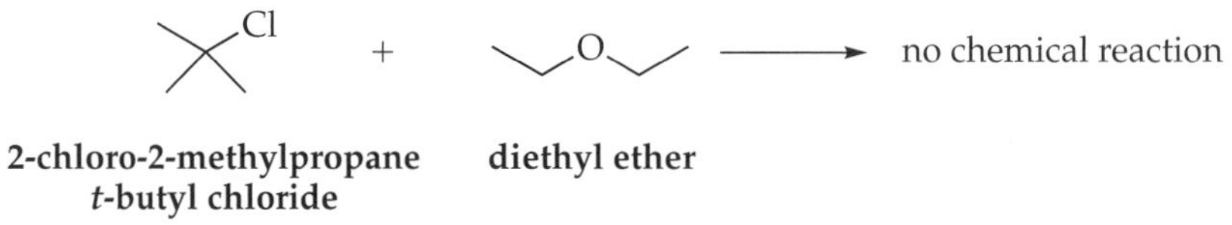

2-chloro-2-methylpropane   **diethyl ether**
t-butyl chloride

The same result occurs if you mix acetone and hexane. The van der Waals forces in the acetone and those in hexane are disrupted and replaced by those between the acetone and the hexane molecules. No bonds in the molecules are broken. You still have acetone molecules and hexane molecules after the substances are mixed.

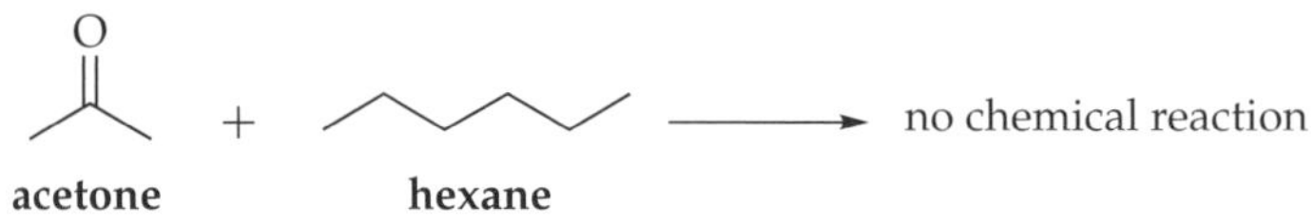

acetone   **hexane**

5.6 PROBLEMS

1. Which of the molecules shown below has a molecular dipole moment? Justify your answer by briefly explaining why the other molecules do not have a molecular dipole moment.

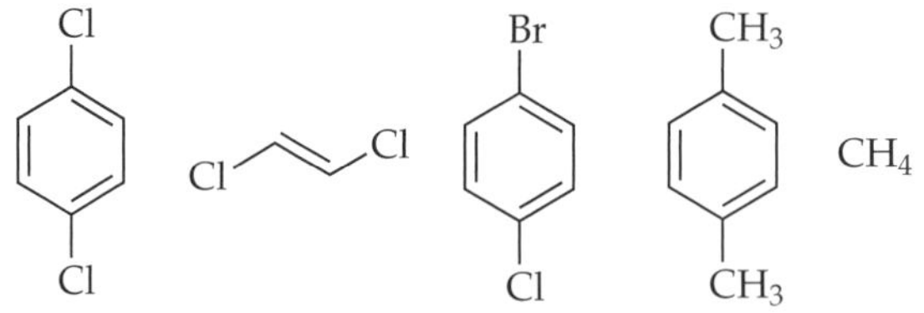

2. The *cis* and *trans* isomers of the dimethyl ester shown below have net molecular dipole moments. The *cis* isomer of 1,2-dinitroethene has a net molecular dipole moment, but the *trans* isomer does not. Explain why the *trans* isomer of the dimethyl ester has a net molecular dipole moment while the analogous isomer of 1,2-dinitroethene does not.

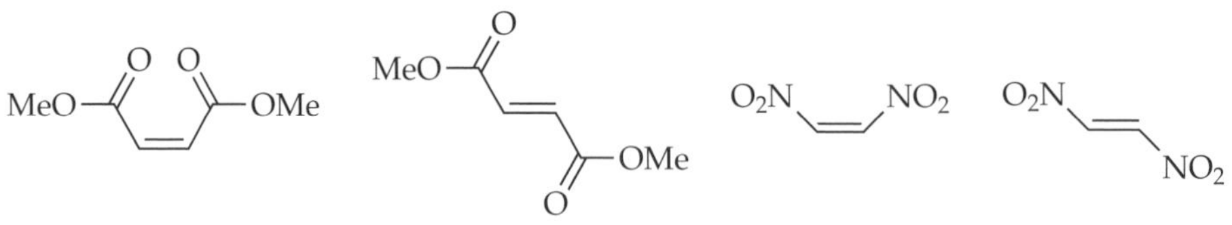

3. Examine the intermolecular forces present in acetic acid and in acetone and explain the differences in their boiling points.

acetic acid   acetone
MW 60 g/mol   MW 58 g/mol
BP 118°C   BP 56°C

4. Because sulfur is larger and more polarizable than oxygen, one might expect that hydrogen sulfide (H_2S) would have a higher boiling point than water (H_2O). Explain why water boils at 100°C while hydrogen sulfide is a gas at room temperature and pressure (bp −60°C).

5. List the intermolecular forces present in Br_2 (159.8 g/mol) and in ICl (MW 162.4 g/mol). How do these forces account for the differences in boiling points (Br_2 bp 58.8°C and ICl bp 97°C)?

6. Compare the boiling points of the straight-chain alkyl chlorides and alkyl bromides using the data in the following table. Consider the effect of molecular weight, dipole moment, and size of the alkyl group on the boiling point. For example, what factors explain the boiling point data if you compare compounds with the same size alkyl group? What forces are important if you compare compounds with similar molecular weights?

	MW g/mol	μ D	BP°C		MW g/mol	μ D	BP°C
CH_3Cl	50	1.89	−24	CH_3Br	95	1.82	4
C_2H_5Cl	64	2.05	12.3	C_2H_5Br	109	2.03	37–40
C_3H_7Cl	79	2.05	46–47	C_3H_7Br	123	2.18	71
C_4H_9Cl	93	2.05	77–78	C_4H_9Br	137	2.08	100–104
$C_5H_{11}Cl$	107	2.16	107–108	$C_5H_{11}Br$	151	2.20	130
$C_6H_{13}Cl$	121		133–134	$C_6H_{13}Br$	165		154–158
$C_7H_{15}Cl$	135		159	$C_7H_{15}Br$	179		178–179

7. Using the data in the following table, explain the difference in the boiling points for the following pairs of compounds.
   a. Dimethyl ether and ethanol
   b. Ethanol and acetonitrile
   c. Ethanol and acetaldehyde
   d. Acetonitrile and acetaldehyde

   Indicate which intermolecular forces predominate.

		MW g/mol	μ D	BP°C
CH_3OCH_3	dimethyl ether	46	1.3	−25
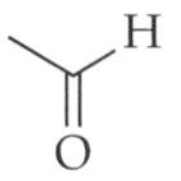	acetaldehyde	44	2.7	21
CH_3CH_2OH	ethanol	46	1.69	79
CH_3CN	acetonitrile	41	3.9	82

8. Explain why the alkenes in the following table have lower boiling points than the aldehydes of comparable molecular weight, but that the effect decreases with increasing molecular weight.

Aldehyde	MW g/mol	BP°C	Alkene	MW g/mol	BP°C
propanal	58	49	1-butene	56	−6
butanal	72	76	1-pentene	70	30
pentanal	86	103	1-hexene	84	63
hexanal	100	128	1-heptene	98	94
heptanal	114	153	1-octene	112	121
octanal	128	171	1-nonene	126	146

9. The following table compares the boiling point data for the straight-chain perfluoro compounds with the hydrocarbons having the same number of carbon atoms. In the case of the five carbon compounds, the boiling point of the perfluoro compound is higher as would be expected because of the significantly higher molecular weight. However, for the six and seven carbon compounds, the boiling points of the hydrocarbons are higher than those of the fluorinated counterparts. Offer a possible explanation for these facts.

	MW g/mol	BP°C		MW g/mol	BP°C
C_5H_{12}	72	36	C_5F_{12}	288	58
C_6H_{14}	86	69	C_6F_{14}	338	57
C_7H_{16}	100	98	C_7F_{16}	388	82

10. Explain the differences in the boiling points for the isomeric hexanes.

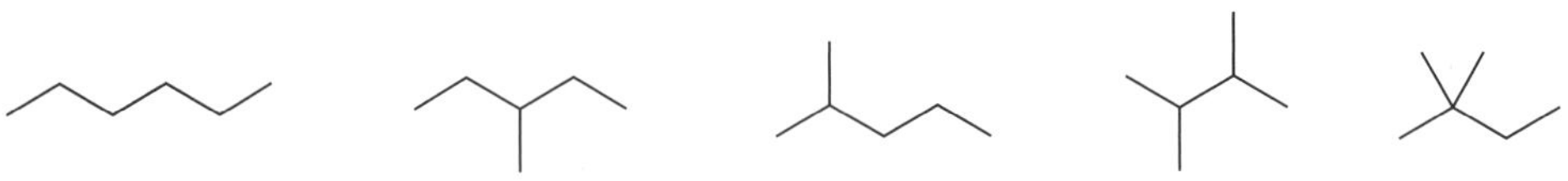

| BP °C | 69 | 63 | 60 | 58 | 50 |

11. On the diagram, indicate which one of the three van der Waals forces best represents the interaction occurring in each of the boxes identified by the letters A, B, C, D: induced dipole-induced dipole (ID-ID), dipole-induced dipole (D-ID), or dipole-dipole (D-D).

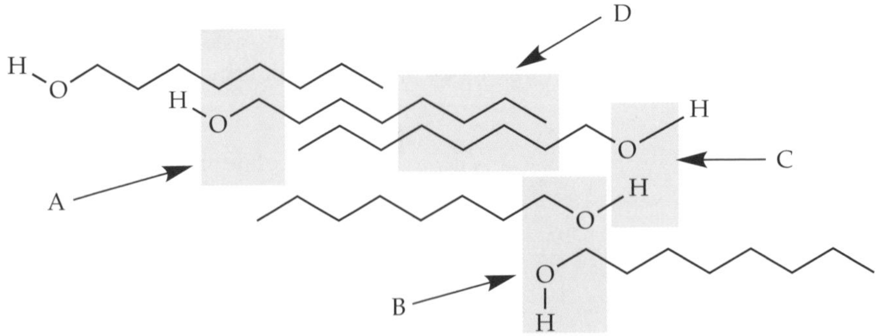

CHAPTER 6

Characteristic Physical Properties of Pure Compounds

The three-dimensional structure of molecules and the intermolecular forces between the molecules of a compound determine its physical properties that are measurable and invariant under specified conditions. Properties such as the molecular weight, density, solubility, melting point, boiling point (at a specified pressure), and specific rotation of a compound have numerical values associated with them that are called physical constants. These constants are characteristic of the pure compound and can be used to identify known compounds and to characterize new compounds. A guide to references that report this data can be found in Chapter 3.

6.1 SOLUBILITY

A **solution** is a mixture of two or more substances such that its composition and properties are uniform throughout. The component present in the lesser amount is called the **solute** whereas the major component is referred to as the **solvent**. Solids, liquids, and gases or any combination can form solutions. For example, air is a solution of gases and soft drinks are solutions of solid or liquid flavorings and carbon dioxide gas in water. In the organic chemistry lab, we normally deal with liquid solvents and either solid or liquid solutes. Two substances that are identified as **miscible** are infinitely soluble in one another in all proportions. When the maximum amount of solute is dissolved in the solvent at any given temperature, we say that the solution is **saturated** with the solute at that temperature. Most substances are not completely insoluble in one another. We measure **solubility** by the amount of solute that dissolves in a given amount of solvent to give a solution. Solubility can be expressed in weight or volume units. In the literature, solubility is also listed as miscible, very soluble, soluble, slightly soluble, and insoluble, but the definitions and limits of these terms are arbitrary and vary with the source.

Two substances form a solution if the free energy associated with mixing them ($\Delta G_{mixture}$) is negative. In other words, a solute is miscible with a

53

solvent if $\Delta G_{mixture}$ is negative and is not miscible (is immiscible) if $\Delta G_{mixture}$ is positive (Equation 6.1).

$$\Delta G_{mixture} = \Delta H_{separation\ A:A} + \Delta H_{separationB:B} + \Delta H_{association\ A:B} - T\Delta S_{mixture} \quad (6.1)$$

As discussed in Section 5.4, energy is required to separate the identical molecules so that $\Delta H_{separation\ A:A}$ and $\Delta H_{separationB:B}$ are positive (endothermic). Energy is released when the dissimilar molecules associate, which means that $\Delta H_{association\ A:B}$ is negative (exothermic). Because the entropy of mixing ($\Delta S_{mixture}$) is positive (increased disorder of the system), mixing is thermodynamically favored if the forces between dissimilar molecules are the same as or stronger than the forces between identical molecules. *When the intermolecular forces between the dissimilar molecules are weaker than those between identical molecules, the entropy term must compensate for the positive enthalpy term in order to achieve a favorable negative free energy of mixing* (Section 5.4).

Note that the entropy contribution ($\Delta S_{mixture}$) is multiplied by the temperature in the equation for the free energy. At higher temperatures, greater positive enthalpy of mixing contributions ($\Delta H_{mixture} = \Delta H_{separation\ A:A} + \Delta H_{separationB:B} + \Delta H_{association\ A:B}$) can be overcome. Recall also that the dipole-dipole intermolecular forces are inversely proportional to the absolute temperature (Section 5.3.2) so that for polar molecules, these intermolecular forces decrease as the temperature is increased. *In most cases, we find that solubility increases with increasing temperature.*

It is not possible to easily quantify these entropy and enthalpy terms, but by analyzing the intermolecular forces present, we can gain an understanding of solubility and formulate some broad generalizations that allow us to predict trends.

For example, all of the van der Waals forces (D-D, D-ID, and ID-ID) contribute to the enthalpy of association of ethanol molecules. Let's examine what happens to the dispersion (induced dipole-induced dipole) forces between the alkyl portions of the ethanol molecules and the hydrogen bonding forces between the polar O-H groups when ethanol is added to water. In a solution of ethanol in water, there is hydrogen bonding between the water and ethanol molecules, but the water molecules interfere with the dispersion forces between the alkyl groups in ethanol. In water molecules, the predominant forces are those from hydrogen bonding. Because there are two hydrogen atoms available for hydrogen bonding, a three dimensional network forms, but the ethanol molecules partially break up this hydrogen bonding network. Any decrease in these intermolecular forces in proceeding from the pure substances to the mixture must be offset by the entropy of mixing. Because it is known that ethanol and water are miscible, the free energy of mixing must be negative (favorable).

Pentane and hexane are also miscible. The induced dipole-induced dipole intermolecular forces between the pentane and hexane molecules are comparable to the forces between the molecules of the pure substances. The entropy term contributes to the negative free energy of mixing.

These examples illustrate the familiar solubility rule: *Like dissolves like*. Substances with similar intermolecular forces dissolve in one another because the intermolecular forces between the identical molecules are replaced by similar intermolecular forces between the dissimilar molecules. Polar molecules dissolve in polar solvents and non-polar substances dissolve in non-polar solvents.

This general rule has its limitations. What happens when we dissolve salt (sodium chloride) in water? The ion-ion intermolecular forces in NaCl are strong as indicated by the high melting point (801°C) and boiling point (1413°C). In addition, there are hydrogen bonding forces (special dipole-dipole interactions) between the water molecules. In the mixture however, these forces are replaced by ion-dipole forces that can partially compensate for the forces in the pure substances. Each Na^+ ion is surrounded by water molecules so that the negative oxygen end of the dipole is pointed toward the cation. Water molecules orient themselves around each Cl^- ion so that the positive hydrogen end of the dipole is toward the anion. We say that the ions are **hydrated**, which means that they are surrounded by water molecules.

Water and sodium chloride are not completely miscible. There is a point at which the entropy of mixing and the intermolecular forces between the water and the ions do not compensate for the loss of the intermolecular forces in the pure substances. The solubility is 1 g of NaCl in 2.8 mL of H_2O at 25°C and 1 g in 2.6 mL at 100°C. The increase in solubility with temperature is expected from the temperature dependence of the entropy contribution to the free energy of mixing (Section 5.4 and Equation 6.1) and the decrease in the strength of dipole-dipole interactions with increasing temperature (Section 5.3.2).

Unlike ethanol, higher molecular weight alcohols are not completely miscible with water. In the pure alcohol, the dispersion forces (ID-ID) between the alkyl portions of the alcohol increase in importance relative to the hydrogen bonding forces as the size of the alkyl group increases (page 40). These induced dipole-induced dipole forces are significantly decreased in the mixture. In addition, the increased size of the alkyl group results in a greater disruption of the hydrogen bonding network of water in the mixture relative to ethanol. The difference in the enthalpy between the unmixed and mixed states is therefore greater than it is in ethanol-water mixtures and the entropy term cannot compensate for this positive enthalpy of mixing. For example, 1-heptanol and water are not completely miscible. The solubility of 1-heptanol is 0.100 g/100 mL of water at 18°C; 0.285 g/ 100 mL at 100°C; and 0.515 g/100 mL at 130°C. This increase in solubility with temperature is due to the increased contribution from the temperature-dependant entropy term (Section 5.4 and Equation 6.1) and the decrease in the strength of the dipole-dipole interactions with increasing temperature (Section 5.3.2). The higher molecular weight straight-chain alcohols, such as 1-octanol, are listed as insoluble in water.

When diethyl ether and water are mixed, an upper layer of ether (density 0.7 g/mL) and a lower layer of water (density 1.0 g/mL) form. Many references state that they are insoluble in one another or immiscible. However, the two substances are not totally insoluble in one another. At 25°C, 7.5 g of diethyl ether dissolves in 100 mL of water and 1.3 g of water dissolves in 100 mL of ether. Looking at the intermolecular forces, this partial solubility is easily understood. Diethyl ether is a polar, aprotic molecule that does not have any partially positive hydrogens and therefore is not capable of hydrogen bonding with itself. It can, however, form hydrogen bonds with water because of the polarity of the carbon-oxygen bond and the partially negative oxygen atom.

$$\delta+ \diagdown \begin{array}{c} \text{O} \\ \delta+ \diagup \end{array} \overset{\delta-}{} \cdots \overset{\delta+}{\text{H}} - \overset{\delta-}{\text{O}} \diagdown \underset{\text{H}}{\overset{\delta+}{}} $$

The loss of the hydrogen bonds between molecules of water and the dispersion forces (ID-ID) and dipole-dipole interactions (D-D) between the diethyl ether molecules is partially compensated for by the formation of hydrogen bonds between the ether and water molecules. Because there is a delicate balance between the change in intermolecular forces and the increase in entropy in forming the mixture, diethyl ether and water are not miscible (infinitely soluble in one another) nor are they totally immiscible.

Pentane is said to be insoluble in or immiscible with water. The London dispersion forces between the pentane molecules are significantly reduced in the mixture as are the hydrogen bonding forces between the water molecules. Pentane is not capable of hydrogen bonding with water so that the only significant intermolecular forces possible in the mixture are dipole-induced dipole and dispersion forces. The entropy of mixing cannot compensate for this change in the strength of the intermolecular forces, therefore pentane and water form two layers when mixed with the less dense pentane (density 0.62 g/mL) forming the upper layer.

6.1.1 Inert vs. Reactive Solvents: Acid-Base Reactions

If we wish to merely dissolve a substance in a solvent, care must be taken to choose a solvent with which the substance does not react. Diethyl ether, hydrocarbons, and chlorinated compounds, such as carbon tetrachloride and dichloromethane, are considered to be **inert solvents** because they do not undergo chemical reactions under the conditions used for dissolving materials in them. Water is an excellent solvent in many cases, but it can react with certain solutes as you will learn as you study chemical reactions. It is important to always consider the reactivity of the solute in the solvent.

There are **reactive solutions** that are deliberately used as solvents in order to test for or to effect solubility. These include, for example, 5% aqueous HCl, 5% aqueous $NaHCO_3$, and 5% aqueous NaOH that react with compounds containing certain functional groups. If a chemical reaction occurs with the solvent, the solubility of the product of the reaction in the solution must be considered. Because the reactions involved are acid-base reactions, the reactivity of a solute is easily predicted from the relative pK_a values of the solute and the solvent. (Refer to the list of pK_a values in Table 4.1.)

Hydrochloric acid is a strong acid (pK_a −2.2). It therefore protonates the **conjugate base** of any acid that has a larger pK_a. For example, dicyclohexylamine is only sparingly soluble in water. The pK_a of its conjugate acid (the protonated form) is 10.4. The amine is therefore protonated by 5% hydrochloric acid to form the amine salt that is soluble in the aqueous solution.

dicyclohexylamine pK_a −2.2 pK_a 10.4

The hydroxide ion is the conjugate base of water, which has a pK_a of 15.7, so that it reacts with acids that have smaller pK_a values. For example, decanoic acid is practically insoluble in water, but is freely soluble in 10% aqueous sodium hydroxide that reacts with it to form the sodium salt of the acid.

decanoic acid
$pK_a \sim 4.8$ pK_a 15.7

A 5% aqueous solution of sodium bicarbonate ($NaHCO_3$) is used to solubilize or test for weak acids. Sodium bicarbonate can act as an acid or a base, so it is important to use the correct pK_a value for the reaction under consideration. Because we are reacting it with a weak acid, the bicarbonate functions as the base and we need to use the pK_a of its conjugate acid, H_2CO_3 or carbonic acid. A solution of $NaHCO_3$ can be used to test for carboxylic acids even if they are soluble in pure water because the chemical reaction that occurs produces carbonic acid that is only slightly soluble in water (10^{-5} M). Carbonic acid decomposes into carbon dioxide and water, and the carbon dioxide gas is observed as bubbles.

pK_a 4–5 pK_a 6.37

Therefore, solubility data can be used to gain information about the functional groups present in a molecule. We know that molecules that are miscible with water are polar, but additional information can be gained by checking the pH of the resulting solution. Many compounds such as alcohols, aldehydes, and ketones result in neutral solutions in water. Low molecular weight carboxylic acids dissolve in water to give acidic solutions and low molecular weight amines result in basic solutions. This simple test can therefore confirm or eliminate the possible presence of functional groups in an unknown solute. Such careful observations in the lab often reveal a great deal of information about the substances with which you are working.

6.2 CHANGES IN PHYSICAL STATE

Recall that interactions between molecules are referred to as **intermolecular forces**. Depending on the strength of these forces, and the temperature and pressure, a compound may exist as a solid, liquid, or gas. For example, we saw that salts such as sodium chloride are solids with high melting points because the molecules are held together by the strong ion-ion interactions. Compounds such as ethanol and acetone exhibit weaker intermolecular forces and exist as liquids at room temperature and pressure. The London dispersion forces between the molecules of low molecular weight alkanes are much weaker and straight-chain alkanes with fewer than five carbon atoms are gases at room temperature and atmospheric pressure (Table 5.4).

As heat is added to a solid, the temperature increases (line AB in Figure 6.1). At point B, the kinetic energy of the molecules becomes

Figure 6.1   The Change in the Temperature of a Compound at Constant Pressure as Heat Is Added

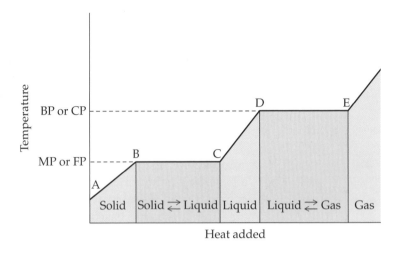

great enough to transform the more ordered structure of the solid into the more random structure of a liquid in which the intermolecular forces are weaker. During this phase change, the heat added is used to disrupt the intermolecular forces between the closely packed molecules so the temperature remains constant (line BC). The solid and liquid phases are in equilibrium at this temperature and the vapor pressures of the solid and liquid are equal. The temperature at which this transformation takes place is called the **melting point** (MP) when the phase change is from solid to liquid or **freezing point** (FP) when the phase change is from liquid to solid. The heat absorbed (melting) or evolved (freezing) per unit mass is called the **heat of fusion**.

After all of the solid has become liquid, the temperature again rises with the addition of heat (line CD in Figure 6.1). At point D, the kinetic energy of the molecules is such that the vapor pressure of the liquid equals the applied pressure. This temperature, called the **boiling point** (BP), remains constant along the line DE as the liquid becomes vapor (boiling) or the gas becomes liquid, the **condensation point** (CP)[1]. The heat required for the liquid to gas phase change (the **heat of vaporization**) is the same as the heat released (the **heat of condensation**) for the liquidization of the gas.

Addition of more heat increases the temperature of the gas (vapor).

The phase of a compound can also be altered by changing the external pressure. A phase diagram (Figure 6.2) can be used to show the relationship between the solid, liquid, and gas states as the external pressure and temperature are altered. The lines show the temperature-pressure combinations at which phase changes occur. The areas represent the temperatures and pressures at which the different phases exist.

Vapor pressure is defined as the pressure exerted when a solid or liquid is in equilibrium with its own vapor at a specific temperature, that is, as many molecules of the solid or liquid are going into the vapor state as there are vapor molecules condensing into the solid or liquid state. If a solid or liquid is placed in a closed container and the container is evacuated, some

---

[1] In the case of water, the condensation point is also called the dew point.

Figure 6.2   A Phase Diagram

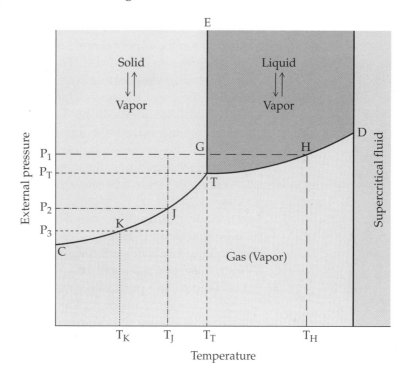

of the solid or liquid vaporizes. As long as some of the solid or liquid is still present, the pressure in the container represents the vapor pressure of the compound at the temperature of the measurement. Vapor pressures vary with temperature but are independent of the external pressure. Remember that the vapor pressure depends on:

- the kinetic energy of the molecules, and therefore is temperature dependent
- the intermolecular forces between the molecules (page 39).

Vapor pressure is independent of external pressure because, for example, as the external pressure increases at a constant temperature, the number of moles in the vapor phase decreases as the volume decreases, so the vapor pressure remains the same (ideal gas law).

SUMMING IT UP!

The vapor pressure of a compound is independent of the external pressure. It depends only on the temperature.

The pressures along line CT in Figure 6.2 are the vapor pressures of the solid at the given temperature. The line represents the pressure-temperature combinations where the phase changes from solid to gas (**sublimation**) or gas to solid (**deposition**). Below line CT, the solid cannot exist. If the temperature of a compound is held constant at T_J and the external pressure is lowered from P_1 to P_3 (Figure 6.2), all of the solid is converted to gas when the external pressure reaches P_2 (point J). Above line CT the vapor pressure of the solid is the pressure at the corresponding temperature along line CT and the solid and vapor exist in equilibrium. For example, the vapor pressure of the compound depicted in Figure 6.2 is P_2 at temperature T_J and external pressure P_1. Because the volume change for the solid-vapor transition is significant, the vapor pressure of the compound changes as the temperature is changed. As illustrated in Figure 6.2, the vapor pressure of the compound is P_3 at T_K (point K).

The pressures along line TD are the vapor pressures of the liquid at the corresponding temperature. The pressure-temperature combinations along this line represent the phase changes for liquid to vapor (**boiling**) and vapor to liquid (**condensation**). The material cannot exist as a liquid in the area below line TD. Above line TD, the vapor pressure of the liquid is the pressure at the corresponding temperature along this line. As in the case of sublimation/deposition, the temperature at which this phase change occurs is pressure dependent because the volume change in going from a liquid to a gas is significant. *The liquid boils when its vapor pressure equals the applied pressure.* In listings of the physical constants for compounds, the value for the boiling point is given as the temperature at which the vapor pressure of the liquid is 1 atm unless a different applied pressure is stated. If P_1 is 1 atm, the vapor pressure of the liquid equals atmospheric pressure at point H and the boiling point of the material is T_H.

At point T, the vapor pressure of the solid equals the vapor pressure of the liquid. Because the solid is in equilibrium with the vapor at this temperature and pressure, and the liquid is also in equilibrium with the vapor at this point, the solid and liquid phases must be in equilibrium. Point T is called the **triple point** and, at pressure P_T and temperature T_T, all three phases, solid, liquid, and vapor exist in equilibrium.

Point D in Figure 6.2 represents the critical point and liquid cannot exist in equilibrium with vapor above this temperature. The phase above this temperature is called a **supercritical fluid** and has properties between a gas and a liquid.

Line TE in Figure 6.2 represents the phase change between the solid and liquid. The vapor pressure of the solid equals the vapor pressure of the liquid along this line and the solid and liquid phases are in equilibrium. The temperature and corresponding pressure along this line represent the melting point (MP) or freezing point (FP), for example point G at external pressure P_1. This transition is relatively independent of pressure because the volume change between the liquid and solid states is usually very small. Although Figure 6.2 shows the melting/freezing point to be constant with changes in pressure, in reality the line TE usually slopes slightly to the right because solids are usually more dense than liquids. An increase in pressure increases the melting point/freezing point. Water is one important exception. Ice is less dense than liquid water, so line TE slopes slightly to the left and the melting point/freezing point decreases with increased pressure.

Melting points, boiling points, and sublimation points are characteristic physical properties of a pure compound. As a result, every compound has a unique phase diagram.

6.3 MELTING POINT RANGE

*The **melting point** of a compound is the temperature at which the solid and liquid are in equilibrium* (line TE in Figure 6.2). Unless stated otherwise, it is assumed that the pressure is 1 atm. Because melting points are physical constants, relatively independent of pressure, and easily measured on very small amounts of material, they are extremely useful in characterizing and identifying compounds. References such as the *CRC Handbook of Chemistry and Physics* (Chapter 3) have indices of melting points that give a list of compounds melting at specified temperatures. In spite of the fact that many compounds have the same melting point range, the melting point can be used to confirm the identity of a compound and to assess its purity because impurities lower the melting point of a solid.

6.3.1 The Effect of Impurities on the Melting Point Range

In the solid phase, a compound and any impurities act as if they are distinct. The impurities do not affect the vapor pressure of the solid, but they affect the melting point range depending upon their effect on the vapor pressure of the liquid.

6.3.1.a Impurities That Are Insoluble in the Liquid Phase

Impurities that are not soluble in the liquid do not form a solution with the liquid compound. They do not affect the intermolecular forces between the molecules of the liquid and therefore they do not affect the vapor pressure of the liquid. *Insoluble impurities have no effect on the melting point/freezing point of the compound.* For example, sand has no effect on the vapor pressure of water and therefore has no effect on its melting/freezing point.

6.3.1.b Impurities That Are Soluble in the Liquid Phase

Impurities that are soluble in the pure liquid compound affect the intermolecular forces between the molecules of the liquid. Soluble impurities contribute to the total vapor pressure of the solution. As long as the impurities (B) form a dilute solution in the liquid compound (A) or the intermolecular interactions between the molecules of the pure compound (A and A) and the molecules of the impurity (B and B) are the same as the interactions between the molecules of the dissimilar compounds (A and B), the solution is said to be ideal and to be described by Raoult's law.

S U M M I N G   I T   U P !

An **ideal solution** is one in which the interactions between the molecules of the same compound (A and A or B and B) are the same as those between the molecules of the dissimilar compounds (A and B). In other words, $\Delta H_{mixture} = 0$ (Equation 5.1).

S U M M I N G   I T   U P !

Raoult's law states that the equilibrium vapor pressure of a liquid compound over a dilute or ideal solution is equal to the mole fraction of the compound in the solution times the vapor pressure of the pure liquid compound: $P_A = X_A P_A^\circ$.

In the presence of soluble impurities, the mole fraction of the compound (X_A) in the liquid solution is decreased ($X_A < 1$) and consequently the vapor pressure of the liquid (A) above the solution is reduced ($P_A < P_A^\circ$). Remember that impurities do not affect the vapor pressure of the solid. In Figure 6.3, line CT represents the vapor pressure of the pure solid as a function of temperature. Line TD is the vapor pressure of the pure liquid as a function of temperature. At point T, the vapor pressure of the solid equals the vapor pressure of the liquid, so temperature T_T

Figure 6.3 The Effect of Soluble Impurities on the Vapor Pressure of a Compound

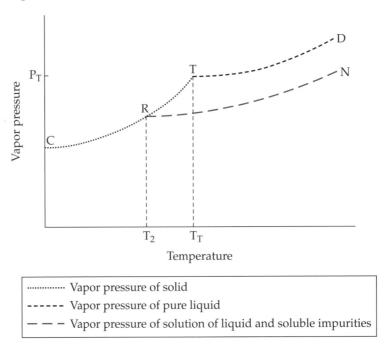

.... Vapor pressure of solid

- - - - - Vapor pressure of pure liquid

— — — Vapor pressure of solution of liquid and soluble impurities

is the triple point where the solid, liquid, and gas phases are in equilibrium. Line RN is the vapor pressure of the liquid solution of the pure compound and the soluble impurities. The temperature at which the solid, liquid solution, and vapor phases are in equilibrium is now at point R. The temperature of the triple point where the impure solid, the liquid solution, and the vapor phases are in equilibrium is decreased to T_2.

Let's look at the phase diagram (Figure 6.4), which shows the phases that exist as a function of external pressure and temperature, to determine the effect of the lowering of the triple point on the melting/freezing point of the impure compound.

The melting/freezing point is the temperature at which the solid and liquid are in equilibrium at 1 atm pressure unless stated otherwise. For the pure compound, this happens at point P, so the melting point of the pure substance is T_P. Because the vapor pressure of the liquid is decreased in the presence of an impurity that is soluble in the liquid phase, the temperature at which the impure solid and the liquid solution are in equilibrium at a pressure of 1 atm is now at point M. The melting point of the impure compound is decreased to temperature T_M. *In other words, impurities that are soluble in the liquid phase lower the melting point of a compound.*

If the amount of impurity is increased, the vapor pressure of the liquid is lowered even further and therefore the melting point is lowered more. For example, when salt is added to water, the freezing point of the solution is lowered by an amount that increases with the amount of salt added.

Figure 6.4 **The Effect of Soluble Impurities on the Melting Point of a Compound**

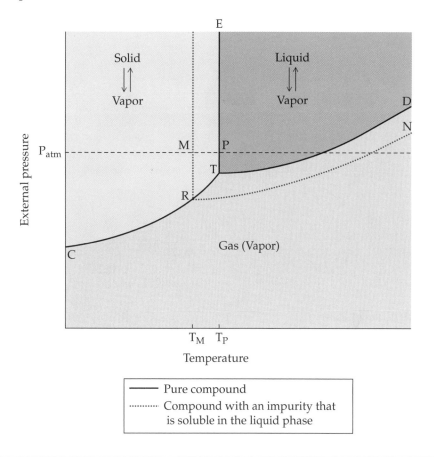

Did you know?

- Salt and sand are both added to roadways in some parts of the country during cold weather to improve driving conditions. Applying salt to cold, wet roads is a deliberate attempt to stop the water from freezing. Because salt is soluble in water, it reduces the vapor pressure and lowers the freezing point. Sand on the other hand is not soluble in water and has no effect on its freezing point. It is simply added to roadways to improve traction.

- Glass made from pure silica (SiO_2) is not common due to the high cost associated with achieving the temperatures necessary to melt the silica (mp 1723°C) so it can be poured. Therefore, most glass found in ordinary items, such as windows and drinking glasses, is made from a mixture of silica and other additives, including sodium, calcium, and magnesium oxides. By adding these soluble impurities, the melting point of silica is reduced dramatically. The resulting glass is more affordable because of the savings in energy associated with melting the mixture.

- A variety of reptiles, amphibians, fish, and insects are capable of surviving in temperatures at which their blood should be freezing. One mechanism for survival is to produce large quantities of water-soluble organic molecules as the ambient temperature drops. Soluble molecules, such as glucose and glycerol, depress the freezing point of the blood and prevent the killer effects of "blood slush."

Figure 6.5   Melting Point – Composition Diagram

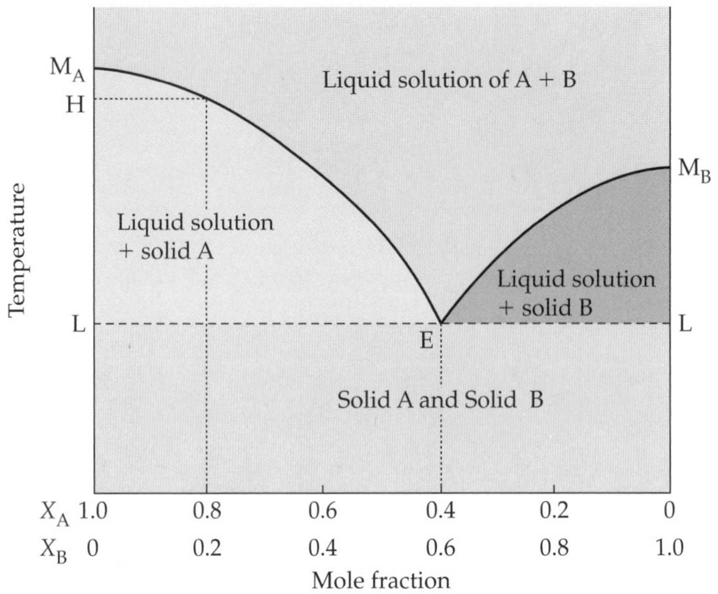

6.3.2 Mixed Melting Points

Figure 6.5 shows the simplest effect on the melting/freezing point when a mixture of two compounds form a homogeneous liquid solution. Pure compound A melts at temperature M_A and pure compound B melts at temperature M_B.

If a liquid solution of composition $X_A = 0.8$ and $X_B = 0.2$ (above curve $M_A - E - M_B$) is cooled, at temperature H solid A is in equilibrium with liquid A (the definition of the melting/freezing point). Therefore at temperature H, $P_{A\,solid} = P_{A\,liquid} = 0.8\,P_A^\circ$ and crystals of A form. The vapor pressure of $P_{B\,solid} \neq P_{B\,liquid}$ and compound B remains in the solution.

As compound A continues to freeze, the mole fraction of A in the liquid solution decreases resulting in a decrease in the vapor pressure of liquid A ($P_{A\,liquid} = X_A\,P_A^\circ$). As a result, the temperature at which the liquid and solid are in equilibrium decreases (Figures 6.3 and 6.4), that is, the melting/freezing point decreases. Solid A continues to crystallize out and the temperature of the mixture continues to decrease with cooling along the curve $M_A - E$ until the composition of the liquid solution for the example in Figure 6.5 is at $X_A = 0.4$ and $X_B = 0.6$ (point E).

At point E and temperature L, the mole fraction of B in the solution has increased, so solid B is now in equilibrium with liquid B ($P_{B\,solid} = P_{B\,liquid} = 0.6\,P_B^\circ$) and B freezes. Because both compound A and compound B are freezing at point E, the composition in the liquid solution stays constant at $X_A = 0.4$ and $X_B = 0.6$. The temperature remains constant at L until all of the material is in the solid state.

Point E, which is called a **eutectic**[2], divides the area where A crystallizes out of a liquid mixture of A and B (area $M_A - L - E$) from the area where B crystallizes out of the mixture (area $M_B - L - E$). At the eutectic point E, $P_{A\,solid} = P_{A\,liquid} = X_{A\,eutectic}\,P_A^\circ$ and $P_{B\,solid} = P_{B\,liquid} = X_{B\,eutectic}\,P_B^\circ$ so that compounds A and B both freeze or melt in a constant ratio called the **eutectic mixture**. A eutectic mixture has a sharp melting/freezing point at temperature L. Eutectics can occur at any ratio between two compounds. It

---

[2] **Eutectic** comes from the Greek word *eutēkos*, meaning easily melted.

should be noted, however, that Figure 6.5 shows a simple case. Some mixtures show more complex behavior in which there is no eutectic or there is more than one eutectic.

Melting is the reverse of freezing, so a solid composed of $X_A = 0.8$ and $X_B = 0.2$ starts to melt at temperature L where $P_{A\ solid} = P_{A\ liquid} = 0.4\,P_A^\circ$ and $P_{B\ solid} = P_{B\ liquid} = 0.6\,P_B^\circ$ (the eutectic mixture). As heat is added, the temperature remains constant at L and compounds A and B continue to melt in the ratio of 4:6 until all of B is in the liquid solution. As more heat is added, compound A melts and the liquid becomes richer in component A (less of the B impurity). As the mole fraction of compound A in the solution increases, the vapor pressure of A increases (Raoult's law); consequently the temperature at which solid A and liquid A are in equilibrium increases (Figures 6.3 and 6.4). At temperature H, the last crystals of A melt to give the liquid solution of composition $X_A = 0.8$ and $X_B = 0.2$, the composition of the original mixture.

Using the melting point equipment commonly found in most laboratories, it is difficult to see the first crystals melt. Therefore, for the example of the 8:2 mixture shown in Figure 6.5, the complete temperature range from L to H is not observed. The beginning of the observed range is always greater than L, but with some practice, the temperature at which the last crystals melt (H) can be more accurately detected. As the purity of the compound increases, the observable melting point temperature increases and the melting point range decreases.

A melting point is always reported as a range of temperatures. The beginning of melting is the appearance of the first visible drops of liquid and the end of melting is the disappearance of the last crystalline material. Even for a very pure compound, the limitations of the equipment are such that a melting point range is observed. Usually a compound with a melting point range of less than 2°C is considered pure. Of course it is possible that the sharp melting point corresponds to a eutectic mixture.

SUMMING IT UP!

A sharp melting point (<2°) indicates a pure compound or a eutectic mixture.

A depressed and broad melting point is an indication that the compound is wet, contains soluble impurities, or is decomposing to form impurities when it is heated. In rare cases, compounds crystallize in more than one crystalline form and a mixture of these forms shows a melting point that is depressed and broad.

SUMMING IT UP!

A depressed and broad melting point can result if:

- the compound is wet
- the compound contains soluble impurities
- the compound decomposes during heating
- the compound crystallized in more than one crystalline form.

Melting points are useful in determining the identity of a compound. Compounds with very different melting point ranges can be excluded as possibilities. However, many compounds have very similar melting point ranges. A **mixed melting point** can be used to determine if compounds with similar melting point ranges are indeed identical. It is best to take the melting point of at least two mixtures of different compositions to avoid the possibility of hitting a eutectic mixture. In this way, one can determine if two samples are identical (no broadening or depression of the melting point range is observed) or different (a broadened and depressed melting point range of the mixture results).

6.3.2.a The Melting Point of Chiral Molecules

Chiral molecules show a unique melting point behavior. Pure crystals of the *R* and *S* enantiomers of an optically active compound and the liquid that forms when they melt have the same vapor pressure and consequently the same melting point range. The intermolecular forces between the molecules of the pure *R* enantiomers are identical to those between molecules of the pure *S* enantiomers because they differ only in that they are mirror images of one another. However, if one mixes crystals of the *S* enantiomer with a pure sample of the *R* enantiomer, the mixture behaves like a compound that has a soluble impurity. The intermolecular forces between the molecules of the *S* enantiomer and its mirror image, the *R* enantiomer, are not the same as those between the pure *R* molecules. Because the *S* enantiomer is soluble in liquid *R*, the melting point range of the mixture is depressed and broadened from that of pure *R*. When the composition of the mixture is 50% *R* and 50% *S*, a minimum, or eutectic, is observed and the melting point is sharp and lower than that of the pure enantiomers. The 50–50 mixture is called a **racemic**[3] **mixture**.

Did you know?

RACEMIC MIXTURE VS RACEMIC COMPOUND
When he was 26 years old, Louis Pasteur was fortunate when he chose tartaric acid to study the shapes of organic crystals. Large quantities of the compound were found in the sediments of fermenting wine. It puzzled Pasteur that two different compounds with identical compositions were found in these sediments. They differed only in that a solution of one of the compounds rotated a beam of polarized light to the right while a solution of the other did not affect polarized light. Pasteur did not accept the belief of other scientists that the two compounds were identical. He examined the crystals of the sodium ammonium salts of both compounds under the microscope. The crystals of the compound that rotated polarized light were all identical. However, the crystals of the compound that did not affect polarized light consisted of two types that were mirror images of one another. Pasteur separated these crystals into two piles and found that one was indeed identical to the compound isolated from wine sediments and its solution rotated the beam of polarized light to the right. The solution of the other type rotated the beam of polarized light to the left by an equal amount. In this way, Pasteur showed that the rotation of polarized light is a property of the molecule. His studies proved that some molecules are asymmetric and exist in two forms that are mirror images of one another.

Louis Pasteur (1822–1895)

© SPL/Photo Researchers, Inc.

---

[3] Racemic comes from the Latin word *racemus* meaning cluster of grapes.

"Did you ever observe to whom the accidents happen? Chance favors only the prepared mind."

—Louis Pasteur

His discovery can be attributed not only to his meticulous observations, but also to chance. Below 26°C, a solution of the sodium ammonium salts of an equimolar mixture of the *R*- and *S*-tartaric acid enantiomers separates into a mixture of the right- and left-handed crystals.

tartaric acid enantiomers

The crystals obtained are referred to as a **racemic mixture**. This is the same as the 50–50 eutectic mixture of the *R*- and *S*-enantiomers and the melting point of the racemic mixture is sharp and always lower than that of either of the pure enantiomers. Above 26°C, the crystals that form are called a **racemic compound** and are a combination of equal numbers of the *R*- and *S*-forms in each crystal that cannot be separated physically. Indeed most equimolar mixtures of the *R*- and *S*-forms of optically active compounds form racemic compounds when crystallized from a solution. The intermolecular forces between the molecules in the racemic compound are different from those in the pure enantiomers and the melting point of the racemic compound can be higher, lower, or the same as that of the pure enantiomers.[4] A racemic compound behaves like a pure compound and has a sharp melting point. The melting point diagram of mixtures of the racemic compound with either of the pure enantiomers shows a eutectic that is not necessarily at the 50–50% composition.

6.3.3 Determining the Melting Point of a Compound

Many different types of instruments are available to determine melting points. Most accept a capillary tube with one end sealed. Finely ground crystals of the solid are introduced into the capillary by pressing the open end into the solid. This is a situation where more is not better because greater quantities of material take more time to melt and therefore give a larger and thus less accurate melting point range. The sample should be just large enough to see, about 1 to 2 mm in the tube. Tight packing of the crystals maximizes heat transfer between the molecules. To accomplish this, the capillary tube should be bounced by dropping it through a drinking straw that is held upright with one end pressed against the countertop. The resulting vibrations move the crystals down to the bottom of the capillary. (Holding the capillary between your fingers and tapping it on the bench rarely does any good because the pressure from the fingers on the glass dampens these vibrations.) The sample should be heated at a rate of about 1 to 2°C per minute once the temperature is close to the melting point so that the thermometer reflects the temperature of the sample. If a second melting point measurement of a given compound is necessary, a new sample should be prepared because the compound may have decomposed during the heating process. Sometimes it is difficult to determine the beginning of melting

---

[4] Mitchell, Alan G. J. Pharm. Pharm. Sci. 1998, **1**, *8*.

because the sample may contract and pull away from the sides of the capillary. Keep in mind that the lower temperature of the melting point range is the temperature at which the first drops of liquid are visible and the upper temperature is the temperature at which the last crystals of solid disappear.

To obtain a mixed melting point, a small amount of the two compounds to be tested is put on a watch glass and the crystals are crushed and mixed with a solid glass rod or a stainless steel spatula. This well-mixed solid is then used for the melting point.

Did you know?

Melting points reported in the literature are usually reported as uncorrected. This means that the thermometer was not calibrated using known samples before the data was collected. Although this may not seem accurate, most good thermometers are rarely off by more than 1°C

6.4 BOILING POINT

*The **boiling point** of a compound is the temperature at which the vapor pressure of the liquid equals the applied pressure (normally 1 atm)* (Section 6.2). It is a characteristic physical constant of the compound and pure samples of the compound exhibit a sharp boiling point. Therefore, like the melting point, the boiling point can be used to identify and characterize materials. However, the boiling point is not as useful as the melting point for this purpose because a larger amount of material is required to obtain a boiling point and its dependence on the pressure must be taken into consideration when comparing literature values.

6.4.1 The Effect of Impurities on the Boiling Point

The effect of an impurity on the boiling point of a liquid varies with the characteristics of the impurity, depending upon its solubility in the liquid and its volatility.

6.4.1.a Insoluble, Non-Volatile Impurities

An insoluble, non-volatile impurity does not affect the intermolecular forces between the molecules of the liquid and does not alter the vapor pressure of the liquid. The boiling point of the mixture is the same as the boiling point of the pure liquid at the given pressure. For example, sand does not affect the boiling point of water.

6.4.1.b Soluble, Non-Volatile Impurities

An impurity that is not volatile, but is soluble in the liquid affects the intermolecular forces and lowers the vapor pressure of the liquid above the solution (Figure 6.3), resulting in an increase in the boiling point of the impure liquid. For this discussion, it is assumed that the solution is either dilute or ideal (page 61) so that Raoult's law (page 61) can be applied.

In Figure 6.6, at atmospheric pressure the boiling point of the pure compound is T_J. In the presence of a soluble, non-volatile impurity, the vapor pressure of the liquid solution is less than that of the pure liquid (according to Raoult's law assuming a dilute or ideal solution). At temperature $T_{J'}$ the vapor

Figure 6.6   The Effect of a Soluble, Non-Volatile Impurity on the Vapor Pressure of a Liquid

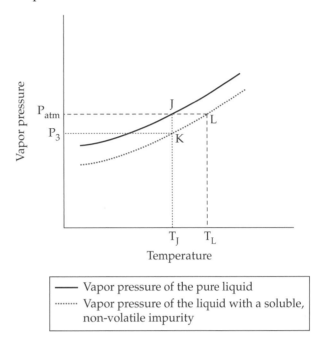

	Vapor pressure of the pure liquid
	Vapor pressure of the liquid with a soluble, non-volatile impurity

pressure of the solution at point K is less than atmospheric pressure and the solution does not boil. The vapor pressure of the impure liquid solution is equal to atmospheric pressure at point L, so that the boiling point of the liquid solution of the pure compound and the soluble, non-volatile impurity is T_L.

If you measure the temperature of the impure liquid when it is boiling, the thermometer registers temperature T_L. However, because the impurity is not volatile, the vapors above the solution are composed of molecules of the pure liquid. The **reflux ring** is that area above a solution where these vapors recondense to liquid and where the vapor is in equilibrium with the pure liquid. The temperature within the reflux ring reflects the boiling point of the pure liquid, T_J. The temperature of the solution and that of the pure liquid in equilibrium with the vapor above the solution within the reflux ring are not the same.

Figure 6.7 shows the phase diagram that illustrates this point. The temperature at which the liquid phase and the vapor phase are in equilibrium at atmospheric pressure is increased from point J to point L in the presence of non-volatile, soluble impurities. Remember that line TD is the vapor pressure of the pure liquid and line RN is the vapor pressure of the liquid solution containing soluble, non-volatile impurities (Figure 6.6).

Did you know?

Why is salt added to water when cooking pasta and vegetables? Salt is a soluble, non-volatile impurity. The boiling point of water is elevated 0.512°C when 58.44 g of sodium chloride is added to 1000 g of water. Adding 1 tablespoon (20 g) of salt to 4 quarts (3785 g) of water would raise the boiling point only 0.046°C, which is not enough to make a difference in cooking time. Salt is added to water when cooking pasta and vegetables for taste only. It does not raise the temperature of the liquid significantly.

Figure 6.7   Phase Diagram Showing the Effect of a Soluble, Non-Volatile Impurity on the Boiling Point of a Compound

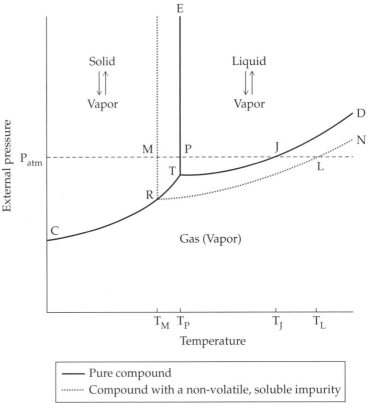

From Figure 6.7, it is apparent that *a soluble, non-volatile impurity lowers the melting/freezing point of a compound (Section 6.3.1.b) and raises its boiling point.*

Did you know?

Why is it important to add antifreeze (1,2-ethanediol) to the water in the radiator of a car in both summer and winter? The boiling point of 1,2-ethanediol is 196–198°C and it can therefore be treated as a non-volatile, soluble additive to the water. As such, it lowers the freezing point of the water to protect the radiator in winter and also raises the boiling point to help prevent boil-over problems in summer.

6.4.1.c Soluble, Volatile Impurities

The boiling point behavior of a liquid with a soluble, volatile impurity depends on the difference in the boiling points of the liquid and the impurity. It is easiest to understand the behavior of mixtures of soluble, volatile components if we examine the dependence of the vapor pressure on the composition of the mixture at constant temperature. Both components of a two-component mixture contribute to the vapor pressure of the solution. Dalton's law and Raoult's law apply to dilute or ideal solutions[5] and relate the vapor pressure of these solutions to the mole fraction of the components in the solution.

---

[5] See page 61 for the definition of an ideal solution.

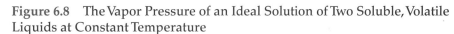

SUMMING IT UP!

Dalton's law: The total pressure of a gas or the vapor pressure of a liquid is the sum of the partial pressures of its individual components, $P_t = P_a + P_b$.

Raoult's law: The partial pressure of a compound in a dilute or ideal solution is equal to the vapor pressure of the pure compound times its mole fraction in the solution, $P_a = X_a P_a^\circ$.

If we look at a two-component system from the perspective of Dalton's law, the total vapor pressure of the solution is the sum of the partial pressures of the two components. Using Raoult's law where $X_{a\,liq}$ and $X_{b\,liq}$ are the mole fractions in the solution and P_a° and P_b° are the vapor pressures of the pure components, the total vapor pressure of the solution is given by Equation 6.2.

$$
\begin{aligned}
P_{total} &= P_{a\,liq} + P_{b\,liq}\\
&= X_{a\,liq}\,P_a^\circ + X_{b\,liq}\,P_b^\circ\\
&= X_{a\,liq}\,P_a^\circ + (1 - X_{a\,liq})P_b^\circ\\
&= P_b^\circ + X_{a\,liq}(P_a^\circ - P_b^\circ).
\end{aligned}
\tag{6.2}
$$

The total vapor pressure of the ideal solution is therefore a linear function of the mole fraction of either component as shown in Figure 6.8.

The composition of the vapor is not the same as the composition of the solution with which it is in equilibrium. The ideal gas law gives us Equation 6.3 for compound A in the gas phase (vapor).

$$
P_{a\,vapor} = n_a \frac{RT}{V}
\tag{6.3}
$$

John Dalton (1766–1844) The law of partial pressures was first stated by Dalton in 1802 and published in 1808.

François-Marie Raoult (1830–1901) His first paper on the depression of the freezing points of liquids by soluble impurities was published in 1878.

Figure 6.8 **The Vapor Pressure of an Ideal Solution of Two Soluble, Volatile Liquids at Constant Temperature**

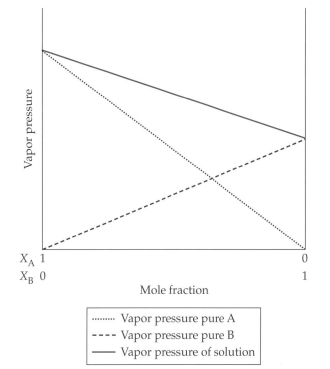

From Equation 6.3 and Dalton's law of partial pressures for a mixture of two gases A and B, we obtain Equation 6.4.

$$P_{total} = (n_a + n_b)\frac{RT}{V} \tag{6.4}$$

By dividing Equation 6.3 by Equation 6.4, we obtain Equation 6.5.

$$\frac{P_{a\,vapor}}{P_{total}} = \frac{n_a}{n_a + n_b} = X_{a\,vapor} \tag{6.5}$$

where $X_{a\,vapor}$ is the mole fraction of compound A in the gas phase (vapor). This is simplified to give Equation 6.6 for the partial pressure of compound A in the gas phase.

$$P_{a\,vapor} = X_{a\,vapor}P_{total} \tag{6.6}$$

Recall that the vapor pressure of A in the solution is given by Raoult's law (Equation 6.7)

$$P_{a\,liq} = X_{a\,liq}P_a^{\circ} \tag{6.7}$$

Because the system is at equilibrium, $P_{a\,vapor} = P_{a\,liq}$, so by combining Equations 6.6 and 6.7, we obtain Equation 6.8.

$$X_{a\,vapor}P_{total} = X_{a\,liq}P_a^{\circ} \tag{6.8}$$

Similarly the partial pressure of compound B in the gas phase (vapor) is $P_{b\,vapor} = X_{b\,vapor}P_{total}$ and the vapor pressure of B in the solution is $P_{b\,liq} = X_{b\,liq}P_b^{\circ}$. Equation 6.9 is valid for the system at equilibrium.

$$X_{b\,vapor}P_{total} = X_{b\,liq}P_b^{\circ} \tag{6.9}$$

If we take the ratio of these two expressions, we obtain Equation 6.10.

$$\frac{X_{a\,vapor}}{X_{b\,vapor}} = \frac{X_{a\,liquid}}{X_{b\,liquid}}\frac{P_a^{\circ}}{P_b^{\circ}} \tag{6.10}$$

The ratio of the mole fractions of the components in the vapor phase is not the same as the ratio in the liquid phase. If the vapor pressure of pure A is greater than the vapor pressure of pure B, then $P_a^{\circ} / P_b^{\circ}$ is greater than 1 and $\frac{X_{a\,vapor}}{X_{b\,vapor}} > \frac{X_{a\,liquid}}{X_{b\,liquid}}$. The vapor is enriched in compound A, which is the component with the higher vapor pressure and lower boiling point.

In Figure 6.9, compound A has a higher vapor pressure (lower boiling point) than compound B. The horizontal lines connect the composition of the liquid and the composition of the vapor that are in equilibrium. The vapor is enriched in the more volatile component. The vertical lines connect the vapor and liquid solution that have the same composition, that is, the liquid that would result from the condensation of the vapor. One vaporization-condensation cycle is called a **theoretical plate**. For example, the vapor from the liquid solution of composition L is enriched in compound A and has the composition at point V. If this vapor is condensed, the composition of the resulting liquid (point M) is the same as that of the vapor. A second vaporization of this liquid (point M) gives a vapor with composition W that has the same composition as the liquid solution at N. In this example, the original solution at L had a composition of 20% A and 80% B. As shown in Figure 6.9, after four vaporization-condensation cycles

Figure 6.9   Vapor Pressure-Composition Diagram for a Solution of Two Soluble, Volatile Compounds at Constant Temperature

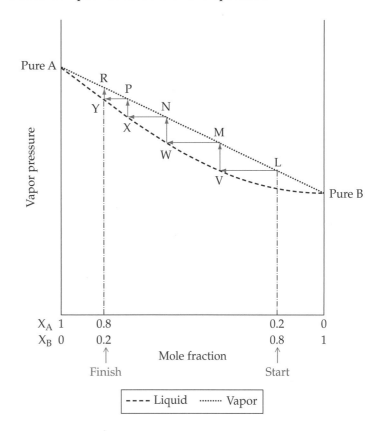

(theoretical plates), the resulting recondensed liquid at R is 80% A (the compound with the lower boiling point) and 20% B.

From a practical standpoint, boiling points are measured at constant pressure and the temperature of the recondensed liquid in equilibrium with the vapor is measured. The relationship between the vapor pressure and the temperature is not linear[6] as shown in Figure 6.6. Figure 6.10 shows the relationships between the composition of the liquid solution containing two soluble, volatile compounds and the vapor in equilibrium with the liquid solution at constant pressure.

The liquid solution of 30% A and 70% B boils when the temperature reaches T_1 at a constant pressure equal to atmospheric pressure. The vapor in equilibrium with the liquid solution at this pressure and temperature T_1 has a composition of 70% A and 30% B, that is, it is enriched in the more volatile component of the mixture. This vapor condenses to a liquid with the same composition as the vapor. The temperature of the liquid in equilibrium with this vapor (70% A and 30% B) is T_2.

Figure 6.10 represents an instantaneous snapshot of the composition and temperature of the liquid solution and the vapor in equilibrium with the solution because the liquid solution becomes richer in component B as component A is removed. Consequently, the temperature of the liquid

---

[6] The Clausius-Clapeyron Equation shows the relationship between the vapor pressure and the temperature, $\ln P = -\dfrac{\Delta H_{vap}}{RT} + C$, where P is the vapor pressure, ΔH_{vap} is the enthalpy of vaporization, T is the temperature Kelvin, and C is a constant characteristic of the compound.

Figure 6.10 Temperature-Composition Diagram for a Liquid Solution of Two Soluble, Volatile Compounds at Constant Pressure

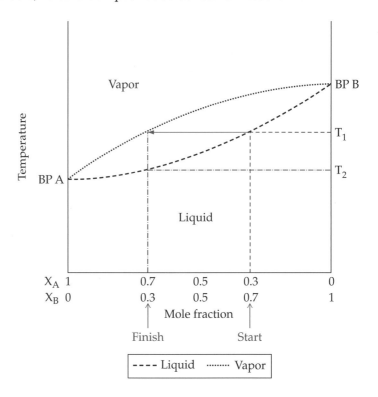

solution increases as the more volatile component (the component with the lower boiling point) is removed. With sufficient vaporization-condensation cycles (theoretical plates), the vapor consists of pure compound A and the temperature of the vapor is the boiling point of compound A. Once all of compound A is removed, the temperature rises abruptly. The vapor consists of pure B and the temperature of the vapor is the boiling point of compound B.

Non-Ideal Behavior for Mixtures of Two Soluble, Volatile Liquids In the previous discussion, we have assumed that the solution of the volatile components was dilute or ideal, that is, that the interactions between identical and dissimilar molecules were the same ($\Delta H_{mixture} = 0$, Equation 5.1, page 61). In many cases, mixtures of liquids do not exhibit ideal behavior and deviations are observed from the vapor pressures predicted by Raoult's and Dalton's laws (page 71). In some cases, the deviations are great enough that azeotropes result.[7] An **azeotrope** is a mixture of two or more components with a fixed composition that boils at a constant temperature.

Minimum-Boiling Azeotropes Positive deviations from Raoult's law are observed if the attractive forces between the identical molecules are greater than those between the dissimilar molecules so that the solutions are not ideal. In these cases, the vapor pressure of the mixture is greater and the boiling point is lower than expected for an ideal solution. For example, in a mixture of methanol and water, the methanol molecules are

---

[7] The word **azeotrope** comes from the Greek "zein tropos" or "constant boiling."

attracted to one another by induced dipole-induced dipole, dipole-induced dipole, and dipole-dipole (including hydrogen bonding) forces. Hydrogen bonding forces between the water molecules are significant. Analogous to the situation with ethanol-water mixtures as discussed on page 54, in a mixture of methanol and water, the induced dipole-induced dipole forces between the methyl groups are disrupted by the presence of the water molecules and the hydrogen bonding network of the water molecules is disrupted by the methanol molecules. Therefore, the forces of attraction between the two components in the mixture are weaker (methanol-water) than those between the identical molecules of the individual components (methanol-methanol and water-water). As shown in Figure 6.11, plots of the vapor pressure vs composition are not straight lines. Compare Figure 6.11 with Figure 6.8 for an ideal solution. The vapor pressures of the components in the mixture are greater than those calculated from Raoult's law at some compositions and the total vapor pressure of the mixture is greater than that calculated from Dalton's law. The boiling points at these compositions are less than predicted assuming an ideal solution.

If the deviations from ideal behavior are large enough, the vapor pressure of the mixture is greater than that of the more volatile component for some compositions and results in a **minimum-boiling azeotrope**. This is

Figure 6.11   Non-Ideal Behavior for a Mixture of Two Soluble, Volatile Compounds at Constant Temperature Exhibiting Positive Deviations from Raoult's Law

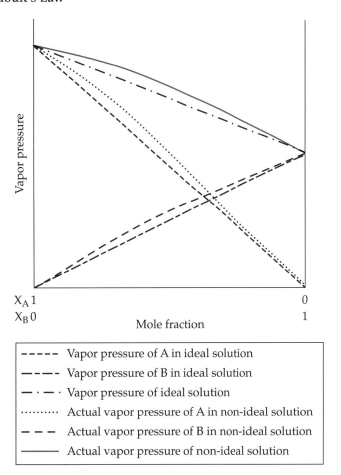

Figure 6.12   Non-Ideal Behavior for a Mixture of Two Soluble, Volatile Compounds at Constant Temperature Showing Positive Deviations from Raoult's Law and a Constant Boiling Mixture

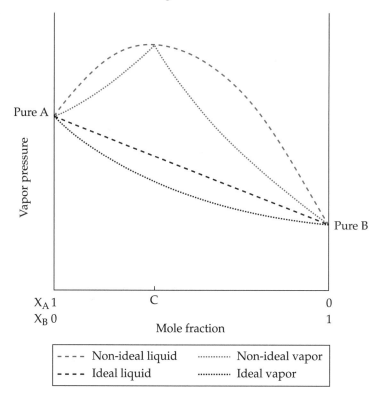

the case with ethanol and water mixtures where the increased size of the alkyl group compared with methanol results in deviations large enough that, at a composition of 95.57 weight % ethanol and 4.43 weight % water, the vapor pressure is greater than that of either pure water or pure ethanol. Consequently, a mixture of this composition boils at a temperature lower than that of either pure component. The boiling point of this mixture is 78.15°C, which is 0.15°C lower than that of the more volatile component, ethanol (bp 78.3°C).

Let's look at a hypothetical case to understand what happens when mixtures forming minimum-boiling azeotropes are repeatedly vaporized and condensed. Figure 6.12 shows a mixture of compounds A and B that exhibit positive deviations from Raoult's law. The vapor pressure of the solution is greater than that for an ideal solution and, at some compositions, is greater than the vapor pressure of the more volatile component (A).

At the compositions where the vapor pressure is greater than that of either pure component, the boiling point is lower than those of the pure compounds as shown in Figure 6.13. There is a minimum in the temperature-composition diagram at composition C.

If a liquid with composition D is heated to its boiling point, the vapor in equilibrium with D has the composition at E. The mole fraction of compound A in the vapor is less than that in the liquid. This vapor condenses to liquid F. If we continue vaporization and condensation, eventually the vaporized and recondensed liquid has the composition at C. The temperature of the vapor in equilibrium with this liquid is T_C. The mixture with composition C is an azeotrope—a constant-boiling mixture with a fixed a composition.

Figure 6.13   Boiling Point-Composition Diagram for a Mixture of Two Soluble, Volatile Compounds Showing Positive Deviations from Raoult's Law and a Constant Boiling Mixture

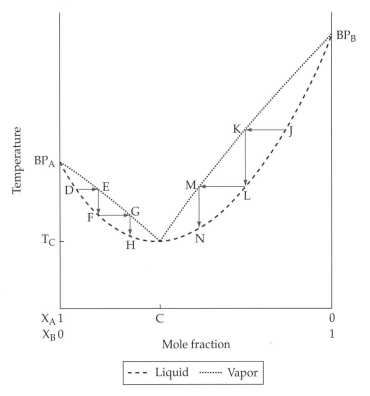

Similarly a mixture at composition J vaporizes to K and recondenses to L. The condensed liquid at L contains less B. Repeated vaporizations and condensations eventually give a liquid in equilibrium with the vapor of composition C, the azeotrope, that boils at temperature T_C. The pure component with the lower boiling point (A) cannot be obtained by beginning with a mixture of composition J.

In Figure 6.14, the temperature-composition diagram from Figure 6.13 has been split into two parts at the mole fraction composition of C. For compositions for which the mole fraction of A is greater than that at composition C (the left side of Figure 6.14), we can treat the liquid as a mixture of pure A and azeotrope C. The vapor E from the liquid of composition D is enriched in the more volatile azeotrope C. Each vaporization-condensation cycle reduces the amount of A in the vapor and enriches it in B until the composition at C is reached. At this point, the temperature remains constant at T_C as both A and B are removed in the ratio of the azeotropic mixture C. All of component B is removed in the azeotropic mixture and only pure A is left behind. With continued heating, the temperature rises abruptly to the boiling point of A and the composition of the vapor in equilibrium with the liquid is pure A. Pure component B cannot be obtained if the mole fraction of B in the solution is less than that at composition C.

If the mole fraction of B in the solution is greater than that at composition C (the right side of Figure 6.14), we can treat the solution as a mixture of pure B and azeotrope C. Again, the most volatile component of the mixture (the one with the highest vapor pressure and the lowest boiling point) is the azeotropic mixture. Each vaporization-condensation

Figure 6.14   Boiling Point-Composition Diagram for a Mixture of Two Soluble, Volatile Compounds Showing Positive Deviations from Raoult's Law and a Constant Boiling Mixture

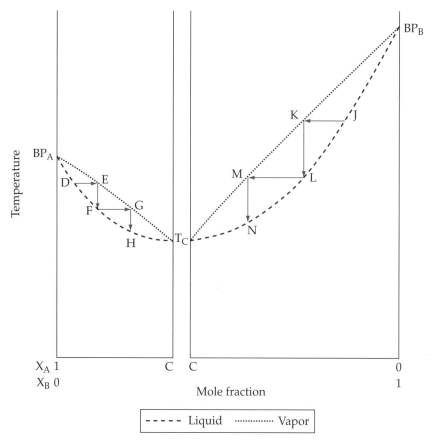

cycle (from J to K to L, from L to M to N, etc.) enriches the vapor in the azeotropic mixture C. Because A and B are being removed in the ratio of the azeotropic mixture C, eventually all of A is removed in the azeotropic mixture and only pure compound B remains. Therefore, after all of C has been removed at the temperature of its boiling point T_C, the temperature rises with continued heating to the boiling point of pure B and the composition of the vapor in equilibrium with the liquid is pure B.

Because ethanol and water form a minimum-boiling azeotrope, it is not possible to obtain 100% ethanol by distillation of the mixture. If you start with 95.57 weight % ethanol, the azeotropic mixture, after many vaporization-condensation cycles, you still have 95.57 weight % ethanol. The ethanol-water azeotrope boils at the constant temperature of 78.15°C. The temperature of the vapor and the temperature of the liquid are always the same and the composition does not change. If you start with a mixture that is 97 weight % ethanol, the azeotropic mixture of 95.57 weight % ethanol and 4.43 weight % water (bp 78.15°C) boils off first. Once all of the water is gone, you are left with pure ethanol. However, it is not possible to obtain 97% ethanol by distillation. If you start with 90 weight % ethanol, all of the ethanol is removed as the azeotropic mixture and you are left with pure water.

Did you know?

World governments are slowly beginning to publicly recognize that our reliance on oil to power engines of all types is a dangerous addiction. Indeed, alternate sources of energy are constantly being sought and existing technologies are becoming more efficient. Ethanol derived from the fermentation of starch and cellulose has long been considered a sustainable fuel alternative to gasoline. In fact, blends of ethanol and gasoline are currently being marketed. One of the problems associated with using ethanol as an alternative fuel has been the economics associated with producing absolute ethanol, which is ethanol free of water. Fuel-grade ethanol must have no water in it, but all methods of making ethanol (e.g. fermentation or hydration of ethylene) employ water as a solvent or reagent. Getting rid of the water is not trivial because of the minimum-boiling azeotrope formed between the water and ethanol. As discussed previously, simple distillation cannot be used to obtain 100% ethanol. There are solutions to this dilemma, but they require extra energy and/or costly reagents. Historically, the most common method is to add benzene to the ethanol-water mixture to form a ternary azeotrope (7.4 weight % water, 18.5 weight % ethanol, and 74.1 weight % benzene) with a boiling point of 64.9°C. Because this azeotrope has a lower boiling point than the ethanol-water azeotrope, all of the water, most of the benzene and some of the ethanol are removed by simple distillation. Water-free ethanol with a trace of benzene is left behind. Because ethanol and benzene form a binary azeotrope (68.3 weight % benzene, bp 67.9°C), a second distillation removes all of the benzene to give absolute or 100% ethanol.

Other methods of removing the water involve drying the ethanol by adding a reagent such as calcium oxide or a molecular sieve. Molecular-sieve technology has advanced considerably in recent years and is becoming the method of choice for producing fuel-grade ethanol.

Maximum-Boiling Azeotropes In some cases, the intermolecular forces between dissimilar molecules are greater than those between the identical molecules, so the vapor pressures of the components in the mixture are lower than those calculated from Raoult's law at some compositions. Therefore, the vapor pressure of the mixture is less than that calculated from Dalton's law at these compositions and the boiling point is higher than expected from ideal behavior. These mixtures are said to exhibit negative deviations from ideal behavior.

When the deviation from ideal behavior is large, the boiling point can be higher than that of either component at some compositions and there is a maximum in the boiling point-composition diagram. The maximum corresponds to a mixture of constant composition that boils at a constant temperature, a **maximum-boiling azeotrope**. For example, in aqueous hydrochloric acid, the hydrogen chloride molecules dissociate and are surrounded by water molecules, resulting in ion-dipole interactions and disrupting the weaker hydrogen bond interactions in the identical molecules. Water (bp 100°C) and hydrogen chloride (bp −111°C) form a constant boiling mixture of 20.22 weight % HCl ($X_{HCl} = 0.111$) that boils at 108.6°C.

Figure 6.15   Boiling Point-Composition Diagram for a Mixture of Two Soluble, Volatile Compounds Showing Negative Deviations from Raoult's Law and a Constant Boiling Mixture

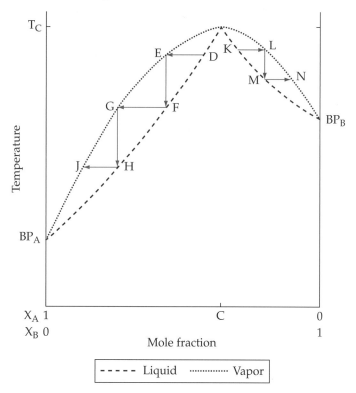

Figure 6.15 shows a general case of a mixture of compounds A and B where the intermolecular forces between A and B are sufficiently stronger than those between A and A or B and B, so that the mixture exhibits a maximum-boiling azeotrope. If a mixture of A and B at composition D is vaporized, the resulting vapor E is richer in A. This vapor condenses to F which is in equilibrium with vapor G and is further enriched in A. Repeated vaporization-condensation cycles give pure compound A. On the other hand, a mixture at composition K gives vapor L that is richer in compound B. Pure B is obtained after many vaporization-condensation cycles.

Figure 6.16 illustrates that we can view mixtures with maximum-boiling azeotropes more simply as mixtures of A and the azeotrope C or as B and the azeotrope C. If the mole fraction of compound A in the mixture is greater than that at composition C (the left side of Figure 6.16), repeated vaporizations and condensations remove compound A at the boiling point of A (BP_A) to leave the azeotrope C behind. Once the composition of the liquid reaches C, the temperature rises abruptly to T_C and the vapor in equilibrium with the liquid condensate has the composition of the azeotrope C.

If we start with a mixture where the mole fraction of compound B is greater than at composition C (the right side of Figure 6.16), repeated vaporizations and condensations remove compound B at its boiling point (BP_B) until the remaining liquid has the composition of the azeotrope C. With continued heating, the temperature rises to the boiling point of azeotrope C (T_C).

In both cases, the more volatile component (the pure component) is removed first. The temperature then rises and the azeotrope C is obtained.

Figure 6.16 Boiling Point-Composition Diagram for a Mixture of Two Soluble, Volatile Compounds Showing Negative Deviations from Raoult's Law and a Constant Boiling Mixture

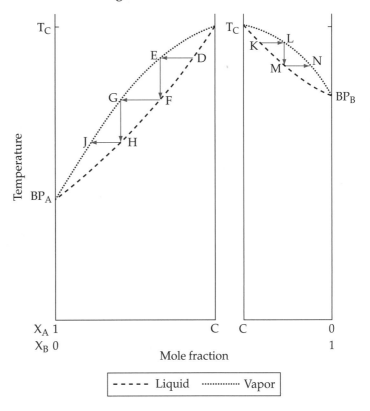

Did you know?

Diethyl ether (ether), an anesthetic first used in surgery in 1846, and halothane, first used in 1958, are historically important anesthetics. When halothane and ether are mixed in a 2:1 ratio, a maximum-boiling azeotrope is formed. The vapor pressure of the mixture (216 mm Hg) is lower and the boiling point (52°C) is higher than either of the constituents: halothane (373 mm Hg, 50.2°C) and ether (243 mm Hg, 34.6°C). The composition of the vapor is the same as that of the liquid when an azeotropic mixture of halothane/ether is distilled. Therefore, an anesthetic vaporizer filled with a 2:1 mixture of a halothane-ether solution and set at any flow rate delivers a 2:1 mixture of the corresponding vapor to the patient. Such reliability is important in clinics and hospitals in developing nations where the personnel administering anesthetics are often less experienced. It has been suggested that this fact, along with the azeotrope's excellent anesthetic profile and its affordability relative to pure halothane, make it an excellent candidate for use in developing countries[8].

halothane
1-bromo-1-chloro-2,2,2-trifluoroethane

diethyl ether

---

[8] Busato, G.A. and Bashein, G. *Update in Anaesthesia*, **2004**, Issue 18, Article 11.

6.4.1.d Insoluble, Volatile Impurities

As the interactions between the identical molecules (A-A and B-B) become increasingly stronger than those between the dissimilar molecules (A-B), liquids A and B eventually become immiscible, that is, insoluble in one another (Section 6.1). This situation can be thought of as an extreme case of positive deviations from Raoult's law. What happens when we heat a mixture of two immiscible compounds to the boiling point of the mixture? *If the liquids are not soluble in one another, each liquid acts independently and does not lower the vapor pressure of the other component in the solution.* The total vapor pressure of the solution is therefore the sum of the vapor pressures of the pure components (Dalton's law, Equation 6.11).

$$P_{total} = P^\circ_{a\ liquid} + P^\circ_{b\ liquid} \tag{6.11}$$

The total vapor pressure of the solution does not depend on its composition and remains constant as long as both A and B are present in the solution. Therefore, the boiling point, which is the temperature at which the total vapor pressure of the mixture equals atmospheric pressure, does not change until one of the components has been removed. In addition, because the total vapor pressure of the solution is the sum of the vapor pressures of the two pure liquid components, the mixture has a higher vapor pressure than either of the components. *The boiling point of the mixture is lower than that of the lower boiling (the more volatile) component.* (Contrast this with Figure 6.10 where the components are soluble in one another.)

All gases are miscible in one another and behave independently, so that the partial pressures of the vapor are given by Equations 6.6a and 6.6b.

$$P_{a\ vapor} = X_{a\ vapor}\ P_{total} \tag{6.6a}$$

$$P_{b\ vapor} = X_{b\ vapor}\ P_{total} \tag{6.6b}$$

At the boiling point, the liquid and the vapor are in equilibrium, so Equations 6.12a and 6.12b apply.

$$P_{a\ vapor} = X_{a\ vapor}\ P_{total} = P^\circ_{a\ liquid} \tag{6.12a}$$

$$P_{b\ vapor} = X_{b\ vapor}\ P_{total} = P^\circ_{b\ liquid} \tag{6.12b}$$

(Compare Equations 6.12a and 6.12b with Equations 6.8 and 6.9 for soluble, volatile impurities.)

The mole fraction is the number of moles of a compound divided by the total number of moles. By dividing Equation 6.12a by Equation 6.12b, we obtain Equation 6.13. This shows the relationship between the moles of compound A and the moles of compound B in the vapor.

$$\frac{Moles_{a\ vapor}}{Moles_{b\ vapor}} = \frac{P^\circ_{a\ liquid}}{P^\circ_{b\ liquid}} \tag{6.13}$$

Substituting the definition that Moles = Weight (g)/Molecular Weight, Equation 6.14 shows the relationship between the weights of A and B that are present in the vapor.

$$\frac{Weight_{a\ vapor}}{Weight_{b\ vapor}} = \frac{P^\circ_a}{P^\circ_b}\frac{MW_a}{MW_b} \tag{6.14}$$

The liquid formed when the vapor condenses consists of compounds A and B in the same weight ratio as given by Equation 6.14. Because compounds A and B are insoluble in one another, they form two layers and can be physically separated.

Steam Distillation If one of the immiscible components is water and the other component has a boiling point greater than 100°C, the mixture boils at a temperature below 100°C and the vapor and the resulting liquid condensate have the composition given by Equation 6.15.

$$\frac{Weight_a}{Weight_{water}} = \frac{P^o_a}{P^o_{water}} \frac{MW_a}{18}$$ (6.15)

This technique, known as steam distillation, allows volatile compounds with high molecular weights that are insoluble in water to be separated from mixtures with non-volatile components. The vapor consists only of water and the volatile, immiscible compound. Because the mixture boils at a temperature less than 100°C, the thermal decomposition of the compound is minimized.

6.4.2 Determining the Boiling Point of a Liquid

Determination of the boiling point of a liquid requires larger quantities of material than that for a melting point determination. If at least 0.3 to 0.5 mL of a liquid is available, a boiling point can be measured by heating the liquid in a reaction tube/small test tube with a boiling chip and then measuring the temperature a few millimeters above the boiling liquid where the vapor recondenses to liquid. This area where the vapor is in equilibrium with the liquid condensate is called the reflux ring (page 69). The temperature in the reflux ring reflects the temperature of the liquid in equilibrium with the vapor above the boiling solution. It is a measure of the boiling point of a liquid if insoluble, non-volatile impurities are present or if the liquid is pure. However as discussed in Sections 6.4.1.b, 6.4.1.c, and 6.4.1.d, the presence of soluble impurities or insoluble, volatile impurities do not allow an accurate determination of the boiling point of the pure liquid.

6.5 SUBLIMATION

As discussed on page 59, line CT in Figure 6.2 gives the vapor pressure of the solid at the corresponding temperature. Along line CT the solid goes directly into the vapor phase without passing through the liquid phase and the pressure-temperature combinations along this line are the sublimation points. The vapor can also be solidified at pressure-temperature combinations along line CT. The volume change in going from solid to gas or the reverse is significant so that the temperature at which these two phases are in equilibrium is pressure dependent. The phase change from solid to gas is known as sublimation and the reverse phase change from vapor to solid is known as deposition. The sublimation point (vapor pressure of the solid equals the external pressure) is analogous to the melting point (vapor pressure of the solid equals the vapor pressure of the liquid) and the boiling point (vapor pressure of the liquid equals the external pressure). The solid phase does not exist below line CT. Above line CT the solid is in equilibrium with the vapor at the vapor pressure corresponding to the pressure along line CT at the corresponding temperature. The liquid phase is not present at the temperatures and pressures along the line CT until the triple point T is reached where all three phases are in equilibrium.

In Figure 6.17, if P_M is atmospheric pressure, the solid melts at point M when the solid and liquid are in equilibrium. On the other hand, if the atmospheric pressure is P_S, the solid sublimes at point S and temperature T_S. The pressure and temperature at point S are less than those at the triple point T, so the liquid phase is not present and the solid becomes a vapor or the vapor becomes a solid.

Let's take carbon dioxide as an example. The triple point of carbon dioxide is at a temperature (T_T) of $-56.6°C$ and a pressure (P_T) of 5.1 atm. At a temperature (T_S) of $-78.48°C$, the vapor pressure of solid carbon dioxide (P_S) is equal to 1 atm. The temperature of a block of solid carbon dioxide at temperatures less than $-78.48°C$ increases until it equals $-78.48°C$. At that point (point S along line CT), the solid sublimes and the temperature remains at $-78.48°C$ until all the solid has vaporized. At 1 atm pressure (P_S in Figure 6.17), carbon dioxide cannot exist in the liquid state and at temperatures greater than $-78.48°C$ (T_S) the solid cannot exist. Carbon dioxide vapor can be solidified at 1 atm pressure (P_S) if the temperature of the vapor is decreased to $-78.48°C$ (T_S).

Figure 6.17 Phase Diagram Showing the Pressure and Temperature of the Sublimation of a Compound

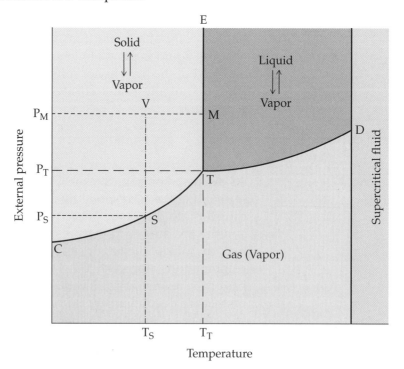

Compounds with triple points below atmospheric pressure can be sublimed at reduced pressure. If the external pressure is reduced until it equals the vapor pressure of the solid (points along line CT), the solid sublimes at the temperature that corresponds to the external pressure. In Figure 6.17, for a compound at temperature T_S and atmospheric pressure P_M (point V), if the external pressure is reduced to P_S (the vapor pressure of the compound), the solid sublimes and the temperature remains constant at T_S until all of the solid has vaporized.

6.5.1 Evaporation vs Boiling and Sublimation

Water doesn't boil at 1 atm pressure and room temperature because the vapor pressure of water is less than atmospheric pressure, but there is still water vapor in the air. This fact is the basis of relative humidity, which is the vapor pressure of water actually present in the air divided by the saturation (equilibrium) vapor pressure (the vapor pressure along line TD in Figure 6.17 that corresponds to the ambient temperature).

If the system is closed, for example, water in a closed container, liquid water evaporates until the liquid and vapor are at equilibrium at atmospheric pressure. The vapor pressure is that pressure along line TD that corresponds to the ambient temperature. Both liquid water and water vapor are in dynamic equilibrium in the container. As many molecules are escaping the liquid surface as are condensing on the surface from the vapor.

However if the system is open to the atmosphere, some of the vapor is removed from the vicinity of the liquid and the liquid keeps evaporating until it is all gone. For example, wet clothing dries at room temperature. We don't have to raise the temperature to the boiling point of water. The water doesn't boil, but rather evaporates from the clothing. Liquid water evaporates at room temperature as long as its actual vapor pressure is less than the saturation vapor pressure at that temperature (the point along line TD corresponding to the temperature).

Vapor also condenses to liquid if the vapor pressure is greater than the saturation vapor pressure at the given temperature. When the relative humidity is 100% the vapor pressure of the water in the air equals the saturation vapor pressure and vapor condenses to liquid if the vapor pressure exceeds this equilibrium vapor pressure.

What happens to the solid at T_S and P_M if P_M is 1 atm (point V in Figure 6.17)? Solid and vapor are still in equilibrium at T_S and the vapor pressure of the solid at that point is P_S (the saturation vapor pressure). Remember that vapor pressure is defined as the pressure exerted when a solid or liquid is in equilibrium with its own vapor at a particular temperature. Vapor pressure is independent of the external pressure so the vapor pressure of the solid is still P_S as long as the temperature is T_S (pages 58–59). Technically, the solid is not subliming because its vapor pressure is not equal to 1 atm, but it vaporizes if its vapor pressure is less than P_S. This is no different than the fact that water has a vapor pressure at room temperature. If we put a solid in a container at a temperature below its melting point, some of the solid vaporizes until the vapor pressure equals its equilibrium vapor pressure (the pressure along line CT) at the temperature of the container. If the system is closed, the solid and the vapor are in dynamic equilibrium and the number of molecules of solid that vaporize equals the number of molecules of vapor that solidify. This can be attained in a closed system, but not in an open system. If the system is open so that the vapor is continually removed, the vapor pressure of the solid in the atmosphere cannot reach the equilibrium vapor pressure and the solid continues to vaporize.

Example

The triple point of iodine is at a temperature of 113°C and a pressure of 0.12 atm. The melting point is 114°C. At room temperature and atmospheric pressure, liquid iodine does not exist, but the solid and its vapor are in equilibrium. Iodine crystals in a closed container vaporize until the vapor pressure of the iodine in the container is equal to the equilibrium vapor pressure of the solid at that temperature (the vapor pressure at the temperature along line CT in Figure 6.17). It doesn't matter what the total pressure in the container is as long as it is above line CT in Figure 6.17. Iodine crystals and iodine vapor are visible in the closed container. Iodine crystals in a closed container can be considered a closed system. The solid and vapor exist in dynamic equilibrium at atmospheric pressure and temperatures below the melting point. Technically this is not sublimation because the vapor pressure of the iodine is not equal to atmospheric pressure. It is evaporation. The dictionary definition of **evaporation** is the process by which the liquid or the solid state is changed into the vapor state. For a liquid, evaporation is vaporization that takes place at a temperature below the boiling point (assuming 1 atm pressure). For a solid, it is vaporization that occurs at external pressures that are greater than the vapor pressure of the solid at that temperature (line CT in Figure 6.17).

What happens if the system is not closed, for example when there is no lid on the container? If you put iodine crystals on a watch glass, they eventually disappear. The solid keeps vaporizing to try to reach the equilibrium vapor pressure, but the vapor is continually removed from the surface of the solid. The solid keeps vaporizing or evaporating.

The same situation occurs with ice cubes in the refrigerator or snow in the mountains. The solid evaporates (vaporizes) at temperatures below the melting point. The triple point of water is at a temperature of 0.01°C and a pressure of 0.006 atm. The recommended temperature for a freezer is 0°F or about −18°C and the vapor pressure of the solid at that temperature is 0.001 atm. We might think that because the vapor pressure is so low we would not lose much solid by vaporization. Why do the ice cubes and the snow keep disappearing? Again, we do not have a closed system, so equilibrium is never established. The ice and snow keep evaporating.

Did you know?

Sublimation is used to remove water and other solvents from solids that have lower vapor pressures than the liquid solvent. In the process known as lyophilization, the solution is frozen and the pressure is reduced. The solid solvent sublimes and the resulting vapor is removed. For example, water is removed from coffee, tea, and foods in this way to reduce their total weight and to prevent spoilage. Microorganisms need water to survive. If the water is removed from the food, spoilage is inhibited. Proteins, enzymes, and nucleic acids that are sensitive to heat can be recovered from frozen aqueous solutions by removing the water by sublimation at reduced pressure. Libraries and museums are able to recover water-damaged documents by freezing them and removing the water by sublimation.

CHAPTER 7

Isolation and Purification of Compounds

When you prepare a compound in the lab or want to isolate a compound from a natural source, it is necessary to purify the particular substance that you are interested in. A reaction in the lab usually proceeds with side-reactions so that the end result is a mixture. Also, natural products contain a wide variety of different compounds. Chemists exploit the differences in the intermolecular forces between like and unlike molecules to obtain a pure substance from a mixture of compounds. The techniques of recrystallization, distillation, sublimation, extraction, and chromatography, which depend on the differences in these forces, are used to separate pure compounds from impurities.

7.1 RECRYSTALLIZATION

Solid products obtained from chemical reactions usually consist of the desired compound and one or more impurities or side-products. Many times the difference in the solubility of these compounds in a liquid solvent is used to purify the desired compound. **Recrystallization** involves dissolving the solid material in the minimum amount of a suitable hot solvent, filtering out insoluble impurities, and allowing the solution to cool slowly to room or ice-water temperature. Ideally, other impurities are more soluble in the solvent and remain in solution. When successful, the result is crystals of a pure compound.

The procedure for recrystallization involves a series of steps you will follow whenever the lab text or your instructor tells you to "recrystallize it."

7.1.1 The Equipment

Recrystallizations are carried out in test tubes or in Erlenmeyer flasks to minimize contamination by dust and evaporation of the solvent, providing that the solvent level is no more than one-half the volume of the vessel. Both vessels promote the condensation of solvent vapors when the solution is heated to the boiling point. Because most organic solvents are flammable, water baths or sand baths should be used to heat solvents and solutions. *Caution: Never use an open flame to heat an organic solvent.*

A boiling stick or chip must be added to the solution *before* it is heated to prevent bumping (superheating). It should never be added to a solution that has already been heated and a new stick or chip must be added before reheating.

THE IMPORTANCE OF BOILING STICKS, CHIPS OR STONES, AND STIR BARS

Heating liquids in the lab is not like heating a can of soup in the kitchen. Before heating liquids in the lab, it is important to add a porous material to prevent superheating and the formation of large bubbles of gaseous solvent within the boiling liquid. Superheating results when boiling does not occur even though the temperature of a liquid is greater than its boiling point. This occurs when there is significant surface tension at the liquid/air interface, which is common in narrow containers. Boiling chips and stones are made out of insoluble, unreactive materials such as calcium carbonate or silicon carbide and, like boiling sticks, introduce a stream of fine air bubbles into the liquid being heated. These air bubbles reduce the tendency of a heated liquid to bump. Bumping occurs when large bubbles of the gaseous solvent form and erupt violently from the solution.

Sticks have an advantage over chips and stones in that they can also be used to stir the solution and are easily removed. The disadvantage of using sticks is that they are more reactive and cannot be used in some cases, for example, in very acidic solutions.

Adding a stick, chip, or stone to a solution that has already been heated can cause the release of large bubbles in the solution and result in sudden and violent bumping. It is also important to remember that when the solution is cooled, the porous spaces fill with liquid and are not effective if the solution is reheated. A new stick, chip, or stone must always be added before reheating.

Alternately, stirring is effective in breaking up large bubbles to prevent bumping. This is most conveniently accomplished using a mechanical device such as a magnetic stirrer and a stir bar.

Pasteur pipettes are useful in transferring solvents and solutions. It is important to hold the pipette correctly in order to direct the stream of liquid into the container. The pipette should be steadied between the middle and ring fingers while squeezing the bulb with the thumb and forefinger as shown in Figure 7.1.

Figure 7.1   Holding a Pasteur Pipette

7.1.2 Choosing the Solvent

Information is available in the literature regarding solvents suitable for the recrystallization of many compounds. However, when the identity of the product is not known or solubility data is not available, analysis of the intermolecular forces involved between solute molecules, solvent molecules, and the mixture is important in the process of choosing a suitable recrystallization solvent. Refer to Section 6.1 for the discussion of solubility.

As discussed in Section 6.1, in most cases solubility increases with increasing temperature because of the temperature dependence of the entropy contribution to the free energy of mixing (Figure 7.2). *A suitable solvent for recrystallization is one in which the desired compound is very soluble in the solvent at its boiling point, but is much less soluble in the solvent at room or ice-water temperatures so that the compound comes out of solution as the mixture cools.* In addition, it is important that the compound does not react with the solvent or decompose at the boiling point of the solvent. *Some of the desired compound is always lost during recrystallization because of its solubility in the cold solvent.*

Impurities should either be very soluble in the solvent so that they remain in solution or very insoluble so that they can be removed by filtration from the hot solution. If an impurity and the desired compound have similar solubilities, recrystallization may still be effective, depending on the relative amounts of the two compounds present. If the desired compound has only a small amount of impurity, that impurity may remain in solution because its solubility is not exceeded.

Most solvents that are useful for recrystallization have boiling points between 50°C and 100°C. There are, of course, exceptions that must be kept in mind when searching for the "right" solvent. Solvents with boiling points below 50°C evaporate readily and are difficult to use as recrystallization solvents. When a solvent has a high boiling point, it is sometimes difficult to remove the residual solvent from the crystals. Ideally, the boiling point of the solvent should be below the melting point of the compound being purified otherwise the compound can melt in the hot solvent instead of dissolving in it. This can result in the formation of an oil, which upon cooling forms a non-crystalline solid with entrapped impurities.

Figure 7.2   The Temperature Dependence of Solubility

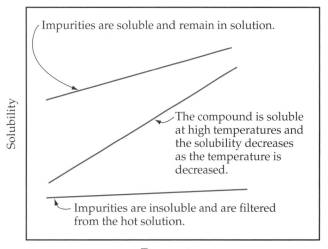

In older literature, you will find solvents such as benzene and carbon tetrachloride used for recrystallization. The use of benzene and chlorinated solvents is now avoided because of their toxicity and the damage they cause to the environment. Diethyl ether, a common laboratory solvent, is not an appropriate recrystallization solvent because it is highly flammable, can form explosive peroxides, and evaporates quickly.

SUMMING IT UP!

Characteristics of a Suitable Solvent for Recrystallization

- The compound is soluble in the hot solvent and much less soluble in the cold solvent.

- Impurities are either very soluble or insoluble in the solvent.

- The solvent is easily removed from the crystals.

- The boiling point of the solvent is below the melting point of the compound.

- The compound does not react with the solvent or decompose at its boiling point.

7.1.2.a Finding a Recrystallization Solvent

In order to find a solvent for recrystallization, a few crystals of the compound are put in a reaction tube/small test tube and a few drops of the solvent are added. Just enough solvent to cover the crystals is used because too much solvent leads to a false positive result in many cases. If the crystals dissolve immediately, the solvent is not useful for recrystallization because the compound will not be recoverable. If the crystals do not dissolve, the mixture is heated to determine if the crystals are soluble in the hot solvent. A few drops more of solvent can be added if necessary. A solvent in which the crystals are soluble in the hot solvent and come out of solution when the mixture is cooled is a suitable recrystallization solvent (Figure 7.3).

7.1.3 Obtaining a Hot, Saturated Solution of the Compound

The weighed impure material is put in a reaction tube, test tube, or an Erlenmeyer flask along with a boiling stick or chip (Section 7.1.1). The solvent is put in a similar container with a boiling stick and heated in a sand bath or hot water bath. The hot solvent is added slowly to the crystals using a Pasteur pipette while the mixture is heated in the sand bath or hot water bath and stirred in the presence of a boiling stick. It is convenient to use a boiling stick rather than a boiling chip or stone because it can be used to stir the solution and it is easily removed when heating is discontinued. The slow addition of the hot solvent is continued while the solution is heated in order to obtain a saturated solution of the compound in the solvent. The desired compound has a finite solubility in the cold solvent, so some loss always occurs. To minimize this loss, it is important to add the *minimum* amount of *hot* solvent needed to dissolve the material (Figure 7.4).

Figure 7.3 Finding a Recrystallization Solvent

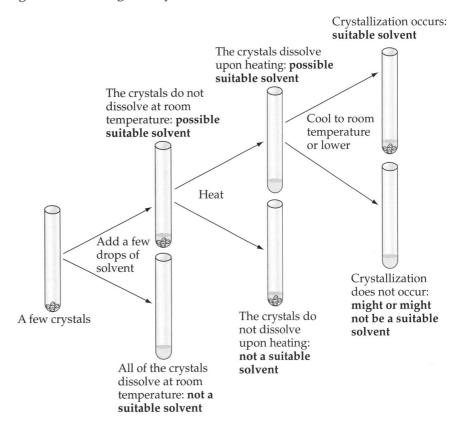

Note: A boiling stick should be used in the solutions that are being heated.
It is not included in the drawings for the purpose of clarity.

Figure 7.4 Obtaining a Hot, Saturated Solution of the Compound

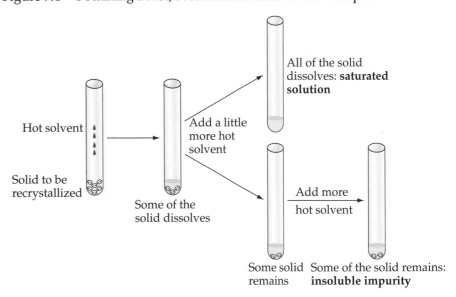

Note: A boiling stick should be used in the solutions that are being heated.
It is not included in the drawings for the purpose of clarity.

Some of the impurities may not be soluble in the solvent and may not dissolve. You have to judge if more material is going into solution as you add small amounts of the hot solvent. Be sure to allow sufficient time for the material to dissolve. Remember that it takes time for the solvent to break down the crystalline structure and dissolve the crystals. If it appears that all of the soluble material has gone into solution, the remaining solid is probably insoluble impurity and will be removed in a later step by filtration (Sections 7.1.5 and 7.2).

7.1.3.a Recrystallization Using a Mixed Solvent (Cosolvent)

There are situations where no single solvent works and miscible solvent pairs, such as ethanol and water or acetone and water, must be used. The compound should be very soluble in one of the solvents and have limited solubility in the second. The crystals are dissolved in the minimum amount of the hot solvent in which they are soluble. If insoluble impurities are present, the solution is filtered. The solvent in which the crystals are less soluble is heated and this hot solvent is added slowly dropwise to the hot solution of the compound. Care should be taken to be certain that the temperature of the solution is not higher than the boiling point of the solvent being added. Once the solution becomes cloudy, a drop or two of the hot solvent in which the crystals are soluble is then added to clear the solution before allowing it to cool to room temperature (Figure 7.5).

7.1.4 Removing Any Colored Impurities

Many crystals of organic compounds are colorless or white. Sometimes highly colored side-products are produced during a chemical reaction and must be removed to effect purification. If the solution of a white or colorless compound is colored by the presence of impurities, activated charcoal is used to adsorb these impurities. Only a small amount of charcoal is needed because of its large surface area. Keep in mind that it adsorbs some of the desirable compound as well as the colored impurity. To prevent foaming and bumping, the solution is removed from the heat source to allow the hot solution to cool slightly. After observing the color of the solution, a small amount of charcoal is added. Then, a new boiling stick is introduced and the solution is boiled for a few minutes. After the charcoal settles, the solution is examined. If the solution is still colored, the steps are repeated and just enough charcoal is added to decolorize the solution.

Figure 7.5   Recrystallization Using a Mixed Solvent

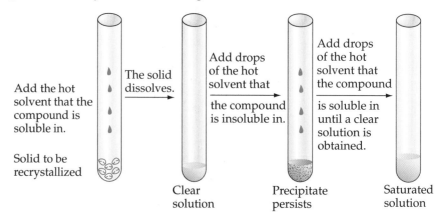

The effectiveness of the decolorization is determined by examining the color of the solution after each treatment.

Did you know?

Activated charcoal, sometimes referred to by the brand name Norit®, is not the kind of charcoal that is used to barbecue. Unlike diamond, graphite, fullerenes, and nanotubes, which have a regular atomic structure, activated charcoal is an amorphous form of carbon. This type of charcoal is obtained by burning wood and is activated by heating with steam to 1000°C in the absence of oxygen. This process results in a finely divided and highly porous material having a very large surface area.

Activated charcoal was used by the Egyptians as early as 1550 BC as an antidote for some ingested poisons. It is still used today for the treatment of poisoning in humans and animals because it is non-toxic and binds to toxic substances in the stomach and intestines. It is also used in water purification and in some gas mask cartridges to remove toxins, such as chlorine, paint fumes, and nerve toxins, such as sarin.

In the laboratory, it is very useful in removing colored impurities from product mixtures. The carbon surface is non-polar and induced dipole-induced dipole (London dispersion forces) and dipole-induced dipole interactions are responsible for the adsorption of molecules. These attractive forces increase with molecular size (Section 5.3.1.b), so that larger molecules are preferentially adsorbed. Colored impurities produced as side-products in chemical reactions are usually large molecules with conjugated double bonds that are more readily adsorbed than smaller molecules. Once all of the active sites are occupied, more activated charcoal must be added to remove additional impurities. The saying, "more is better," does not apply in decolorization because the desired compounds are also adsorbed.

7.1.5 Removing Undissolved Impurities and Charcoal

Sometimes during a recrystallization all of the material does not dissolve because of the presence of insoluble impurities. These impurities and any charcoal that was used are removed by filtering the hot solution (Section 7.2).

If the particles are large enough and the solution is in a reaction tube/small test tube, a square-tip Pasteur pipette is used to remove the hot solution from the solids (Section 7.2.1).

To remove activated charcoal or small particles that are suspended in solution, the solution is passed through **glass-fiber filter paper** wadded into the constriction of a short-stemmed Pasteur pipette (Section 7.2.2). To prevent crystallization of the product in the pipette, a slight excess of hot solvent can be added to the solution before filtration.

The hot solution can also be filtered by gravity using a Hirsch funnel with a frit or a piece of filter paper in place (Section 7.2.3).

Note: The hot solution should not be vacuum filtered. Remember that reducing the external pressure reduces the boiling point of a liquid (page 60). The boiling solvent can bump, resulting in a loss of product. Because the solution is saturated any loss in solvent due to evaporation will cause the product to precipitate on the filter or in the collection vessel.

If extra solvent was added to facilitate filtration it is evaporated first to obtain a hot, saturated solution of the solute before proceeding to the next step.

7.1.6 Allowing the Crystals to Form

The goal of recrystallization is to deposit pure crystals of the desired material. To accomplish this, the solution is slowly cooled to room temperature undisturbed. Avoid the temptation to pick it up and admire it! You want the molecules to fit together like spoons in a drawer so that no impurities are entrapped in the crystalline structure—don't disturb them. Crystallization is an equilibrium process in which the molecules in solution and those in the crystal are in equilibrium. Impurities don't fit in the crystalline lattice and return to the solution, but the molecules of the compound do fit and slowly deposit. Fast cooling or agitation may cause impurities to be occluded into the crystalline structure. You may wish to put the container with the solution in a beaker and surround it with cotton so that it cools more slowly. Slow cooling encourages the growth of large crystals that are much easier to filter and more pleasing to the eye.

If you started with a saturated solution of the material in the hot solvent, you should observe the slow growth of crystals in the test tube or flask. In some cases, the outcome is not what is expected. If no crystals appear, it may be that the solution is supersaturated and it is necessary to introduce a nucleation center to induce crystal formation. First, try *gently* scratching the inside of the container below the surface of the solution with a glass rod. It is not understood why, but many times this induces crystallization. Swirling the flask helps crystal formation if some crystals have deposited above the surface of the liquid as a result of evaporation of the solvent. If these methods don't work, try adding a seed crystal of the compound to encourage crystallization. You can get a seed crystal from another student in the class who is working with the same material or by introducing a boiling stick into the solution. Remove the boiling stick and allow the solvent to evaporate. There should be crystalline material on the boiling stick which will provide the nucleation center when it is reintroduced into the solution. If crystals still do not appear, an excess of solvent may be present. If this is the case, some of the solvent needs to be boiled off. (Remember to add a new boiling stick before heating the solution.)

After the solution has cooled to room temperature, the recrystallization vessel is usually put in an ice-water bath to further decrease the solubility of the compound and maximize the amount of material recovered. Remember that the compound is soluble to some extent in the solvent even at low temperatures so some material is always lost in the filtrate, which is referred to as the **mother liquor**.

Did you know?

A common mantra of candy makers is, "Don't stir the pot." The pot to which they are referring is the boiling solution of sugar (sucrose) in water. As the water evaporates, a supersaturated solution of sugar in water results. This solution is just "waiting" to crystallize at the first sign of a seed crystal. Often that crystal is just above the surface of the solution, clinging to the side of the pot. If the mixture is stirred, the crystal can be knocked down into the liquid leading to rapid and uncontrollable crystallization of the entire mixture. This is a candy maker's worst nightmare!

Did you know?

One mechanism some reptiles, amphibians, fish, and insects use to avoid freezing to death is discussed on page 63. A second approach most of the same animals use is to produce so-called antifreeze proteins. These proteins have evolved to "latch" onto ice crystals in the blood and either completely thwart further crystallization or force it to proceed in a controlled fashion. Either way, the goal of these protein molecules is to stop the spontaneous crystallization of super-cooled blood.

7.1.7 Collecting the Crystals

The mother liquor is removed from the crystals by filtration. The method of filtration depends on the amount of solvent and the nature of the crystals (Section 7.2). A square-tip Pasteur pipette can be used to remove the solvent from crystals in a reaction tube/small test tube (Section 7.2.1). If the amount of solvent is more than about 2 mL, it is better to use a Hirsch funnel to collect the crystals (Section 7.2.3). The mother liquor should be placed in a clean receptacle in case it is desirable to collect a second crop of crystals.

7.1.8 Washing the Crystals

The crystals are rinsed with a small amount of ice-cold recrystallization solvent to remove any soluble impurities adhering to the surface of the crystals. Remember that the crystals are slightly soluble in the ice-cold solvent so more is definitely not better. If the crystals are in a reaction tube or small test tube, the ice-cold rinses are added to the reaction tube/small test tube, the mixture is stirred, and the liquid is removed using the square-tip Pasteur pipette method (Section 7.2.1). The ice-cold solvent is added directly to the crystals filtered on a Hirsch funnel or short-stemmed funnel and the mixture is stirred with a stainless steel spatula. If the crystals have been vacuum filtered on a Hirsch funnel (Section 7.2.5), the vacuum must be released first. The ice-cold rinse is then added directly to the crystals on the Hirsch funnel. The mixture is stirred with a stainless steel spatula before reapplying the vacuum.

7.1.9 Drying the Crystals

When you are working on a microscale, the most difficult part of recrystallization may be removing the crystals from the reaction tube/small test tube. It is best to use the stainless steel spatula and to rotate the reaction tube/small test tube while gradually pulling the spatula out of the tube. Put the wet crystals on a piece of filter paper. You will probably have to repeat this process several times.

Crystals on the Hirsch funnel are easily scraped onto filter paper.

The wet crystals can be blotted with the filter paper and then allowed to air dry on a tared watch glass. When the weight of the crystals and the watch glass is constant after weighings separated by several minutes, the crystals are sufficiently dry.

7.1.10 Obtaining a Second Crop

If the yield of crystals is very small based on the original amount of material recrystallized, a second crop can be obtained by evaporating some of the filtrate (mother liquor) and cooling it slowly to room temperature followed by cooling in an ice-water bath. After collecting and drying the crystals from the second or third crop, their purity should be assessed before combining them with the first crop because they are sometimes less pure.

Summary of the Steps for a Successful Recrystallization

- Choose the appropriate equipment (size and shape).
- Find an appropriate solvent.
- Prepare a hot, saturated solution of the compound in the chosen solvent.
- Remove colored impurities, if necessary.
- Remove undissolved impurities and charcoal (if used).
- Allow the crystals to form by cooling the solution slowly and undisturbed.
- Collect the crystals by filtration, saving the mother liquor.
- Wash the crystals.
- Dry the crystals.
- Obtain a second crop if necessary.

7.2 FILTRATION

Solid materials are removed from solutions using different techniques depending on the amount of liquid and the size of the particles.

If a hot solution is being filtered to remove solid impurities when recrystallizing, the solution and the receiving container should be kept warm to prevent crystallization of the desired material during the filtration process. This is accomplished by placing the solution to be filtered and the receiving vessel in a water bath or other medium that is at an appropriate temperature. In addition, it is sometimes necessary to add a small amount of additional solvent to the solution being filtered to prevent premature crystallization of material in solution. After filtration, the added solvent is removed by evaporation (Section 7.3.1.a).

After filtering a recrystallized compound, the crystals are washed (Sections 7.1.8 and 7.6.3) with a small amount of cold solvent and the same filtration method is used to separate the solid material and the liquid. The wash liquid is added to the filtrate (mother liquor).

7.2.1 Square-Tip Pasteur Pipette Filtration

If the solution is in a reaction tube/small test tube, depending on the size of the crystals, it is sometimes convenient to use a Pasteur pipette to separate the solution from the solid. The crystals must be large enough to enable them to be excluded when the solution is drawn up into the Pasteur pipette.

Choose a pipette with a square end, that is, not chipped. The air is expelled from the pipette bulb as the tip approaches the bottom of the reaction tube/small test tube. The square tip of the Pasteur pipette is placed firmly on the bottom of the tube and the solution is slowly drawn up into the pipette leaving the crystals in the tube (Figure 7.6). The solution is expelled into a clean tube. This process takes practice and you may have to try to seat the pipette on the bottom of the tube a couple of times so the crystals are not drawn up along with the solution.

Figure 7.6   Square-Tip Pasteur Pipette Filtration

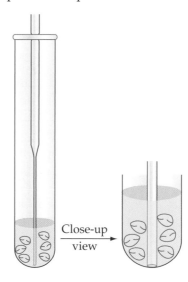

7.2.2 Glass-Fiber Pasteur Pipette Filtration to Remove Small Particles or Charcoal

To remove activated charcoal or small particles that are suspended in solution, a *quarter-piece* of glass-fiber filter paper is shaped into a small ball and the wad is eased into the constriction of a short-stemmed Pasteur pipette using a boiling stick (Figure 7.7). It is best to use a short-stemmed pipette to minimize the problems caused when the material precipitates during filtration. In some cases, a small amount of additional solvent is added to prevent the crystallization of the solid material during the filtration (Section 7.2).

The solution is transferred to the filter pipette using a second Pasteur pipette (long- or short-stemmed). If the solution does not pass readily through the glass-fiber filter plug, a pipette bulb can be put on the Pasteur pipette being used as the filter so that pressure can be applied to speed the filtration. If all of the air has been expelled from the pipette bulb and it is necessary to exert more pressure to complete the filtration, it is important to *first* break the seal between the bulb and the pipette *before* releasing the pressure on the bulb. This is necessary in order to equalize the pressure above

Figure 7.7   Removing Activated Charcoal or Small Particles

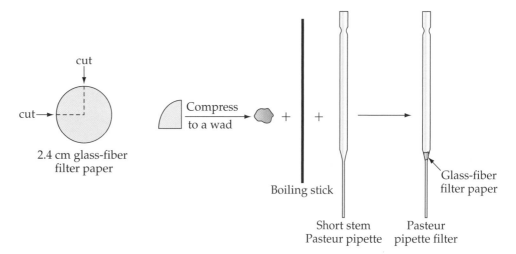

and below the filter plug. If the pressure is not equalized, air is drawn in from the bottom of the pipette and dislodges the filter. If this happens, the remaining solution passes through without being filtered.

7.2.3 Filtration Using a Hirsch Funnel or a Short-Stemmed Funnel

If the amount of solvent is more than about 2 mL, it is better to use a Hirsch funnel or a short-stemmed funnel to remove the solid materials.

If a Hirsch funnel is used, the frit or a piece of filter paper should be firmly in place (Figure 7.8). If a piece of filter paper is used, it should be wet with a small amount of solvent before adding the solution to be filtered so that the filter paper is tightly seated in the funnel. If the crystals are finely divided, a piece of glass-fiber filter paper should be used even with the frit in place.

If a short-stemmed funnel is used, a piece of filter paper that is the correct size for the funnel is folded as shown in Figure 7.8.

A small Erlenmeyer flask, centrifuge tube, or test tube large enough to hold the solution can be used to collect the filtrate. The collection vessel should be secured with a clamp. It may be necessary to put a paper clip or other object between the filter funnel and the collection vessel to prevent an air-lock and to allow the solution to filter readily. The solution is swirled or stirred and the suspension of the crystals is poured onto the filter. A stainless steel spatula is used to scrape out any crystals remaining in the flask and they are added to the crystals on the filter. If you have trouble transferring the crystals, a small amount of ice-cold solvent can be poured into the original container, swirled to get the remaining material in suspension, and filtered.

7.2.4 Decanting the Solvent

If the solution is in a reaction tube, a test tube, or a centrifuge tube, it can be centrifuged to pack the solid material in the bottom of the tube. The solution can then be slowly and carefully **decanted** from the solid into an appropriate clean container (Figure 7.9).

Figure 7.8 Gravity Fitration Using a Hirsch Funnel or Short-Stemmed Funnel

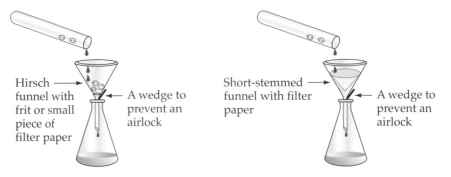

Folding the filter paper used in the short-stemmed funnel.

Figure 7.9   Centrifuging and Decanting a Solution from Solid Materials

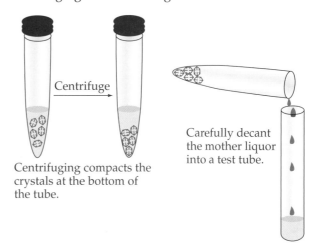

Centrifuging compacts the
crystals at the bottom of
the tube.

Carefully decant
the mother liquor
into a test tube.

If the material is in an Erlenmeyer flask and the particles are compact and dense so that they are not suspended in solution, the liquid can be slowly and carefully poured into a clean vessel leaving the solids behind.

7.2.5 Vacuum Filtration

Vacuum filtration can be used to collect crystalline materials and to remove traces of solvent from them. A hot solution should not be vacuum filtered. Remember that reducing the external pressure reduces the boiling point of a liquid (page 60). The boiling solvent can bump, resulting in loss of product. Because the solution is saturated, any loss in solvent due to evaporation can cause impurities to precipitate on the filter or in the collection vessel.

The Hirsch funnel is placed on the filter flask, the flask is secured with a clamp, and a piece of filter paper is placed in the Hirsch if a frit is not in place. If necessary, the filter paper is wet with a few drops of solvent to achieve a tight seal and the side-arm of the filter flask is attached to a vacuum source (Figure 7.10). The solution is swirled or stirred and the cold suspension of the crystals is poured onto the filter. The stainless steel spatula is used to scrape any crystals remaining in the flask onto the Hirsch funnel. It is important to not leave the vacuum on after filtration is completed because this causes evaporation of the solvent. Release the vacuum by removing the vacuum tubing or, if there is a three-way stopcock between the vacuum source and the flask, by opening the stopcock. If you have trouble transferring the crystals, a small amount of ice-cold solvent can be poured into the original container, swirled to get the remaining material in suspension, and filtered by reapplying the vacuum.

Figure 7.10   Vacuum Filtration

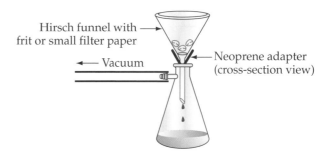

Hirsch funnel with
frit or small filter paper

← Vacuum

Neoprene adapter
(cross-section view)

7.3 DISTILLATION

Distillation is sometimes effective in purifying a liquid product. The liquid is heated in a container that is connected to a downward-sloping cooling tube leading to a collection container. There are many different types of distillation set-ups depending on the available equipment. Examples are shown in Figure 7.11. As the liquid is vaporized, it rises in the container, which is usually a round-bottom flask. The top of the container is fitted with a distillation head so that a thermometer can be inserted to observe the temperature of the vapor in equilibrium with the recondensing liquid—the boiling

Figure 7.11   Set-Up for Simple Distillation

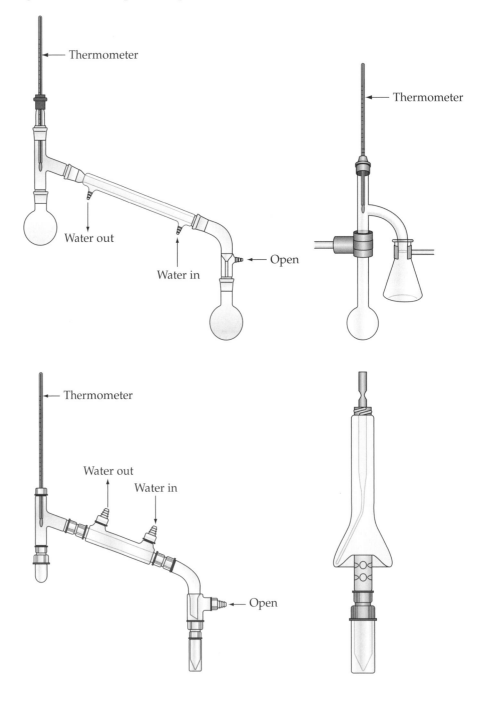

point of this liquid. The distillation head is attached to a downward-sloping cooling tube in which the vapor is condensed to a liquid that has the same composition as the vapor. This tube or condenser can be cooled by ambient air or circulating water. Air condensers are generally used when the liquid being distilled has a boiling point such that the hot vapors are easily condensed at room temperature. These condensers can be made more efficient by wrapping them with a wet paper towel, pipe cleaner, or cotton to aid in the cooling process. In other cases, the cooling tube is surrounded by a second separate jacket through which a constant flow of water is maintained. The water inlet is connected to the nozzle of the condenser closest to the receiving container. The water outlet is connected to the top nozzle near the thermometer. The end of the cooling tube is open to the atmosphere and directs the recondensed liquid into the receiving flask. A vial, test tube, round-bottom flask, or Erlenmeyer flask can serve as the collection container. With the exception of vacuum distillation, it is important that the system is open to atmospheric pressure at a point past the cooling tube where the vapor is recondensed.

As discussed in Sections 6.4.1 and Sections 6.4.1.a through 6.4.1.d, the differences in the volatility and solubility of the components of a mixture affect the boiling point of the liquid and the composition of the vapor. The type of equipment used and the success of the purification depend on these differences. It is assumed in the following discussion that the mixture is an ideal solution that follows Raoult's and Dalton's laws (pages 61 and 71).

7.3.1 Simple Distillation

Simple distillation can be used to achieve purification if the product and its impurities have significantly different boiling points or if there is only a small amount of a soluble impurity.

7.3.1.a Volatile Solvent or Impurities; Soluble, Non-Volatile Product

If the solvent and the impurities have a much lower boiling point than the desired product, heating the mixture removes the solvent and impurities and the product remains behind. For example, diethyl ether can be removed from a product with a much higher boiling point than 35°C by boiling away the ether. Technically this process is referred to as evaporation instead of as a distillation when the vapor is not recondensed and collected. However, the principles of distillation apply. *Care should always be taken in heating the liquid and containing the vapors if the solvent or impurity is flammable.*

7.3.1.b Soluble, Non-Volatile Impurities; Volatile Product

Purification of a volatile product by distillation is successful if the impurities are non-volatile and soluble in the product. As discussed in Section 6.4.1.b and shown in Figures 6.6 and 6.7, the temperature of the impure liquid is greater than that of the vapor in equilibrium with the purified liquid compound. However, the temperature of the thermometer at the top of the distillation head is constant as long as the bulb is immersed in the recondensing pure vapor so that the thermometer records the boiling point of the pure liquid. The pure product is collected in the receiving vessel and the impurities remain behind.

7.3.1.c Soluble, Volatile Impurities; Volatile Product

If the boiling point of the product and the impurities are significantly different, the mixture can be treated as if one of the components is not volatile, as discussed previously. That is, if the impurity boils at a much higher temperature than the product, the product can be distilled away from the impurity. On the other hand, if the product is the compound with the higher boiling point, the impurity can be removed by distillation leaving the product behind.

If the boiling points of the product and impurities are similar, simple distillation is effective if the amount of the impurity is small. For example, if it is assumed compound A represents a small amount of impurity, then $X_{A\ liquid}$ is very small and Equation 6.2 becomes $P_{total} = P_b^\circ$. Pure compound B can be obtained by distillation.

7.3.2 Fractional Distillation

The success of purification by simple distillation depends on the amount of the impurity and the difference in the boiling points of the product and the soluble impurity. When the amount of the impurity is more significant and the boiling points of the product and soluble impurity are similar enough that simple distillation is not effective, the product can sometimes be purified using fractional distillation.

7.3.2.a Soluble, Volatile Impurities; Volatile Product

As discussed in Section 6.4.1.c and shown in Equation 6.10, when a mixture of two soluble, volatile components is heated, the vapor does not have the same composition as the original liquid. Through repeated vaporization-condensation cycles (Figures 6.9 and 6.10), the vapor and its liquid condensate become enriched in the component with the higher vapor pressure (lower boiling point). In Figure 7.12, the boiling point of a liquid containing compounds A and B with a composition at D is T_0. If liquid with the composition at D is vaporized, the vapor has composition H and the mole fraction of compound A in the vapor is greater than that in the original liquid D. Vapor H is condensed to liquid E, which has the same composition as vapor H and a lower boiling point (T_1) than the original liquid D (T_0). By distilling liquid E, vapor of composition J is obtained, which is further enriched in the component with the lower boiling point (compound A). This vapor condenses to liquid F whose boiling point is T_2. By successively redistilling the condensed liquids and collecting the liquid resulting from the condensation of the vapor, pure A is eventually achieved.

The step from liquid D to vapor H and then to liquid E is a theoretical plate and represents a **simple distillation** where the liquid is vaporized and the resulting vapor is condensed and collected. A series of these simple distillations or steps leading to pure compound A is called a **fractional distillation**. Rather than collecting and redistilling each liquid condensate, the process is more easily carried out using a **fractionation column** between the container with the impure material and the distillation head. The repeated vaporization-condensation cycles occur within the column as the material rises up the column. Examples of the equipment required are shown in Figure 7.13.

The fractionation column needs to have a large vertical surface area so that the vapors can condense and revaporize many times before reaching the distillation head. Materials such as stainless steel sponge, glass beads, or glass helices provide a large surface area and are often used as packing

Figure 7.12 Temperature–Composition Diagram for a Liquid Solution of Two Soluble, Volatile Components at Constant Pressure

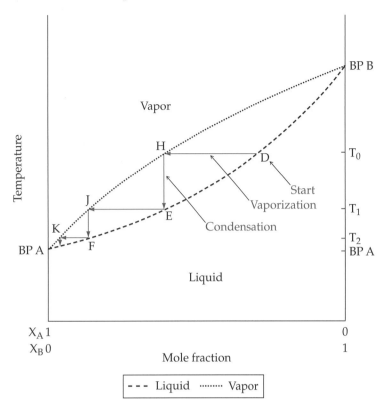

material in the column. The number of theoretical plates achieved in the column depends on the length of the column and the packing material used. In an ideal situation, the complete separation of compounds A and B is accomplished. The number of theoretical plates necessary to achieve this depends on the difference in the boiling point of the two components and their relative amounts.

Figure 7.12 shows a snapshot of what happens as the first vapors rise in the distillation column. As the revaporized material rises in the column, the temperature within the column decreases because the vapor becomes enriched in the lower boiling component as it condenses and revaporizes. Ideally, the vapor at the top of the column is pure A so the thermometer in the distillation head records the temperature of this vapor in equilibrium with pure liquid A, which is the boiling point of compound A. Because the vapor entering the column is richer in the lower boiling component, the mole faction of A left in the container decreases as the distillation proceeds. The composition of the undistilled liquid moves to the right in Figure 7.12 along the curve that indicates the liquid. Consequently, although it is not measured, the temperature of the liquid in the container increases during the distillation. Eventually only pure B is left in the flask and it starts its way up the distillation column. When all of compound A has distilled, the temperature recorded by the thermometer rises sharply to the boiling point of the higher boiling material when vapor B and liquid B are in equilibrium at the top of the column. At this point, one needs to change the receiving flask to obtain pure compound B.

Figure 7.13 Equipment for a Fractional Distillation

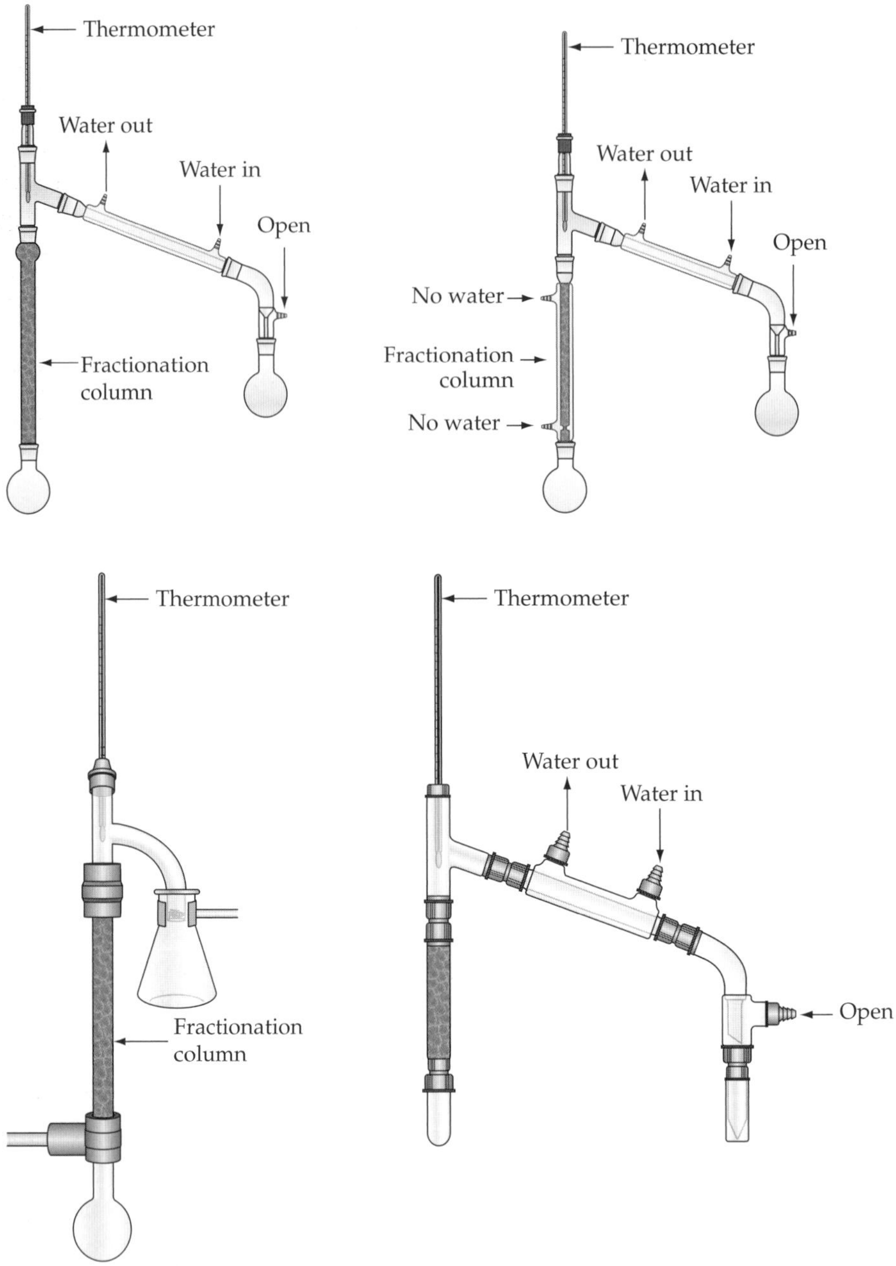

The drawback to fractional distillation is that the packing material retains significant amounts of the material, resulting in losses. Consequently, it is difficult and often impossible to use fractional distillation for small quantities of material. Specialized equipment is available to accomplish fractional distillations on microscale quantities of material, but it is very expensive. Ultimately column chromatography (Section 7.8.4) or gas-liquid chromatography (Section 7.8.6) are more suitable for the separation and purification of small quantities of liquids.

7.3.3 Vacuum Distillation

The boiling point of a liquid depends on the external pressure as discussed on page 60 and shown in Figure 6.2. As the pressure decreases, the temperature at which the liquid and vapor are in equilibrium decreases, that is, the

Figure 7.14 Equipment for Vacuum Distillation

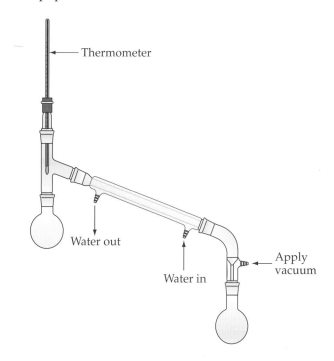

boiling point decreases. Therefore, compounds can be distilled at lower temperatures if a vacuum is applied to the distillation apparatus. This technique is particularly useful for compounds with high boiling points and for those that decompose at their normal boiling point (atmospheric pressure). The equipment for simple distillation is modified (Figure 7.14) to include a connection to a vacuum source and must not be open to the atmosphere.

7.3.4 Steam Distillation

There are situations where a compound with a high boiling point can be distilled at temperatures below its boiling point.

7.3.4.a Insoluble, Volatile Impurities; Volatile Product

In the case of steam distillation, an impurity, water, is deliberately added to an immiscible compound with a boiling point at atmospheric pressure that is greater than 100°C. The set-up for simple distillation (Figure 7.11) is used to distill the aqueous mixture. As discussed in Section 6.4.1.d, the mixture distills at temperatures below the boiling point of either component. The boiling point of a mixture of a high boiling compound and water is less than 100°C at atmospheric pressure. Equation 6.15 gives the ratio of the weight of the compound to the weight of water in the vapor. Because the compound is not soluble in water, it is easily separated from the water in the liquid distillate.

7.4 REFLUX

Reflux literally means to flow back. In the lab, reflux is used to describe a technique in which a liquid is distilled and then recondensed back into the original solution. A liquid solution is heated in a round-bottom flask that has a water or air-cooled condenser attached to it. It is critical that the apparatus

Figure 7.15   Equipment Used for Reflux

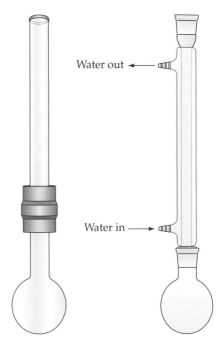

Water out ←

Water in →

be open to the air to avoid heating a closed system (Figure 7.15). If it is necessary to exclude water, a drying tube filled with calcium chloride is attached to the top of the condenser. A boiling stone is always added if the solution is not being stirred to prevent superheating and bumping (Section 7.1.1).

As the liquid boils, the vapors rise up in the condenser until they are cooled so that they recondense to liquid. The area where the condensation occurs, called the reflux ring (page 69), is visible in the condenser. The rate of heating should be controlled so that the reflux ring is no more than a third to halfway up the condenser. All vaporized liquid is returned to the solution, so that the composition of the solution does not change. Consequently, the temperature remains constant at the boiling point of the solution.

Reflux is an important and useful technique because solubility and the rate of a reaction both increase with temperature. Higher temperatures aid in dissolving solids and also increase their solubility. By heating a solution under reflux, a reaction can be carried out at the temperature of the boiling point of the solvent with no change in the composition of the mixture.

7.4.1 Collecting Noxious Gases Using a Gas Trap

In some reactions, noxious gases are given off and must be trapped. On a microscale level, the amount of these noxious gases is usually sufficiently small so that they can be collected in a gas trap attached to the reaction apparatus. For water soluble gases, such as HCl or NH_3, the fumes can be directed into a piece of damp cotton.

Figure 7.16 shows detailed instructions on preparing the gas trap. A piece of small-diameter tubing (approximately 1.7 mm) is threaded through a rubber septum and the threaded septum is attached to the reaction vessel. Wet cotton is inserted into the end of a test tube. The test tube is inverted and clamped to the rack. The end of the tubing is threaded through the wet cotton into the inverted test tube. Gases are therefore directed through the tubing into the inverted test tube and the wet cotton absorbs the small amounts of noxious gas generated.

Figure 7.16   Instructions for Setting Up a Gas Trap for Water-Soluble Noxious Gases

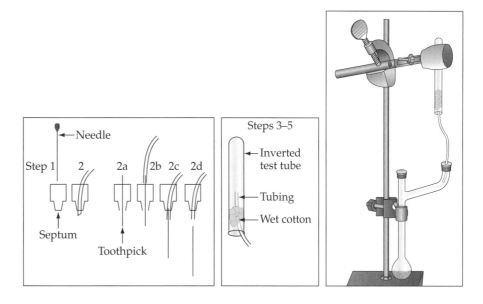

7.4.1.a Instructions for Setting Up a Gas Trap

Step 1. Use a clean syringe needle and pierce the top of a new septum several times in approximately the same area. Do not reuse a septum.

Step 2. Be sure that the end of your tubing is cut at an angle to form a point. Try to force the tubing through the septum. If this doesn't work, pierce the septum several more times with the needle and try again. If this still doesn't work, try the following steps.

Step 2a. Insert a toothpick through the septum.

Step 2b. Place the tubing on the toothpick.

Step 2c. Hold the tubing near where the toothpick and tubing meet, and push the tubing through the septum.

Step 2d. Pull the toothpick out of the tubing.

Step 3. Insert a piece of wet cotton into a test tube, invert the test tube, and clamp it to a rack.

Step 4. Thread the long end of the tubing through the wet cotton in the inverted test tube.

Step 5. Attach the septum to the reaction tube/small test tube or other apparatus containing the reaction mixture.

7.5 SUBLIMATION/EVAPORATION

Compounds that have a significant vapor pressure below their melting point can be purified by sublimation/evaporation[1] if the impurities have significantly higher or lower vapor pressures at the same temperature. Symmetrical, nonpolar molecules such as naphthalene with only induced dipole-induced dipole interactions (London dispersion forces) between the molecules are ideal candidates for this technique. Other compounds, such

---

[1] Refer to the discussion of sublimation vs evaporation in Section 6.5.1.

Figure 7.17   Examples of Molecules that Sublime

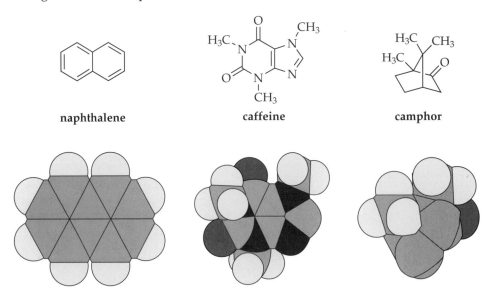

as caffeine and camphor that have structures that allow them to escape the surface of the solid easily, can also be successfully purified by sublimation (Figure 7.17).

At temperatures below the line TME and pressures greater than those along CST in Figure 6.17, the solid and vapor are in equilibrium (Section 6.5). If pressure P_M is atmospheric pressure and the temperature of the solid is T_S (point V in Figure 6.17), the vapor pressure of the solid is P_S and some of the compound exists in the vapor phase. As shown in Figure 6.17, the vapor pressure of a solid (line CST) decreases as the temperature decreases. Therefore, if the vapor hits a surface with a temperature less than T_S, some of the vapor solidifies. Because the vapor pressure of the compound in the container is therefore reduced, the solid and vapor cannot reach equilibrium at temperature T_S and the solid at temperature T_S continues to vaporize (evaporate), depositing the solidified vapor on the cooler surface.

If the temperature is above the melting point of the material (point M in Figure 6.17 if P_M is atmospheric pressure), the process is not a sublimation/evaporation. In this case, the compound melts to liquid that is in equilibrium with the vapor at the vapor pressure along line TD in Figure 6.17 corresponding to the temperature of the liquid. The vapor then condenses to liquid or solid depending on the temperature of the cooler surface on which it is deposited (Figure 6.17).

To prevent the compounds from melting and/or decomposing thermally, solids are sometimes purified using vacuum sublimation (Section 6.5). By reducing the pressure below the pressure at the triple point (point T in Figure 6.17), the solid vaporizes without melting at temperatures below the triple point temperature (T_T) and deposits as a solid on a surface that is at a lower temperature.

The advantage of sublimation/evaporation over recrystallization as a purification technique is that no solvent is needed. There is less waste and less loss due to the solubility of the compound in the solvent used for recrystallization and in the steps necessary to separate the compound from the solvent. It also avoids the occlusion of solvent in the crystals. However, it is less effective than recrystallization in cases where the impurities have vapor pressures that are similar to the compound being purified and it cannot be used for compounds with very low vapor pressures.

To purify a solid by sublimation/evaporation, the solid is heated in a container that has a second, cooler surface inside it. The vaporized/resolidified solid deposits on the outside surface of the inner cooler container. The system should be open to the atmosphere except for situations where vacuum (reduced pressure) sublimation is necessary. Examples of set-ups are shown in Figure 7.18. In each case, ice is put in the inside of the inner collection container.

For a vacuum sublimation, a side-arm test tube or filter flask is used as shown in Figure 7.19. The vacuum is applied at the side arm.

Be creative in assembling a sublimation apparatus. Take care, however, because water vapor from the atmosphere can condense on the outside of

Figure 7.18 Sublimation/Evaporation Set-Ups

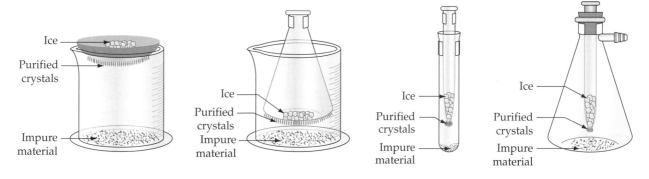

Figure 7.19 Set-Up for Vacuum Sublimation/Evaporation

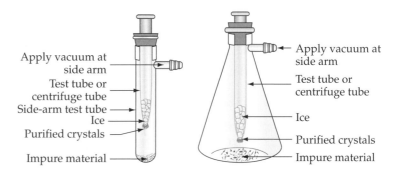

the cold collection container similar to the way drops of liquid deposit on the outside of a glass of ice water (Section 6.5.1). Be careful that the purified solid does not become contaminated with this moisture. In addition, if vacuum sublimation is used, the apparatus should be repressurized gradually to avoid dislodging the purified material.

7.6 EXTRACTION AND WASHING

Isolating and purifying a compound using extraction and washing are simple and quick techniques that require only basic laboratory equipment. However, they are not always effective in removing all of the contaminants.

7.6.1 Liquid-Liquid Extraction

In discussing solubility (Section 6.1), we found that molecules with similar intermolecular forces are more soluble in one another. For example, compounds with hydrogen bonds dissolve in solvents with hydrogen bonds more readily than in solvents with polar, aprotic bonds. Therefore, ethanol and water are miscible while diethyl ether and water form two layers with small amounts of ether dissolved in water and small amounts of water dissolved in the ether layer. Substances such as pentane, with only London dispersion forces, are less soluble in water. What happens if we take a third compound, either a solid or a liquid, and add it to a mixture of two immiscible solvents, one of which it is preferentially soluble in? One of the solvents must be polar and the other less polar or non-polar so that they are not soluble in one another. After shaking the mixture so that the phases are well mixed, the two solvents separate and form two layers or phases. Where is the third compound found? *The technique of **extraction** uses the differences in the solubility of a compound between two immiscible phases to partition compounds between the two phases.* The difference in the interaction of the compound with the two immiscible phases allows us to isolate it or to separate it from other materials that have different solubilities in the two phases.

If a compound is put into a mixture of two immiscible liquid solvents and the phases are then thoroughly mixed, the compound is distributed between the two phases according to its solubility in each. Therefore, polar compounds are found in the polar liquid and less polar or non-polar substances are found in the less polar/non-polar layer. Water is usually used as the polar solvent and diethyl ether, hydrocarbons such as pentane or hexane, or a halogenated solvent such as dichloromethane are commonly used as the less polar/non-polar solvents. To better understand the uses and limitations of extraction, it is useful to look at a quantitative example. However, in practice these calculations are not practical or necessary in order to perform an extraction.

A constant called the **distribution coefficient** or the **partition coefficient** is defined as the ratio of the concentrations of the compound in the two solvents. This is a constant for a particular compound partitioned between the two specified solvents at a given temperature. By convention, because the polar solvent is usually water and we usually want to purify a less polar or non-polar organic compound, the ratio used is usually the concentration in the non-polar solvent (nps) divided by the concentration in water (ps) so that the ratio is greater than one (Equation 7.1).

$$K_p = \frac{concentration_{nps}}{concentration_{ps}} \tag{7.1}$$

The solubility in each phase is used to estimate the concentration (Equation 7.2). This is an approximation because water and the non-polar solvents are soluble in each other to varying degrees, which changes the solvent properties. Also, the presence of other substances in the solvents such as salts affects the solubilities.

$$K_p = \frac{solubility_{nps}}{solubility_{ps}} \tag{7.2}$$

Example

To illustrate the process of extraction, assume we have 100 mg of a non-polar organic compound (NPO) mixed with some more polar substances in 10 mL of an aqueous solution and we wish to isolate the non-polar compound. If the solubility of NPO in diethyl ether divided by the solubility in water is 3, we can use this to approximate the partition coefficient according to Equation 7.2. Remember that the partition coefficient is not strictly the ratio of the solubility in each of the individual solvents because there is ether dissolved in the water layer and water dissolved in the ether layer, but this approximation serves our purpose. If we add 10 mL of ether to the mixture and shake it vigorously to intimately mix the ether and water layers, we can use Equation 7.1 to estimate the amount of NPO in the diethyl ether layer, [A]. There would be (100 - A) mg of NPO in the water layer, so that Equation 7.1 becomes:

$$K_p = \frac{A \text{ mg} / 10 \text{ mL}}{(100 - A) \text{ mg} / 10 \text{ mL}} = 3$$

$$A = 75 \text{ mg}$$

We have recovered 75 mg or 75% of our desired material, NPO, in the diethyl ether layer.

If instead we use 5 mL of diethyl ether and do the extraction twice, we will recover even more in the 10 mL of combined ether extracts. From the first extraction:

$$K_p = \frac{A_1 \text{ mg} / 5 \text{ mL}}{(100 - A_1) \text{ mg} / 10 \text{ mL}} = 3$$

$$A_1 = 60 \text{ mg}$$

Now there is (100–60) mg or 40 mg left in the aqueous layer. A second extraction of this layer with 5 mL of fresh diethyl ether gives the following:

$$K_p = \frac{A_2 \text{ mg} / 5 \text{ mL}}{(40 - A_2) \text{ mg} / 10 \text{ mL}} = 3$$

$$A_2 = 24 \text{ mg}$$

By using two extractions with 5 mL of fresh diethyl ether, we have extracted 84 mg or 84% of the material.

Using one-third the amount of diethyl ether (3.3 mL) and repeating the extraction three times gives us an 87.5% recovery. You may wish to go through these calculations to prove this to yourself.

Multiple extractions using a fraction of the extraction solvent each time increases the amount of the desired material recovered.

The value of K_p depends on the number of carbon atoms in a molecule and also the type and number of functional groups present because these factors affect solubility. Keep in mind that *"like dissolves like"* (Section 6.1).

If two compounds have K_p values that are relatively close, they cannot be separated by extraction.

Example

Let's assume that in addition to NPO in the first example, there is 100 mg of a more polar organic compound MPO in the mixture with a Kp between diethyl ether and water of 1. Using Equation 7.1, we can estimate the amount of MPO in the ether layer, [B].

$$K_p = \frac{B \text{ mg}/10 \text{ mL}}{(100 - B) \text{ mg}/10 \text{ mL}} = 1$$
$$B = 50 \text{ mg}$$

The diethyl ether layer contains 75 mg of NPO and 50 mg of MPO or 60% NPO and 40% MPO.

Extraction is useful only for separating compounds with very different solubilities in the extraction solvent.

Even if they have similar K_p values, it is possible to use extraction to remove organic acids from neutral organic compounds by utilizing the acid-base properties of the acids. The mixture is dissolved in an organic solvent. This non-polar phase containing the neutral compounds and the carboxylic acid can be extracted with an aqueous solution of the conjugate base of an acid with a pK_a value that is greater than that of the carboxylic acid (pK_a 3 to 5). The base converts the carboxylic acid into a salt, which is soluble in the aqueous layer. From Table 4.1, we see that aqueous solutions of $NaHCO_3$ (pK_a of H_2CO_3 6.37), Na_2CO_3 (pK_a of $HCO_3^{\ominus}$ 10.3), or NaOH (pK_a of H_2O 15.7) are all effective in converting the carboxylic acid into the sodium salt. Organic compounds called phenols have pK_a values of 8 to 10. Therefore, an aqueous solution of $NaHCO_3$ does not convert them into sodium salts, but aqueous solutions of Na_2CO_3 or NaOH do (Section 4.1.1.b). If it is necessary to separate a carboxylic acid and a phenol from one another, extraction with an aqueous solution of $NaHCO_3$ removes the carboxylic acid as the salt leaving the phenol in the organic layer. If there are other non-acidic compounds present along with the phenol, the organic layer is then extracted with NaOH to remove the salt of the phenol into the aqueous layer (Figure 7.20). Reversing the order of the extractions by extracting first with NaOH does not accomplish separation of the carboxylic acid and the phenol because both of them are converted into their salts and removed into the aqueous NaOH layer in the first extraction.

a carboxylic acid   **phenol**

Recovering the carboxylic acid or phenol from their sodium salts is accomplished by acidifying the aqueous extracts with hydrochloric acid whose pK_a is –2.2 (Table 4.1, Figure 7.20).

Note that alcohols with pK_a values of 16 to 18 are not removed from the organic layer by aqueous NaOH (pK_a of H_2O 15.7).

An organic base, for example an amine, RNH_2 (pK_a 10) or $ArNH_2$ (pK_a 4), can be removed from neutral or acidic compounds in a non-polar solvent by converting the base to the water soluble hydrochloride salt using aqueous hydrochloric acid (pK_a –2.2). After separation of the aqueous layer, the organic base is recovered by adding NaOH to this aqueous solution.

To better understand the above discussion, you may wish to refer back to the information on acid-base reactions found in Section 4.1.

In macroscale preparations, extractions are carried out in a separatory funnel that allows the layers to be easily separated. The stoppered separatory funnel is shaken with frequent venting by turning the funnel upside down with the stopcock end up pointing away from anyone's face. The stopcock should be opened slowly to equalize the pressure. This process is repeated to ensure complete mixing. After the layers have separated, the stopper is removed and the *lower* layer is drained into a clean container.

When working with small quantities, however, the large surface area of a separatory funnel results in the loss of a significant amount of material. A centrifuge tube is preferred because its conical shape makes

separatory funnel

Figure 7.20 **Separation of Benzoic Acid, Phenol, and Toluene by Acid–Base Extraction**

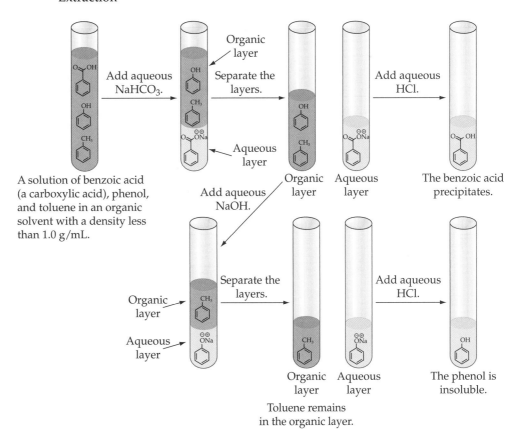

removing the lower layer easier, but a test tube is also adequate. The two layers must be thoroughly mixed by shaking or by drawing up one of the layers into a Pasteur pipette and expelling it into the remaining layer several times (Figure 7.21). The lower layer is then removed with the Pasteur pipette and placed in a clean centrifuge or test tube, taking care not to remove any of the upper layer. The small amount of the lower layer remaining in the original tube can be taken up into the Pasteur pipette. If a small amount of the upper layer is also removed, the phases are allowed to separate in the Pasteur pipette and each of the layers is carefully expelled into the appropriate tube.

It is important to know if the desired compound is in the lower or the upper layer. Knowledge of the polarity of the compound helps in determining if it is in the non-polar or the polar solvent. The density of the solvents provides a good guideline for determining which solvent is the upper and the lower phase. Diethyl ether (density 0.71 g/mL) is

Figure 7.21   Microscale Extraction

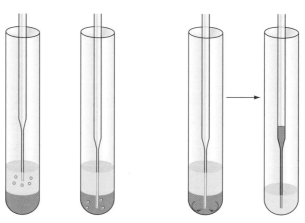

Step 1. Expel the air from the pipette as it passes through the layers.

Step 2. Draw up some or all of the bottom layer into the pipette.

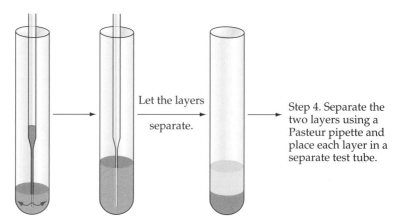

Let the layers

separate.

Step 4. Separate the two layers using a Pasteur pipette and place each layer in a separate test tube.

Step 3. Expel the solution forcefully into the remaining liquid to ensure effective mixing. Let the layers separate and then repeat Steps 1–3 again at least once.

less dense than water and should be the top layer. If dichloromethane (density 1.33 g/mL) is used as the extraction solvent, it is more dense than water and should be the bottom layer. It is important to check, however, because a saturated aqueous salt solution can be more dense than dichloromethane so that it constitutes the lower layer. The identity of the phases is checked by putting a small amount of what is thought to be the aqueous layer in a test tube. The addition of a small amount of water should give a homogeneous solution. If two layers form, the solution must be the organic layer.

To be safe, all of the layers should be saved until the desired compound is obtained. If the compound was not adequately extracted or is not in the expected layer because of an acid-base reaction, it is still possible to recover the material. *Don't throw any of the layers away!*

If the partition coefficient is small, it can be altered to favor the extraction of a non-polar substance into the organic layer by using an aqueous layer that is saturated with sodium chloride. Salts make the aqueous solution more polar so that organic compounds are less soluble in the salt solution than in pure water. The denominator in Equations 7.1 or 7.2 becomes smaller and the K_p is increased. For example, ethanol is completely miscible in water, but it is only slightly soluble in a saturated solution of aqueous sodium chloride.

We have used diethyl ether (bp 35°C, density 0.7 g/mL) to illustrate the extraction solvent, but other organic solvents are also used, including petroleum ether (a mixture of alkanes bp 30–60°C, density 0.7 g/mL), pentane (bp 36°C, density 0.6 g/mL), hexane (bp 69°C, density 0.7 g/mL), ethyl acetate (bp 77°C, density 0.9 g/mL), and dichloromethane (bp 41°C, density 1.3 g/mL). All of these solvents are immiscible with water.

SUMMING IT UP!

The Characteristics of a Suitable Extraction Solvent

- It is immiscible with the solvent that contains the desired compound.

- The compound is more soluble in the extraction solvent, for example, the K_p is greater than 1 for the desired substance and less than 1 for the impurities so that the impurities are not removed by the extraction solvent.

- The extraction solvent is easily removed from the compound after extraction. For example, it should have a relatively low boiling point.

- It does not react chemically with any of the compounds. The exception to this is the use of an aqueous base to remove an organic acid into the aqueous layer or an aqueous acid to isolate an organic base in the aqueous layer.

7.6.2 Liquid-Solid Extraction

Compounds in some solid mixtures are isolated from insoluble materials by adding a solvent to the solid in which the desired compounds are soluble. When you make a cup of coffee from ground coffee beans or a cup of tea from tea leaves, you are using hot water to dissolve the flavor elements and remove them from the grounds or tea leaves. This process is a **liquid-solid**

extraction. Many natural products are isolated using this method. If the compounds are very soluble and the solid is finely divided, stirring with cold solvent is sometimes effective. In most cases, however, the solid material is refluxed (Section 7.4) with the solvent. Simple examples of the equipment used for liquid-solid extractions are shown in the Figure 7.15. After filtering off the solid residue, the desired compounds are isolated from the solvent.

7.6.3 Washing

Extraction involves adding a second, immiscible solvent in which the desired compound is soluble to remove it from a mixture of compounds. Whenever you make coffee using coffee beans or a cup of tea from tea leaves, you are extracting the coffee or tea flavors from the solids. You don't want the grounds or tea leaves. *Washing,* on the other hand, uses a second immiscible solvent that does not dissolve the desired compound in order to remove impurities that are more soluble in the second solvent. Decaffeinated coffee is made by washing the coffee beans to remove the caffeine. You keep the coffee beans.

Did you know?

There are lab experiments in which the objective is to isolate caffeine from coffee. The caffeine is *extracted* from the coffee using solvents like dichloromethane. The goal is to obtain the caffeine. On the other hand, to obtain decaffeinated coffee, the objective is to *wash* the caffeine from the coffee beans while retaining the flavors present in the beans. The goal is to obtain the coffee beans. At least 97% of the caffeine must be removed in order for the coffee to be labeled decaffeinated.

Several methods have been developed to preserve the taste of the coffee while ridding it of caffeine, which lacks aroma and which has a slightly bitter taste. In the direct methods, the coffee beans are exposed to solvents in which the caffeine is soluble. In some cases dichloromethane is used to wash the caffeine from the green coffee beans. The dichloromethane solution of the caffeine is removed and the beans are heated to remove residual solvent. Its use is authorized by the FDA because only 0.1 ppm of residual dichloromethane remains in the coffee after roasting. The disadvantage of the method is the exposure of the individuals involved in the decaffeination process to the solvent. Another method uses carbon dioxide at high temperatures and pressure so that it is in the supercritical state and behaves like a liquid and a gas (Figure 6.2 and page 60). The supercritical fluid carbon dioxide combines with the caffeine selectively and both are removed from the green coffee beans.

There are also indirect methods where the green, unroasted beans are soaked in hot water under pressure. The caffeine is water-soluble and is removed along with the compounds that give coffee its flavor. These beans lack the flavor elements important to coffee and are discarded. The caffeine is removed from the aqueous solution by filtering it through activated charcoal in what is called the Swiss Water Process. Alternately, the solution is washed with dichloromethane or ethyl acetate. When ethyl acetate is used to remove the caffeine from the aqueous solution, the coffee can be called "naturally decaffeinated" because this solvent is found in some fruits. The aqueous solution is then used to wash the caffeine from a new batch of green coffee beans. Because this solution is saturated with the flavor elements, it does not remove any of these from the new beans. It does not contain any caffeine, however, so it removes this compound from the beans, which are then ready for roasting. The aqueous liquid that is saturated with the coffee flavors can be reused many times with new batches of the green, unroasted beans.

Example

Does washing work to separate compounds with similar K_p values? Let's go back to the example in which we had a mixture of 100 mg of compound NPO ($K_p=3$) and 100 mg of compound MPO ($K_p=1$) in 10 mL of an aqueous solution. We extracted with 10 mL of diethyl ether and found 75 mg of NPO and 50 mg of MPO in the ether layer. That means the extracted material is 60% NPO and 40% MPO. What if we now washed that diethyl ether layer with 10 mL of fresh water? Remember that this is washing because now we are trying to remove impurities from the desired compound. Let A_w and B_w represent the amounts of NPO and MPO respectively that remain in the ether layer after it is washed with the water.

$$K_p = \frac{A_w \text{ mg}/10 \text{ mL}}{(75 - A_w) \text{ mg}/10 \text{ mL}} = 3$$
$$A_w = 56.25 \text{ mg}$$

$$K_p = \frac{B_w \text{ mg}/10 \text{ mL}}{(50 - B_w) \text{ mg}/10 \text{ mL}} = 1$$
$$B_w = 25 \text{ mg}$$

We now have 81.25 mg of material that is 69% NPO and 31% MPO. This is a slight improvement over our original extraction, but not sufficient for purification and we have recovered only 56% of our desired compound NPO.

Washing is useful, however, to remove compounds from the organic layer that have significantly smaller K_p values, such as additional polar impurities. By adding a fresh aqueous solution or a saturated sodium chloride solution, more of the polar impurities are removed. Also, as discussed previously, washing with an aqueous solution of a base removes acid impurities and washing with an aqueous solution of an acid removes basic impurities because of the formation of salts that are soluble in the aqueous layer.

After recrystallization, a solid product is washed with a very small amount of the cold solvent used for recrystallization. This procedure removes the impurities that are soluble in the cold solvent from the surface of the crystals.

7.7 DRYING

An organic solvent that has been used for an extraction contains water in varying amounts. This water must be removed before the desired compound is isolated. If there is a lot of water in the organic phase, washing it with a saturated aqueous sodium chloride solution removes some of the water. This is important for ether extracts that contain quite a bit of water, but not, for example for dichloromethane extracts because water is much less soluble in this solvent.

More complete drying is accomplished by adding an anhydrous inorganic salt to the organic layer. These salts form insoluble hydrates that are removed by decanting or filtering them from the solution (Section 7.2.3 and 7.2.4). Many different compounds are effective for drying organic solvents. These vary in the speed of drying, their capacity for water, and their effectiveness in removing all of the water. It is important that they do not react with the desired material and that they are easy to separate from the solution. For microscale work, granular anhydrous sodium sulfate (Na_2SO_4) that slowly forms the decahydrate, $Na_2SO_4 \cdot 10H_2O$, is preferred even though it cannot be used in a heated solution because the hydrate breaks down when heated above 32°C. In addition, the hydrate forms slowly and it is not as efficient in removing the last traces of water as other drying agents. Its advantage is that it is easily separated from the organic solution because of its granular nature. Anhydrous calcium chloride ($CaCl_2$) that forms $CaCl_2 \cdot 6H_2O$ is also very easily removed, is fast, and efficient, but it loses water above 30°C, so cannot be used for heated solutions. It has the disadvantage or, in some cases, the advantage that it reacts with compounds containing oxygen and nitrogen such as alcohols and amines and also removes these compounds. Magnesium sulfate, which is commonly used in macroscale preparations, is not as useful for microscale chemistry because it is sold as a fine powder. This means it has a large surface area and is faster than other drying agents, but it results in the loss of more product and it is difficult to remove. In addition, it is a strong Lewis acid and can complex with or cause rearrangements of the desired material.

Five to ten minutes of drying time is usually sufficient. If the organic solution is clear with no visible signs of water and the drying agent is not clumped, but flows freely, a sufficient amount of the drying agent is present and one can assume that the solution is dry.

7.7.1 Removing the Drying Agent From the Solution

The solution can be removed from the drying agent by decanting the solution from the solid (Section 7.2.4). If the solution is in a test tube or centrifuge tube, the contents are centrifuged to pack the drying agent in the bottom of the tube. The liquid is decanted from the solid drying agent into a clean container (Figure 7.9). If the mixture is in an Erlenmeyer flask and the drying agent has settled to the bottom, the solution is slowly and carefully poured into a second, clean Erlenmeyer flask leaving the drying agent behind. If it is not possible to remove the solution by decanting, it is filtered using a Hirsch funnel with the frit or a piece of glass-fiber filter paper in place or using a short-stemmed funnel fitted with a piece of filter paper (Section 7.2.3 and Figure 7.8). The Hirsch funnel has the advantage that the frit or small piece of filter paper does not absorb as much of the solution as the larger filter paper used in the short-stemmed funnel.

The drying agent is rinsed with a small amount of clean solvent and the rinse is added to the solution after decanting or filtering.

Vacuum filtration should not be used for separating the solution from the drying agent.

7.8 CHROMATOGRAPHY

Just as runners, boats, cars, and swimmers become separated because they move at different rates in a race, mixtures of compounds can be separated by a process called **chromatography**.[2]

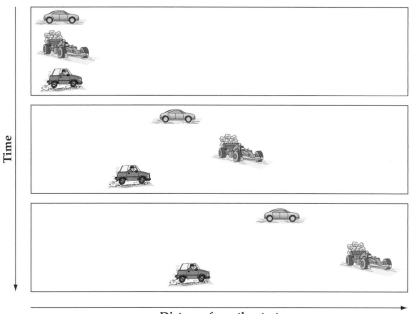

Distance from the start

The separation of mixtures of compounds into the individual components depends on differences in their physical properties and their interaction with the two phases used in chromatography: the stationary phase, which does not move, and the mobile phase, which moves through the stationary phase. We have examined the intermolecular forces between identical and dissimilar molecules (Sections 5.3 and 5.4), solubility (Section 6.1), and vapor pressure (Section 6.4). These are the concepts that we must consider in order to understand chromatography. Components of a mixture interact with the stationary phase by being adsorbed on it, dissolving in it, or reacting with it. The **stationary phase** can be a solid, as in column chromatography or thin-layer chromatography, or a liquid film coated on a solid, as in paper chromatography or gas chromatography. The **mobile phase** that flows through the stationary phase is a liquid with the exception of gas chromatography where it is a gas, usually helium or nitrogen. If the mobile phase is a liquid, the solubility of each component in the mobile phase is an important factor in determining the rate at which a component moves. *The stronger the interaction between a component and the stationary phase, the slower the component is moved by the mobile phase.*

---

[2] In 1903, Mikhail Semenovich Tswett, a Russian botanist applied a petroleum ether solution of plant pigments to a tube of calcium carbonate. He found that by adding more petroleum ether, the pigments in the tube were separated into colored zones or bands, "Like the light rays of the spectrum." He called the preparation a *chromatogram* and the technique the *chromatographic method* after the Greek words Croma or Khromatos for color and Graphein for written.

Separation of components of a mixture using chromatographic techniques depends on differences in their adsorption on the surface of the stationary phase, in their solubility in the stationary and mobile phases, and in their vapor pressures. These differences all depend on the intermolecular forces between the molecules and the stationary phase and those forces between the molecules and the mobile phase. In Figure 7.22, molecules of compounds A, B, and C are put on a support that is the stationary phase. The mobile phase is then allowed to flow through the support. Molecules of compound A are less attracted to or less soluble in the support compared with the mobile phase and are therefore carried through the support by the mobile phase more quickly than those of compounds B and C. Molecules of compound C are more strongly held or more soluble in the support and their transport through the support is much slower than those of compounds A and B. As the mobile phase passes through the stationary phase, the three components of the mixture are separated.

Another way to look at the separation shown in Figure 7.22 is that compound A prefers the mobile phase more than compounds B or C. Compound C prefers the stationary phase more than compounds B or A. The stationary and the mobile phases compete for the compounds resulting in an equilibrium distribution of each compound between the two phases. Depending on the difference in the strengths of the interactions with the stationary phase and the mobile phase, the components in a mixture move through the stationary phase with the mobile phase at different rates.

Figure 7.22   The Separation of Compounds by Chromatography

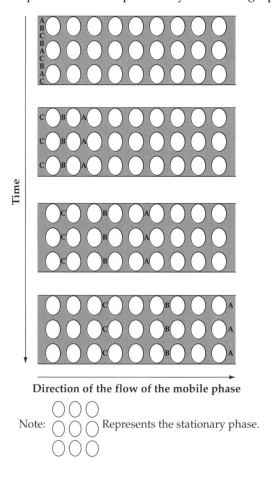

Direction of the flow of the mobile phase

Note: Represents the stationary phase.

As long as two substances do not have identical interactions with the two phases and the number of distributions or adsorption/desorption cycles is large enough, they can be separated. Recall in the discussion of extraction, two compounds with similar distribution coefficients were not separated by extraction and a large amount of material was lost (Section 7.6.1). In the case of chromatography, a very large number of distributions occur as the materials move along the stationary phase. Consequently, compounds with small differences in their distribution coefficients can be separated. In addition, while extraction depends only on solubility differences in two phases, separation of compounds by chromatography can depend on differences in their adsorption on the stationary phase and differences in their vapor pressures, as well as differences in their solubility in the two phases depending on the type of chromatography being used.

The choice of the stationary and mobile phases is important. If the interaction between the stationary phase and the compounds is too great, the compounds are not moved by the mobile phase. If the interaction is too weak, all the compounds in the mixture move with the mobile phase without being separated.

The chromatographic techniques used in the organic chemistry lab differ in the physical states of the mobile and the stationary phases as shown in Table 7.1.

7.8.1 Common Types of Chromatography

The terms adsorption, partition, normal-phase and reverse-phase are used to distinguish different types of chromatographic techniques.

7.8.1.a Adsorption Chromatography

If the stationary phase is a solid, it **ad**sorbs the components of the mixture on its surface as a monolayer. It is important to distinguish **ad**sorption and **ab**sorption. If you drink a cup of coffee, you **ab**sorb it. It becomes part of you. If you spill a cup of coffee on your shirt, it is **ad**sorbed on the surface of the shirt and can be removed by washing.

In **adsorption chromatography,** the intermolecular interactions between the stationary phase and the components of the mixture determine the strength of adsorption. Compounds are adsorbed and desorbed selectively

Table 7.1 Some Chromatographic Techniques

Type of Chromatography	Mobile Phase	Stationary Phase
Paper	liquid	liquid (water) adsorbed on paper
Thin Layer (TLC)	liquid	solid
Column	liquid	solid
Gas (GC)	gas (He or N_2)	liquid adsorbed on an inert support
High Performance Liquid (HPLC)	liquid	solid or liquid adsorbed on an inert support
Reverse-phase	polar liquid	non-polar liquid or waxy film attached to or adsorbed on an inert support

from the solid stationary phase by the mobile liquid phase as it flows through the solid phase. As with the solid phase, the intermolecular forces between the liquid and the components determine the rate at which this desorption occurs and, consequently, the rate at which the individual components move through the stationary phase. There is a competition between adsorption by the solid phase and desorption by the liquid phase and an equilibrium is established. Depending on the differences in the interactive forces between the components of the mixture and these two phases, the rates of desorption of the components are different. The different compounds move through the stationary phase at different rates and become separated.

With a polar stationary phase, dipole-dipole interactions (including hydrogen bonding), dipole-induced dipole interactions, and induced dipole-induced dipole interactions between the components and the stationary phase must be considered, depending on the characteristics of the components (Sections 5.3 and 5.4). If the stationary phase is non-polar, only induced dipole-induced dipole and dipole-induced dipole interactions are possible with the components (Sections 5.3.1, 5.3.3, and 5.4). These same van der Waals forces occur between the liquid mobile phase and the sample depending on their polarities.

7.8.1.b Partition Chromatography[3]

In **partition chromatography**, the stationary phase is a thin film of liquid adsorbed on the surface of an inert solid support. The mobile phase is an immiscible liquid or a gas, such as helium or nitrogen. If the mobile phase is a liquid, the separation of a mixture depends on the differences in the intermolecular interactions of the components of the mixture between the stationary and mobile phases. As with extraction (Section 7.6), the compounds are partitioned between the two liquid phases according to their relative solubilities in each of the phases. In the case of chromatography, however, one of the liquid phases is immobilized and the number of distributions is very large. As long as the partition coefficients of the components are different, they move through the stationary phase at different rates and are separated. This is equivalent to doing many extractions. If the mobile phase is a gas, the separation of components depends on differences in their solubility in the liquid stationary phase and in their vapor pressures.

7.8.1.c Normal-Phase Chromatography

In adsorption and partition chromatography, the typical stationary phase is polar and the mobile phase is less polar than the stationary phase (**normal-phase chromatography**). Therefore, polar compounds are more strongly attracted to the stationary phase and are moved along by the mobile phase

---

[3] In 1952 A.J.P Martin and R.L.M. Synge were awarded the Nobel Prize in chemistry for their contributions to the separation of complex mixtures using partition chromatography (<http://www.nobel.se/chemistry/laureates/1952/>). In the presentation speech (<http://www.nobel.se/chemistry/laureates/1952/press.html>), Professor A. Tiselius stated that, "The novelty in Martin and Synge's method is thus not the "chromatographic column" or "filter paper analysis", but rather concerns the fundamental chromatographic process itself. This can now be formulated as the partition of a substance between two liquids, instead of — as previously — entirely as its concentration at the surface of a more or less poorly defined active powder. Thus we have a rational basis for the method and enormously larger possibilities for choosing the experimental conditions that will be most suitable in any particular case. The almost explosive development of chromatography since the discovery of Martin and Synge's principles shows the power and scope of their invention."

more slowly than non-polar substances. Non-polar compounds are eluted[4] more rapidly. If the stationary phase is too polar or the mobile phase is not polar enough, the components do not move but remain on the stationary phase. If, on the other hand, the stationary phase is not polar enough or the mobile phase is too polar, the compounds move through the stationary phase without being separated. A mobile phase that is too polar may also compete with the components in binding to the stationary phase.

Two polar stationary phases commonly used in adsorption chromatography are alumina and silica gel.

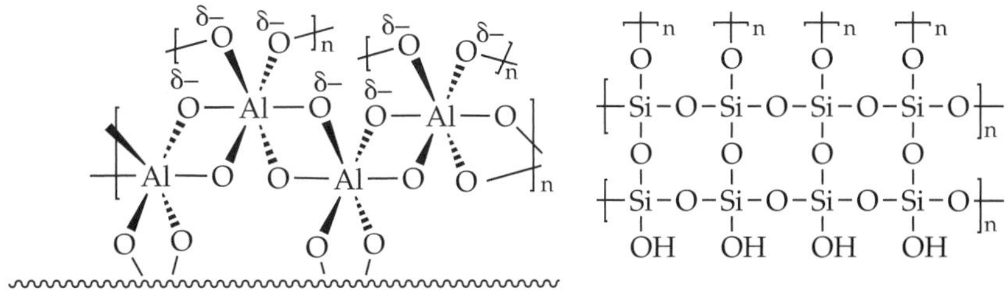

alumina   **silica gel**

Both of these are polar so that dipole-dipole (including hydrogen bonding) and dipole-induced dipole interactions with compounds occur.

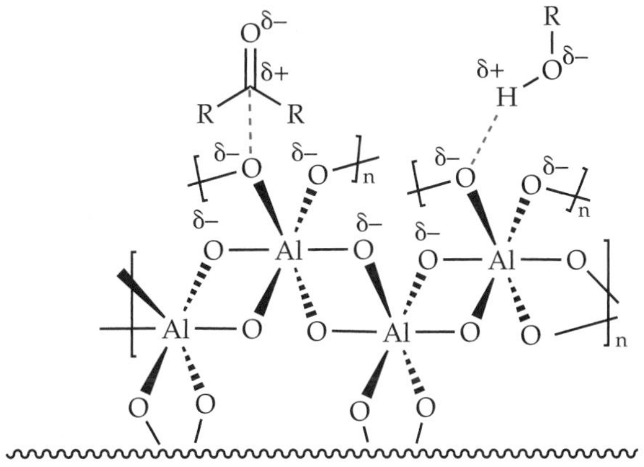

In addition to interactions analogous to those of alumina, silica gel is a hydrogen bond donor as well as acceptor.

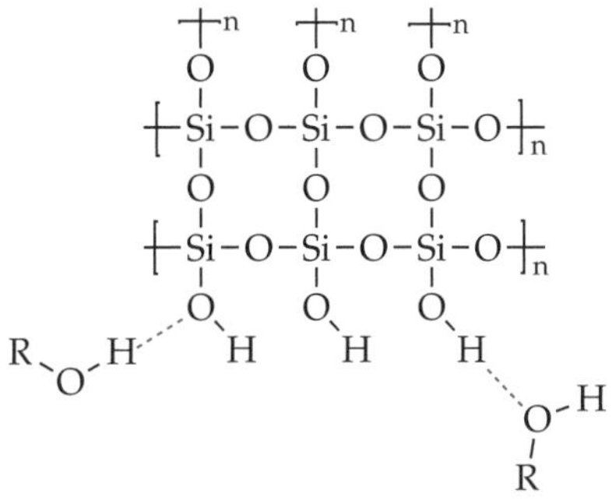

---

[4] Elute means to wash out; to extract one material from another, usually using a solvent.

The polarity of alumina and silica gel can be decreased by the addition of small amounts of water.

Did you know?

Silica gel is a matrix of silicon-oxygen-silicon bonds. There are hydroxyl groups on the surface of the matrix. The hydroxyl groups give silica gel its polar properties and the ability to hydrogen bond to other compounds. Perhaps you have opened a pair of shoes, a piece of electronic equipment, or a bottle of vitamins and found a little packet inside labeled, "Silica Gel, Do Not Eat". The material inside the package is beads of silica gel that have a high surface area and they hydrogen bond to water in the atmosphere. This adsorption of water serves to reduce the humidity of the surrounding environment. Hydrated silica gel is dehydrated by heating, making it a recyclable water adsorbent.

Silica gel was developed during World War II to keep penicillin dry. In addition to its use for consumer goods, it is used by museums to protect art work. Although the packages say do not eat, it is actually inert, chemically stable, and non-toxic. It is not known to be particularly tasty and it is recommended that any inadvertent ingestion be followed by a large quantity of liquid.

There are many other adsorbents used. The only requirements are that they be insoluble in the mobile phase, have a good adsorptive capacity, and have a large surface area. For example, a cationic or anionic residue attached to the stationary phase (an ion-exchange resin) allows charged materials to be separated. If a chiral molecule, such as a naturally-occurring carbohydrate or amino acid, is attached to the stationary phase, the stationary phase can distinguish between the enantiomers of an optically active compound. In this way, racemic mixtures can be separated into the individual enantiomers (**chiral chromatography**).

In choosing an adsorbent and a mobile phase, the relative polarity of the sample, the adsorbent, and the mobile phase must be considered.

7.8.1.d Reverse-Phase Chromatography

In **reverse-phase chromatography**, a non-polar stationary phase is used. For example, if silica gel is modified with a long-chain hydrocarbon, the stationary phase behaves as if it were coated with a thin, non-polar liquid or waxy film. The components are partitioned between the liquid or waxy film layer of the stationary phase and the liquid mobile phase.

For example, $R = C_4H_9$, C_8H_{17}, or $C_{18}H_{37}$

A solvent that is more polar than the stationary phase is used for the mobile phase. For example, a mixture of water and a miscible organic solvent, such as methanol or ethanol, can be used. Non-polar compounds are more strongly held to the stationary phase and are eluted more slowly than polar substances. Polar substances prefer the polar mobile phase and are eluted first.

7.8.2 Limitations of Chromatography

In general, chromatography is useful for separating mixtures of compounds and is valuable as both an analytical and a preparative technique. However, it is not possible to determine if a substance is pure using chromatography alone. Meaningful information is provided if more than one component is detected by chromatography, but a mixture of compounds may exhibit the same behavior toward the stationary and mobile phases and the individual components may not be distinguished.

For the same reasons, it is not possible to definitively identify a compound using chromatography. If a known sample of the compound is available and subjected to the same conditions as an unknown, different behavior of the two substances is proof that they are not the same. However, if they exhibit identical behavior, it merely means that they might not be separable by the method used.

7.8.3 Thin Layer Chromatography

Thin layer chromatography (TLC) is a simple, fast, and inexpensive technique that requires only a small amount of material.

7.8.3.a The Stationary Phase

In thin layer chromatography, a thin film of finely divided solid adsorbent on an inert support, such as a glass microscope slide, a plastic sheet, or aluminum foil, serves as the stationary phase. The adsorbent is usually alumina or silica gel, which may include a fluorescent compound such as manganese-activated zinc silicate to aid in the visualization of compounds using ultraviolet light. Because the film is so thin, there is not a large amount of adsorbent available and the quantities of material to be separated (chromatographed) must be small, on the order of <1 mg or 10 μl. Consequently, this is an analytical tool, not a preparative method.

Did you know?

In 1938, Russian chemists N.A. Izmailov and M.S. Shraiber used microscope slides coated with calcium, magnesium, and aluminum oxides to analyze plant extracts. They applied a spot of the compound to the plate and added a drop of solvent to produce concentric rings. They referred to the process as spot chromatography. In 1941, M. Crowe of the New York State Department of Health adapted the technique to determine solvents for column chromatography, but it wasn't until 1951 that J.G. Kirchner of the U.S. Department of Agriculture used adsorbent held on the plate with a binder. He was investigating the flavor components of citrus juices and developed the plates by allowing the solvent to ascend up the plate. Egon Stahl from Germany perfected methods for applying a uniform layer of silica gel to a glass plate in 1956 and first used the term *thin layer chromatography*. The technique became widely used when, in 1963, Eastman Kodak developed methods for precoating flexible sheets of plastic with a variety of adsorbents.

7.8.3.b The Mobile Phase

Pentanes or hexanes, dichloromethane, ethyl acetate, acetone, 2-propanol, ethanol, or methanol are used as the mobile phase. Miscible mixtures of these, for example, ethyl acetate and hexane, can also be employed to achieve good separation of the components. A sufficient amount of the mobile phase is added to a jar or beaker (the developing chamber) to cover the bottom to a depth of about 0.5 cm. A piece of filter paper is placed in the liquid and pressed against the inside of the chamber to aid in keeping the atmosphere in the chamber saturated with the mobile phase that is known as the developing solvent.

7.8.3.c Applying the Sample to the Stationary Phase and Developing the Plate

The compounds being investigated must be relatively non-volatile so they do not evaporate from the plate. A capillary pipette is used to apply a small spot of a compound or mixture dissolved in a volatile solvent on a pencil line or mark, called the **spotting line**, about 1 cm above the bottom of the chromatography plate. The plate is put into a developing chamber taking care that the spots of the compounds are above the solvent level and that the plate does not come in contact with the filter paper. The chamber is kept covered with a lid or piece of aluminum foil so that the atmosphere inside remains saturated with the developing solvent (the mobile phase). The solvent rises on the plate due to capillary action. As the mobile phase moves up the plate, it competes with the solid adsorbant for the compound(s) being analyzed. Depending on the polarity of the compound and the solvent, materials move up the plate at different rates. If a mixture is spotted, the components of the mixture *may* move at different rates and be separated.

7.8.3.d Visualizing the Plate and Analyzing the Data

The plate is removed from the developing chamber after the developing solvent has migrated to about 0.5 cm from the top of the plate. The distance that the solvent migrated is immediately marked by drawing a pencil line before the solvent evaporates. This line is known as the **solvent front**.

The areas containing the compounds are usually not colored unless the compounds themselves are colored, but they can be visualized using various methods. If the adsorbent on the plate contains a fluorescent indicator, compounds that have conjugated double bonds absorb ultraviolet (UV) light and typically show up as dark spots on a colored background when the plate is viewed under an ultraviolet lamp. *Warning: Never look into a UV lamp.*

An additional method is to place the plate in an iodine chamber (page 86). Iodine forms reversible complexes with many organic compounds so that dark spots reveal the position of the compounds on the plate. Visualization with UV light and iodine are very convenient and often complimentary techniques. It is important that the plate be visualized with UV and the spots marked before it is placed in the iodine chamber.

There are also a variety of reagents that have been designed for identifying compounds with specific functional groups. These reagents are typically applied by dipping the plate into the reagent or by spraying the reagent on the plate. One such reagent is dinitrophenylhydrazine (DNPH) that reacts with aldehydes and ketones to give a class of compounds known as dinitrophenylhydrazones (Equation 7.3). The key to the success of this visualization method is that dinitrophenylhydrazones are always dark yellow or orange in color. Therefore, if one dips a developed TLC plate into an acidic solution of DNPH, the plate looks pale yellow, but any aldehydes or ketones will show up as dark yellow or orange spots. The rate at which these spots appear is a function of the reactivity of the carbonyl compound. In general, ketone hydrazones take longer to appear than aldehyde hydrazones.

2,4-dinitrophenylhydrazine   a 2,4-dinitrophenylhydrazone

A **retention factor** or R_f **value** is defined as the distance the compound moves from the spotting line divided by the distance the mobile phase travels along the plate (the solvent front) (Equation 7.4).

$$R_f = \frac{\text{distance the compound moves from the spotting line}}{\text{distance from the spotting line to the solvent front}} \qquad (7.4)$$

These distances are measured from the spotting line to the center of the spot and from the spotting line to the solvent front respectively. The retention factor is not a physical characteristic of the compound, but is specific to the amount of material spotted, the adsorbent, the solvent, and the temperature. Although exact R_f values cannot be compared on different plates and with different solvents, their relative values provide useful information about the relative polarity of compounds.

7.8.3.e Factors Affecting Separation

Both alumina and silica gel are polar adsorbents and the solvent that serves as the mobile phase must be chosen to desorb the components of the mixture selectively in order to separate them. The preferred solvent results in the greatest difference in the R_f values for the compounds in a mixture. If the liquid mobile phase is not polar enough (plate A in Figure 7.23), the compounds do not move from the spotting line and are not separated. If the solvent is too polar (plate B in Figure 7.23), the compounds move with the mobile phase and no separation is accomplished. The mobile phase in plate C has a polarity between that used for developing plates A and B and separates the mixture into three spots. Because the stationary phase is polar and spot Z has the largest R_f, it is the least polar component of the mixture. It prefers the mobile phase more than the more polar stationary phase. Compound X is the most polar material because it is held most tightly by the stationary phase resulting in the smallest R_f.

7.8.3.f Applications of Thin Layer Chromatography

Thin layer chromatography can be used as a preliminary check on the purity of a compound, to determine the minimum number of substances in a mixture, to follow the course of a reaction by tracking the disappearance of starting material or the appearance of product, and to provide a guide to establish suitable conditions for and to monitor column chromatography. It can also be used as an initial check of the identity of a compound

Figure 7.23   The Effect of Solvent Polarity on the Separation of a Mixture of Compounds

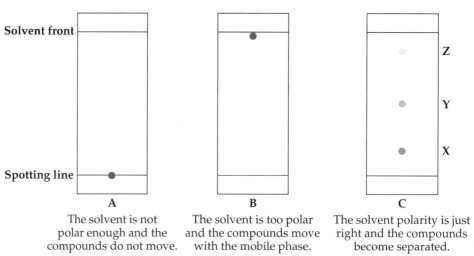

A	B	C
The solvent is not polar enough and the compounds do not move.	The solvent is too polar and the compounds move with the mobile phase.	The solvent polarity is just right and the compounds become separated.

if a sample of known material is available. Spots of the unknown and known compounds are **cospotted**, which means they are placed on top or next to one another on the same plate. If the two compounds show different R_f values, the compounds are not identical. If, however, the R_f values are the same, it does not prove that the compounds are the same. Just as many compounds have the same melting points, it is possible that different compounds will exhibit the same behavior toward the stationary and mobile phases and therefore have the same R_f values. A single spot on a TLC plate does not prove that a substance is pure (Section 7.8.2). Also, the size of the spot is not correlated with the amount of material present, so that thin layer chromatography is not a quantitative technique in most circumstances. In fact, the spots diffuse and become larger as they move up the plate.

7.8.3.g General Procedure for Thin Layer Chromatography

You will use thin layer chromatography frequently in the organic chemistry laboratory. The procedure is very fast and simple.

Prepare the Micropipettes *Caution: Make sure that there are no flammable solvents on either side of the lab bench.*

Use the capillary tubes that have two open ends (not the melting point capillaries) and the microburners to draw micropipettes to use for spotting solutions on the TLC plate. Heat the middle portion of the capillary in the hottest part of the flame. This is the area just above the dark blue cone of the flame. Rotate the tube slightly until the flame in the center is orange-red and the tube is softened. Quickly remove the tube from the flame and immediately pull the two ends gently in opposite directions to get a constricted length. Do not try to pull the tube while it is in the flame because it will form two closed-end pieces. Break apart the two ends to make two micropipettes. It is usually best to remove all but approximately 1.5 to 2 cm of the constricted length. With practice you will be able to draw micropipettes with very small diameters for use in TLC.

Figure 7.24   Filter Paper for the TLC Developing Chamber

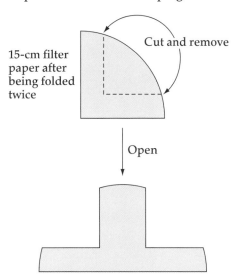

15-cm filter paper after being folded twice

Cut and remove

Open

   When breaking the drawn capillary to make the two micropipettes, hold the capillary and the drawn area close to where you wish to break it. Then push down slightly, exerting a downward motion near the point of breakage and lifting the undrawn ends slightly up.

The micropipettes are very sharp and can cause serious injury if proper precautions are not exercised. To store extra pipettes, place them in a thick-walled sealable plastic bag or other container that can be closed. The sharp tip should be pointed toward the bottom of the bag or other container. Do *not* place them with the sharp tips pointing upward in the container because this can lead to injury.

Prepare the Developing Chamber It is convenient to use a 150-mL beaker for the developing chamber. Fold a piece of 15-cm filter paper twice to form a quarter of a circle and cut it as shown in Figure 7.24. Unfold it once to give the shape shown in the figure. Place the long edge down in the beaker and press the filter paper to the walls so that it clings to the side of the beaker. Add 10 mL of developing solvent to the beaker. Cover the beaker with aluminum foil. Allow time for the chamber to equilibrate so that the filter paper is saturated with the developing solvent. It is important to keep the aluminum foil on the beaker so that the chamber remains saturated with the solvent vapor and the composition of the solvent mixture does not change due to evaporation of the more volatile components.

Spot the Plates Avoid touching the surface of the plates. Handle the plates only by the edges. Do not use plates that have chips of silica gel missing. Use a pencil to lightly mark the position for spotting the compounds (the spotting line) about 1 cm from the bottom of the plate. Make sure the line where the compounds are spotted is above the solvent level. All of the spots must be above the solvent level in the beaker and at the same distance from the bottom of the plate. Do not use a pen because ink contains dyes that will migrate on the TLC plate. You can put more than one spot per plate, but do not put them too close to the edges. Refer to the full-sized drawing of the TLC plate shown in Figure 7.25.

Figure 7.25 Spotting the TLC Plate

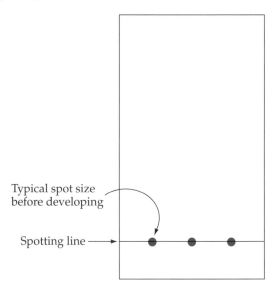

Dip the micropipette into the solution of the compound you are analyzing, touch the plate lightly on the spotting line with the micropipette, and quickly remove the pipette. Be careful not to gouge the plate. Make each spot as small as possible. A spot 2 to 2.5 mm in diameter is a good size. Do not spot the sample too heavily. To obtain small spots, it is helpful to blow gently on the plate as the sample is applied. Sometimes it is useful to examine the plate by ultraviolet light before developing it to assure that the spots are visible. Be aware, however, that not all compounds show up under ultraviolet light.

Develop the Plates Place the plate in the developing chamber. It is important that the TLC plate does not touch any part of the filter paper. If it does, the solvent from the paper will begin to migrate onto the plate from the point of contact, leading to erroneous results.

Allow the solvent to rise to within 0.5 cm of the top of the plate. Mark the solvent front (Section 7.8.3.d) with a pencil immediately upon removing the plate from the developing chamber before the solvent evaporates. Make sure that you replace the cover on the developing chamber.

Visualize the Spots Allow the solvent to evaporate from the plate. This process can be accelerated by waving the plate like a fan. Use the UV lamp to detect the position of the compounds that absorb UV light. Circle the spots with a pencil. After visualization with the UV lamp, put the plate in the iodine chamber unless instructed differently. Mark the spots that are stained by the iodine. Look for differences between how spots stain with iodine compared with how they appeared under the UV light. Alternately, you may be instructed to apply a reagent to the plate that reacts with the compounds on the plate. These reagents may be corrosive so that it is important to follow the instructions for how plates stained with these reagents are to be marked.

Warning: Do not look into the UV lamp because serious injury to the eyes can result.

Record the Data in the Laboratory Notebook Make full-size sketches of the plates in your notebook, marking what was spotted in each lane. Clearly indicate which spots were detected under the UV lamp and which were stained by other methods. Also record the solvent used for developing the plate. Calculate the retention factor (R_f value) for all spots. (Equation 7.4.)

7.8.4 Column Chromatography

The same principles are involved in **column chromatography** as in thin layer chromatography. In order to achieve a reasonable flow rate, however, the particle size of the adsorbent is not as small and therefore the surface area is not as large. In addition, the stationary phase is packed in a column, such as a Pasteur pipette (0.5 to 2 g adsorbent, 10 to 100 mg of sample to be separated), a titration burette, or a glass chromatography column with a larger diameter. The size of the column is determined by the amount of material to be separated. The amount of adsorbent used is much greater than that in thin layer chromatography, so larger quantities of samples can be separated. A longer column allows a larger number of adsorption/desorption cycles and better separation of compounds, but it increases the time and the volume of the solvent necessary to elute the components. Because column chromatography is much slower than TLC, it is not an analytical, but rather a preparative technique. A suitable adsorbent and solvent for column chromatography are established using TLC. In normal-phase chromatography, the stationary phase is polar and the mobile phase is less polar (Section 7.8.1.c).

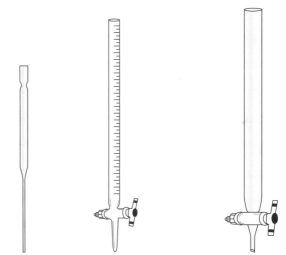

Pasteur pipette   Titration burette   Glass chromatography column
Note: The images are not to scale.

The sample can be a non-volatile liquid or a solid. It can be dissolved in a small amount of the solvent to be used as the mobile phase and applied directly to the column. Alternatively the solution of the sample in the solvent can be mixed with a small amount of the adsorbent. After the solvent is evaporated, the adsorbent and sample mixture is placed on top of the stationary phase packed in the column.

The mobile phase descends through the adsorbent by gravity. As the mobile phase flows through the column, the components are adsorbed by the stationary phase and desorbed by the solvent in a process called **elution** and are separated into bands or zones. Because the mobile phase is added to the top of the column, the composition of the solvent can be changed gradually during the elution of the components of the mixture. Solvents are classified in order of their polarity. Hydrocarbon solvents such as pentane and hexane are non-polar. Dichloromethane, ethyl acetate, and acetone are polar, aprotic solvents. Ethanol, methanol, and water are polar, protic solvents. If, for example, the sample consists of both polar and non-polar compounds, a non-polar solvent can be used to remove the non-polar components. The polarity of the solvent can then be gradually increased by mixing increasing amounts of a miscible and more polar solvent with the original non-polar solvent in order to elute the more polar components. As is the case with TLC, if a solvent that is too polar is used as the mobile phase, all of the components of the mixture are transported through the stationary phase with the solvent and no separation is achieved. On the other hand, if the solvent is not polar enough, the components remain adsorbed on the stationary phase and are not eluted. If a strongly polar solvent is required, it is usually preferable to use a less polar adsorbent for the separation.

The **eluent** (solvent) from the column is collected in fractions that are analyzed for the presence of components of the mixture using TLC or other analytical methods. The solvent is then evaporated from these fractions to recover the separated materials. As Tswett observed in his 1906 paper, "It can be supposed that two substances in a solvent might be adsorbed to the same degree . . . It can happen that any one zone is not absolutely pure." (Section 7.8.2)

In addition to the relative polarity of the stationary phase, the solvent, and the components to be separated, the flow rate of the mobile phase (solvent) affects the success of the separation. If the flow rate is too rapid, not enough time is allowed for an equilibrium to be established in the adsorption/desorption of the compounds and they are not well separated. On the other hand, a flow rate that is too slow allows the separate bands of components to diffuse and broaden.

Did you know?

Although Tswett succeeded in separating plant pigments by column chromatography in 1903, his work was rejected when Willstatter and Stoll in Germany failed to replicate the separation in 1913. They did not heed his warnings not to use adsorbants that were too "aggressive" and the components of the plant pigments were decomposed by the stationary phase. It was not until 1931 that the technique was rediscovered and revived.

©Chemistry of Tallinn University of Technology.

M.S. Tswett 1872–1919

7.8.4.a Flash Chromatography

To allow the use of an adsorbent with smaller particle size and therefore a larger surface area, the technique of **flash chromatography** has been developed. A glass column is used and moderate air or nitrogen pressure is applied to the top of the column. For example, if the column is a Pasteur pipette, a pipette bulb is placed on the top of the pipette and gently squeezed to force solvent through the packing material. The rate of elution is controlled by the amount of pressure applied to the top of the column. The separation of the components is faster than that achieved with ordinary column chromatography and there is less diffusion of the components. Separation is effective, however, only if the difference in the retention factors determined by TLC is greater than or equal to 0.15.

7.8.5 Paper Chromatography

In **paper chromatography,** the cellulose paper is not the stationary phase. Cellulose adsorbs about 22% water from the air so that the stationary phase is water. The separation depends on the competition between the water adsorbed on the cellulose and the mobile solvent for the components of the sample. The result is **liquid-liquid partition chromatography** (Section 7.8.1.b) because the compounds partition between the stationary water phase and the mobile solvent in a process identical with extraction. Compounds that are water soluble or that have the ability to hydrogen bond are more soluble in the stationary phase and do not migrate as readily as less polar molecules. The technique is used to separate highly polar or polyfunctional compounds such as amino acids and carbohydrates. The mobile phase usually contains water with an added miscible polar solvent such as methanol or ethanol.

Did you know?

Paper chromatography was first used in 1938 by British scientists R. Consden, A.H. Gordon and A.J.P. Martin. Martin was later the co-recipient of the Nobel Prize with Synge for his contributions to partition chromatography[5].

7.8.6 Gas-Liquid Chromatography (GC)

Gas chromatography is very useful for the qualitative and sometimes the quantitative analysis of mixtures of thermally stable, volatile compounds.

Did you know?

In 1941, Martin and Synge suggested using a gas as the mobile phase for partition chromatography. It wasn't until 1952, however, that the technique was first used.

7.8.6.a The Stationary Phase

The column is a coiled metal tube from 1 to 15 m long and usually about 2 to 4 mm in diameter. The length of the column determines the number of partitioning cycles and therefore the efficiency of the separation. It is packed with an inert solid support such as crushed fire brick or clay that is coated with a non-volatile liquid. This immobilized liquid serves as the stationary phase. There are also columns called capillary columns that have no packing material. The liquid phase is applied directly to the wall of the column. Capillary columns have less surface area than a packed column and the sample size therefore must be smaller (less than 0.1 μl) when compared with that for a packed column (0.1 to 2 μl). The liquids used as the stationary phase are classified on the basis of their polarity. Long-chain hydrocarbons, such as $C_{30}H_{62}$, have a low polarity as does silicone oil. Carbowax™ (polyethylene glycol) is moderately polar and diethylene glycol succinate is very polar.

---

[5] See footnote 3 on page 124.

Examples of Liquids Used as Stationary Phases

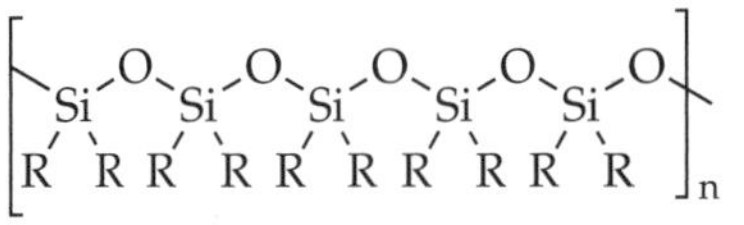

silicone oil 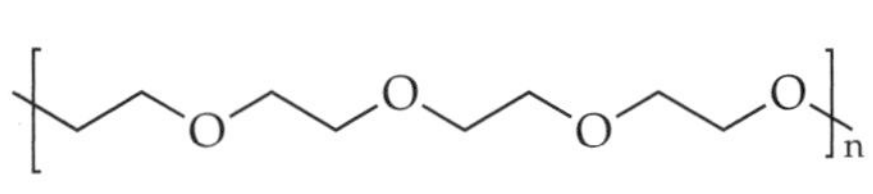   Carbowax™ (polyethylene glycol)

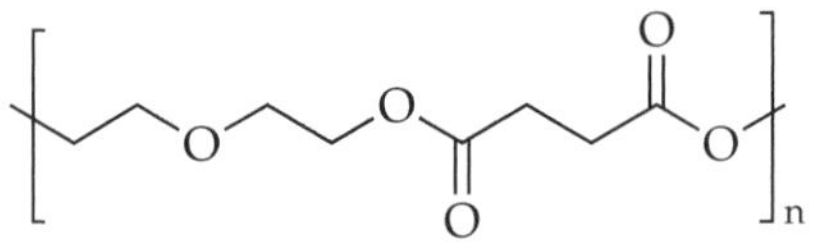

diethylene glycol succinate

The column is contained in a temperature-controlled oven. In some instruments, the temperature can be programmed to increase gradually during the elution of the compounds. If the column temperature is too high, there is no separation because the equilibrium distribution of the components between the liquid phase and the gas phase is not established. The temperature must be high enough, however, so that the components remain in the gas state. The boiling point of the liquid stationary phase determines the maximum temperature that can be used. If the column temperature is too high, the liquid coating bleeds off of the inert support. Many times, two columns—one polar and one non-polar—are connected side-by-side in the oven to enable the analysis of mixtures with different polarities. The sample is injected into only one of the columns, the one with the polarity most appropriate for the components in the sample.

7.8.6.b The Mobile Phase

The eluting solvent is an inert carrier gas, usually helium or nitrogen. Because all gases are infinitely miscible, the mobile phase does not interact with the molecules of the sample, but merely transports them through the column. Only the liquid stationary phase is involved in the separation of the components of the sample. The compounds are not adsorbed on the surface of the stationary phase, but dissolve in it so that the process is partition chromatography (Section 7.8.1.b) and is analogous to extraction. The flow rate of the carrier gas contributes to the success of the separation. If the flow rate is too high, the compounds are swept through the column without attaining equilibrium. A flow rate that is too slow increases the time necessary to elute the compounds.

7.8.6.c Applying the Sample to the Stationary Phase

The sample can be a gas, volatile liquid, or a volatile solid dissolved in a small amount of a volatile solvent. It is injected into a chamber where it is vaporized and mixed with the carrier gas that transports the compound(s) onto the column. The temperature of this inlet chamber is normally slightly higher than that of the column so the compounds in the sample must be thermally stable at this temperature. As the gaseous sample is carried through the column by the carrier gas, the components are partitioned between the stationary liquid phase and the carrier gas. The separated components concentrate in bands as they are dissolved and revaporized.

7.8.6.d Factors Affecting Separation

The success of the separation depends on the nature of the stationary phase, the column length, the flow rate of the carrier gas, the column temperature, and the amount of material being separated. The difference in the partition coefficients of the components depends only on their relative vapor pressures and the differences in their interaction with the liquid stationary phase. The compounds are eluted in the order of their vapor pressure with those with the higher vapor pressure moving more quickly through the stationary phase. If there is little or no interaction between the compounds and the stationary phase, one can view the sample as having an insoluble, non-volatile impurity (the stationary phase). As discussed in Section 6.4.1.a, insoluble, non-volatile impurities do not affect the intermolecular forces between the molecules of the substance. They do not alter its vapor pressure and consequently have no effect on its boiling point. Therefore, if the components of the sample do not interact with the liquid stationary phase, they are eluted from the column in the order of their boiling points, the component with the highest vapor pressure (lowest boiling point) coming off first.

On the other hand, if the functional groups of the sample interact with the liquid stationary phase, the vapor pressure of the sample in solution is affected. As discussed in Section 6.4.1.b, soluble, non-volatile impurities (the liquid stationary phase) lower the vapor pressure of the sample and raise its boiling point. Therefore, if a component interacts with the stationary phase, its vapor pressure is lowered. For example, if a polar compound and a non-polar compound have the same boiling point, the intermolecular forces between the non-polar substance and a non-polar stationary phase are greater than those for the polar compound. The vapor pressure of the non-polar component is therefore lowered more than that of the polar substance. It is eluted from the column after the polar molecule that has less affinity for the stationary phase and a higher vapor pressure. On the other hand, the polar compound has a greater affinity for a polar liquid stationary phase and its vapor pressure is lowered more so that the non-polar substance is eluted first if the stationary phase is polar.

It is not possible to predict the exact effect of the stationary phase on the vapor pressure of the components and therefore it is sometimes difficult to predict the order of elution of the compounds. A lower boiling material (higher vapor pressure) can be polar and more soluble on a polar adsorbent (more lowering of its vapor pressure) and a high boiling material (lower vapor pressure) can be non-polar. The order of elution depends on how much the vapor pressure of the low boiling material is lowered by the stationary phase. With non-polar stationary phases, the intermolecular attractions between the components and the stationary phase are much weaker and the order of elution usually follows the boiling points, so the material with the lowest boiling point (highest vapor pressure) comes off first followed by successively higher boiling compounds. Polar compounds that have the same functional group are usually separated on polar stationary phases in the order of their boiling point because the vapor pressures of the compounds are affected by the stationary phase to relatively the same degree.

To summarize, if compounds have similar polarities, the stationary phase has similar effects on their vapor pressures and they are moved through the column in the order of their boiling points. If the components have different polarities, the time they spend on the stationary phase depends on their vapor pressure, their relative polarity, and the polarity of the stationary phase. Consequently, the order of elution is more difficult to predict.

7.8.6.e Detection of Compounds

As the components leave the column, they pass through either a thermoconductivity detector or a flame ionization detector or into a mass spectrometer as described in Section 8.2.1.d.

With the thermoconductivity detector, the carrier gas used is helium, which has a higher thermal conductivity than most organic compounds. When a component flows over the detector wire, the wire is less efficiently cooled and heats up, increasing the resistance of the wire. The signals are recorded on a chart recorder to give a chromatogram that is the record of the resistance of the wire as a function of time. The signals show up as peaks and the **retention time** for a compound is the time from the injection of the sample to the maximum intensity of its peak (Figure 7.26).

In the case of the flame ionization detector, nitrogen is used as the carrier gas and the gas eluted from the column is mixed with oxygen or air and hydrogen and burned. The conductivity of the flame, which depends on the ionic content of the flame, is measured and the signals are plotted as a function of time. Instruments with flame ionization detectors are more sensitive than those with thermoconductivity detectors.

Figure 7.26   A Gas Chromatogram of a Mixture of Monoterpenes

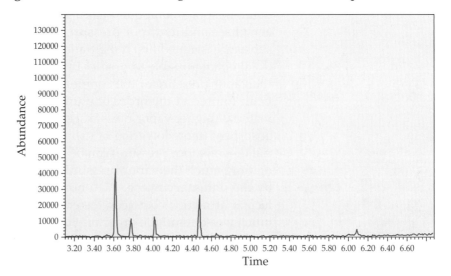

7.8.6.f Applications of Gas Chromatography

The number of signals in the gas chromatogram indicate the number of components in the mixture that have different retention times. As is the case with all types of chromatography, some compounds may not be separated under the conditions used (Section 7.8.2). A known compound can be coinjected with the sample to indicate that a given signal is different from the known if two signals result. However, a single signal for the compound and the known does not prove that the two are identical.

In instruments with either a thermoconductivity or flame ionization detector, the signals produced are proportional to the amount of material in the carrier gas. The areas under the peaks indicate the relative amounts of each of the components if the compounds have the same response to the detector. This is usually the case if the compounds are structurally similar. Otherwise, for quantitative determination, known amounts of compounds mixed with a third compound are injected and the response to the detector is determined. A factor is then applied to the ratio of the peak areas of the components to determine the relative amounts of the components in the sample.

7.8.7 High Performance or High Pressure Liquid Chromatography (HPLC)

At the same time that Martin and Synge[6] suggested using gas as the mobile phase, they also proposed that chromatographic separations would be improved if smaller particles were used for the stationary phase in order to increase the surface area of the adsorbent. The rate of flow of the mobile phase could be increased by applying high pressure to the column. The development of **high performance or high pressure liquid chromatography** was delayed until the mid 1960's, however, by the introduction of gas-phase chromatography.

HPLC can be used for analytical or preparative work and for adsorption or partition chromatography. The stationary phase is in a metal column that is shorter than a GC column and that can withstand pressures of 1000 to 6000 psi. The process is similar to GC in that the sample is injected into the mobile liquid phase. The eluent is forced through the stationary phase by high pressure applied by a pump. HPLC has the advantage that the compounds do not have to be vaporized so that it can be used for compounds that are not volatile or are not thermally stable. The only restriction is that the compounds must be soluble in a solvent. Whereas separations by GC depend only on the stationary phase, in HPLC the stationary phase and the mobile phase affect the selectivity. As with column chromatography, the polarity of the mobile phase can be varied by the gradual addition of a solvent with a different polarity. HPLC is useful for the separation of natural products such as carbohydrates, steroids, nucleosides, amino acids, alkaloids, peptides, and proteins. It has the disadvantage that it is usually slower than GC, not as sensitive, uses large volumes of solvent, and the instrumentation is more expensive.

The stationary and mobile phases used in HPLC are similar to those used in column chromatography. Silica gel is used extensively for normal-phase

---

[6] Nobel Prize in Chemistry in 1952. See footnote 3 on page 124.

chromatography (adsorption chromatography, Sections 7.8.1.a and 7.8.1.c). For the separation of compounds such as amino acids by reverse-phase chromatography (partition chromatography, Sections 7.8.1.b and 7.8.1.d), the silica gel is modified by replacing the hydroxyl functions with ether groups where the R is a linear alkyl group (page 126). In this case, a mobile phase is used that is more polar than the stationary phase, so the polar compounds are eluted first.

The components of the separated mixture are detected by an instrument that measures the conductivity, ultraviolet light absorption, or some other physical property of the eluent as it leaves the column. The resulting signals are recorded as a function of time to give a trace called a chromatogram similar to that obtained in gas chromatography.

CHAPTER 8

Structure Determination

The development of methods for determining the connectivity of atoms in a molecule has played a significant role in the advancements made in organic chemistry. The primary methods for structural analysis currently used by organic chemists are X-ray crystallography, nuclear magnetic resonance (NMR) spectroscopy, and mass spectrometry (MS).

X-ray crystallography is the most definitive technique because it provides direct information about bonding in a molecule. The final result of an X-ray crystal structure is a ball-and-stick drawing of the molecule (Figure 8.1). However, X-ray crystallography is not a routine form of analysis because the collection and analysis of data can take anywhere from one to several days. Furthermore, it is expensive. Finally, as the name implies, X-ray crystallography is limited to crystalline analytes. Organic chemists generally turn to X-ray crystallography only when a combination of other techniques does not provide definitive results.

NMR spectroscopy and mass spectrometry are the two most commonly used methods for determining the structure of organic molecules. Both methods are capable of analyzing solids, liquids, and gases and the normal time required for such analyses is on the order of minutes to hours.

There are volumes of work dedicated to the understanding and use of both NMR spectroscopy and mass spectrometry. These treatises range from very sophisticated treatments of the underlying principles behind these experimental techniques to the basic explanations found in undergraduate textbooks. NMR spectroscopy and mass spectrometry are important enough to be considered disciplines and each has several scientific journals

Figure 8.1 An X-Ray Crystal Structure of an Organic Molecule

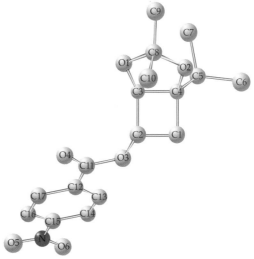

dedicated to disseminating the most up-to-date research in their field. The primary purpose of the following sections on NMR spectroscopy and mass spectrometry is to introduce you to how these methods can be used to determine the structures of organic molecules.

8.1 NUCLEAR MAGNETIC RESONANCE SPECTROSCOPY

NMR spectroscopy is an analytical method used by chemists to help determine the connectivity and spatial orientation of atoms in a molecule. The NMR experiment has its roots in the physics community and the three pioneers of this field, Isidor Rabi[1], Felix Bloch[2], and Edward Purcell[2], were awarded Nobel prizes in physics for their contributions. In 2002, one half of the Nobel Prize in Chemistry was awarded to Kurt Wüthrich[3] for his work in applying NMR to the study of structurally complex biomolecules. In 2003, Paul Lauterbur and Sir Peter Mansfield[4] were awarded the Nobel Prize in Medicine or Physiology for their contributions to making magnetic resonance imaging (MRI), a type of NMR experiment, an important tool in medical diagnostics.

The theory behind an NMR experiment is well beyond an introduction to interpreting NMR data, but beginning students in organic chemistry can make great strides in obtaining important structural information from NMR spectra without a comprehensive understanding of the theory.

Nuclear magnetic resonance is an appropriate description of the phenomenon used to obtain structural information about a molecule because the technique involves nuclei resonating in a magnetic field. Table 8.1 lists some relevant information about nuclei commonly encountered in organic molecules. The only nuclei that can be detected in an NMR experiment are those possessing a nuclear spin not equal to zero. Hydrogen and carbon can both be detected using NMR even though the most abundant isotope of carbon (^{12}C) does not have a nuclear spin. The carbon NMR signal comes from

Table 8.1 **Common Nuclei in Organic Molecules**

Nucleus	Nuclear Spin	% Natural Abundance	Nucleus	Nuclear Spin	% Natural Abundance
^1H	1/2	99.98	^{31}P	1/2	100
^2H	1	0.02	^{32}S	0	94.8
^{12}C	0	98.9	^{33}S	3/2	0.8
^{13}C	1/2	1.1	^{34}S	0	4.4
^{14}N	1	99.6	^{35}Cl	3/2	75.5
^{15}N	1/2	0.4	^{37}Cl	3/2	24.5
^{16}O	0	99.6	^{79}Br	3/2	50.5
^{17}O	5/2	0.04	^{81}Br	3/2	49.5
^{18}O	0	0.2	^{127}I	5/2	100
^{19}F	1/2	100			

[1] See http://nobelprize.org/nobel_prizes/physics/laureates/1944/.
[2] See http://nobelprize.org/nobel_prizes/physics/laureates/1952/.
[3] See http://nobelprize.org/nobel_prizes/chemistry/laureates/2002/index.html.
[4] See http://nobelprize.org/nobel_prizes/medicine/laureates/2003/.

the 1.1% of the carbon–13 (^{13}C) isotope occurring naturally in compounds. The following discussion focuses on both ^{1}H and ^{13}C NMR spectroscopy.

When the nuclei in a molecule with a nuclear spin of 1/2 are subjected to a strong magnetic field (B_0), they orient themselves either with or against the field resulting in two energetically different **spin states** referred to as alpha (α) and beta (β) (Figure 8.2).

The energy separation (ΔE) between the two spin states α and β is dependent on the magnetic field strength. In the absence of an external magnet ($B_0 = 0$), the difference in energy (ΔE) is zero. As the strength of the field increases, ΔE increases. The presence of this energy difference between the two spin states is the basis of the NMR experiment. Nuclei in the lower energy α state can absorb energy and be excited to the β state when electromagnetic radiation of the correct frequency is applied. This frequency is known as the **resonant frequency** and is determined by $\Delta E = h\nu$, where h is Planck's constant and ν is the frequency of electromagnetic radiation. The energy differences are very small (10^{-6} kcal/mol) and correspond to frequencies in the radio wave region of the electromagnetic spectrum. The excitation to a higher energy spin state is known as a **spin flip** or **resonance**, and it is measured and plotted in a graphical form known as an **NMR spectrum** (Figure 8.3). The resonant frequencies for carbon and hydrogen at different magnetic field strengths are shown in

Figure 8.2   The Influence of an External Magnetic Field (B_0) on Spin-State Splitting

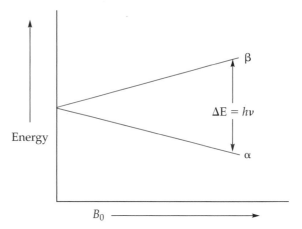

Figure 8.3 ^{1}H NMR Resonance at 7 Tesla

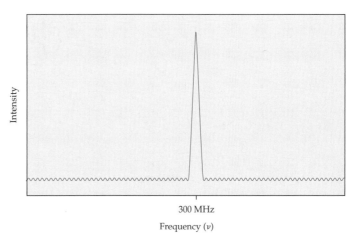

Table 8.2. It is important to note that the resonant frequencies of these nuclei are different, allowing us to obtain separate NMR spectra for the carbon and hydrogen atoms in an organic molecule.

Because of the strength of the magnet required, NMR instruments are very expensive. The cost would be prohibitive if they were capable only of determining if a compound contained hydrogen and carbon.

8.1.1 Determining the Number of Different Types of Hydrogen and Carbon

In addition to the differences in the resonance frequencies of hydrogen and carbon (Table 8.2), hydrogen and carbon atoms that are located in different electronic environments within a molecule resonate at slightly different frequencies. These differences are small and are measured in hertz (Hz, 10^{-6} MHz), but they are detectable. Consequently, NMR spectra provide an important tool for determining the number of different or nonequivalent hydrogens or carbons in a molecule.

For example, the ^{1}H and ^{13}C NMR spectra of dimethyl ether each show only one resonance. A mirror plane can be drawn through the oxygen atom and its two lone pairs of electrons (Figure 8.4). Because of this symmetry, the two methyl groups are equivalent so that the two carbon atoms are identical and they appear as only one resonance in the carbon spectrum. To understand the single resonance in the hydrogen NMR, one must examine a Newman projection looking down the oxygen–carbon bond as shown in Figure 8.4. It is evident in the conformations shown that, at any

Table 8.2 Resonant Frequencies of ^{1}H and ^{13}C at Different Magnetic Field Strengths

Magnetic Field Strength (Tesla)	^{1}H Resonant Frequency (MHz)	^{13}C Resonant Frequency (MHz)
2.35	100	25.3
7	300.13	75.47
11.7	500	125.7

Figure 8.4   Conformers of Dimethyl Ether Looking Down an Oxygen-Carbon Bond

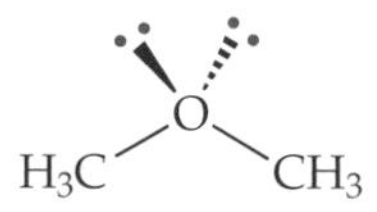

dimethyl ether

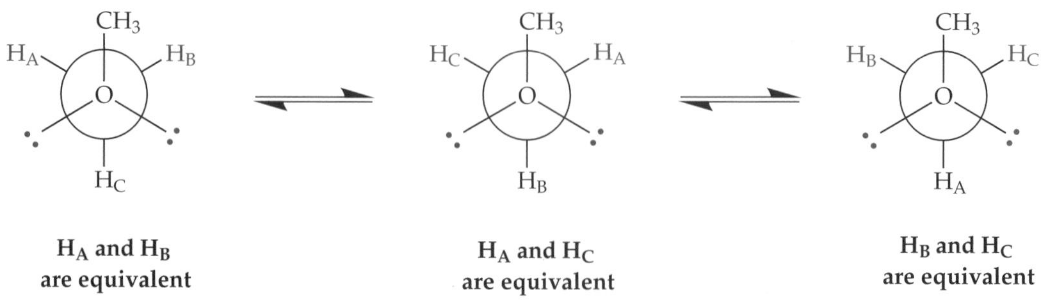

H_A and H_B
are equivalent

H_A and H_C
are equivalent

H_B and H_C
are equivalent

given time, two of the hydrogen atoms are equivalent to each other but the third is not. If the time it took to run the NMR experiment were on the same time scale as bond rotation, we would expect to see two different hydrogen resonances—one from the two hydrogens that are adjacent to the second methyl group and one from the hydrogen that is across from the second methyl group. However, unless we label each hydrogen, there is no way to distinguish them if there is rapid bond rotation so that the three hydrogen atoms of each of the methyl groups appear to be equivalent. Because bond rotation is much faster than the time scale of the NMR experiment and the two methyl groups are equivalent, a single resonance representing the average of all conformations is observed for the six hydrogens in dimethyl ether.

The ^{1}H NMR and ^{13}C[^{1}H] NMR spectra of *tert*-butyl methyl ether are shown in Figure 8.5. (The notation [^{1}H] after the ^{13}C will be discussed in Section 8.1.4.c.) This molecule contains twelve hydrogens and, unlike dimethyl ether, the two alkyl groups are not the same so we expect to see more resonances (signals) in the ^{1}H NMR spectrum (Figure 8.5a). The spectrum shows only two resonances for *tert*-butyl methyl ether. As discussed for dimethyl ether, the hydrogens in the methyl group are equivalent and are responsible for one resonance. Similarly, the hydrogens in each of the methyl groups of the *tert*-butyl substituent are equivalent because of free rotation. Free rotation also explains why the three methyl groups of the *tert*-butyl group are equivalent. Consequently, the nine hydrogens of the *tert*-butyl residue all resonate at the same frequency, resulting in one signal for all of these hydrogens. In the ^{13}C[^{1}H] NMR spectrum (Figure 8.5b), three resonances account for all five of the carbons. The tertiary carbon of the *tert*-butyl group and the methoxy carbon are unique and the three methyl carbons of the *tert*-butyl group appear equivalent due to rapid bond rotation.

Let's examine how many hydrogen and carbon resonances we would expect to see in the NMR spectra of the molecules shown in Figure 8.6. For 1,2–dichloroethane, there is a point of symmetry in the molecule so that the two halves of the molecule are equivalent. Therefore, we would expect to see only one hydrogen resonance and one carbon resonance.

For 1,1,2-trichloroethane, we would see two carbon and two hydrogen resonances. Because the molecule has a mirror plane containing the two

Figure 8.5a ^{1}H NMR Spectrum of *tert*-Butyl Methyl Ether

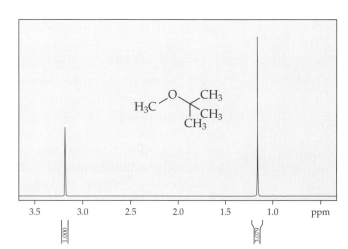

Note: The numbers in the brackets will be explained in Section 8.1.3.b.

Figure 8.5b $^{13}C[^1H]$ NMR Spectrum of *tert*-Butyl Methyl Ether

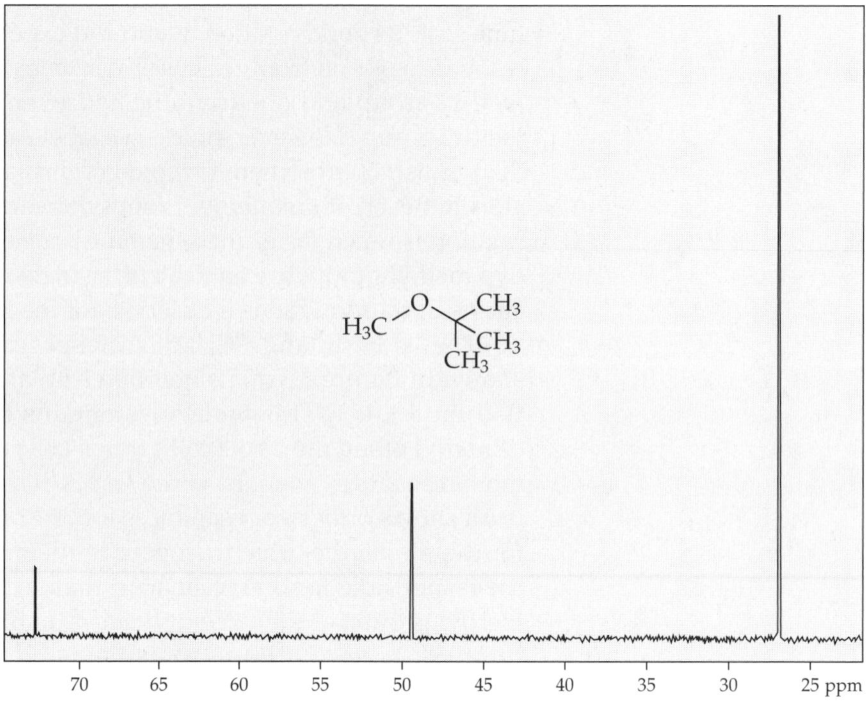

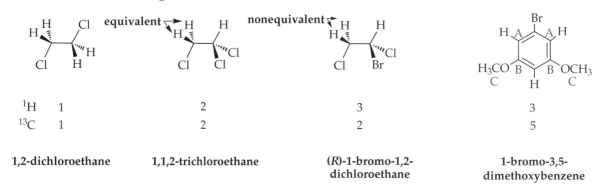

Figure 8.6   The Number of ^1H and ^{13}C NMR Resonances Expected in Some Organic Molecules

	1,2-dichloroethane	1,1,2-trichloroethane	(R)-1-bromo-1,2-dichloroethane	1-bromo-3,5-dimethoxybenzene
^1H	1	2	3	3
^{13}C	1	2	2	5

carbons and bisecting the chlorines on carbon 1 and the hydrogens on carbon 2, the two hydrogens on carbon 2 are equivalent.

Changing one of the chlorines in 1,1,2-trichloroethane to bromine to give 1-bromo-1,2-dichloroethane results in a chiral molecule. Again, there would be two carbon resonances. However, there is no plane of symmetry in this molecule and even rapid rotation about a carbon-carbon bond does not make the two hydrogen atoms on carbon 2 equivalent. Therefore, we would expect to see three different hydrogen resonances for 1-bromo-1,2-dichloroethane.

In 1-bromo-3,5-dimethoxybenzene, there is a plane of symmetry that is perpendicular to the plane of the ring and contains the bromine atom, the carbon bonded to it, and the carbon-hydrogen bond across the ring. This makes the two methoxy groups equivalent as well as the two sets of remaining ring carbons, resulting in five carbon resonances. (The equivalent carbons are labeled with letters.) Three hydrogen resonances arise from the six equivalent hydrogens on the methoxy groups, the two equivalent hydrogens *ortho* to the bromine, and the single hydrogen *para* to the bromine.

Cyclohexane has six carbons and twelve hydrogens and it might be thought of as being symmetrical at first glance. However, it is important to consider that the molecule exists predominantly in the chair conformation. In the case of carbon, the chair conformation has a three-fold axis of symmetry as can be seen by rotating it by 120° around the center of the ring. (This is most easily observed by building a model.) All six carbons are therefore equivalent and only one resonance is seen in the carbon spectrum. There are two distinct types of hydrogens, the axial and the equatorial. Again, relative to the NMR time scale, the two chair forms are rapidly interconverting at room temperature so that we observe an average of the axial and equatorial hydrogens and only one resonance is recorded. However, if the solution is cooled to a temperature of less than −89°C, the process of ring flipping slows down sufficiently so that two signals are observed, one for the axial hydrogens and one for the equatorial hydrogens.

**chair conformation
of cyclohexane**

8.1.2 The Chemical Shift in ¹H and ¹³C NMR Spectra

Nonequivalent hydrogens and carbons in a molecule resonate at different frequencies depending on the electronic environment of the atom. In order to standardize the differences between these frequencies, the resonances of all hydrogens and carbons in a molecule are measured by the distance of their signal from that of tetramethylsilane (TMS), $(CH_3)_4Si$. This molecule has the advantage that both its hydrogen and carbon spectra show single resonances that are at lower frequencies and do not overlap with the signals from those in the majority of organic molecules. Therefore, when an NMR spectrum is recorded, a small amount of TMS is added to the sample and the distance between a resonance and the TMS peak is measured.

Because the frequency of a resonance is dependent on the external magnetic field strength (Figure 8.2 and Table 8.2), the distance of the resonance from the TMS signal measured in hertz (Hz) depends on the magnetic strength of the instrument used to record the spectrum. To standardize the reporting of the frequency of a signal, a ratio called the **chemical shift** or **δ** is used. *The chemical shift, δ, is defined as the difference in the distance of the signal from that of TMS in Hz divided by the frequency of the NMR spectrometer in megahertz (MHz = 10^6 Hz). The resulting chemical shift value in parts per million (ppm) is therefore independent of the magnetic field strength (Equation 8.1).*

$$\text{Chemical shift } \delta \text{ (ppm)} = \frac{\text{Resonance frequency of sample (Hz)} - \text{Resonance frequency of TMS (Hz)}}{\text{Frequency of NMR spectrometer (MHz)}} \tag{8.1}$$

Table 8.3 ^{1}H NMR Chemical Shifts for Common Functional Groups

Chemical Shift Range (ppm)	Type of Hydrogen	Chemical Shift Range (ppm)	Type of Hydrogen
0.2–0.8	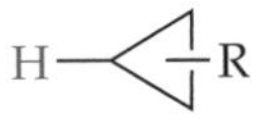 H—⊲—R	2.7–4.0	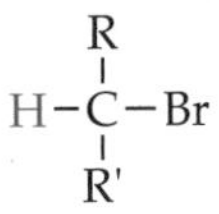 H—C(R)(R')—Br
0.8–1.2	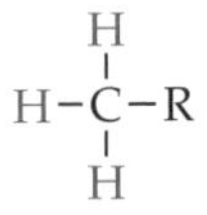 H—C(H)(H)—R	2.8–3.8	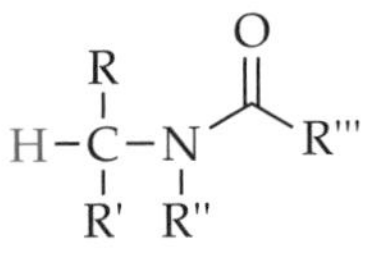 H—C(R)(R')(R'')—N—C(=O)R'''
1.2–1.8	 H—C(H)(R')—R	3.1–4.0	 H—C(R')—Cl
1.4–1.8	 H—C(R')(R'')—R	3.2–3.6	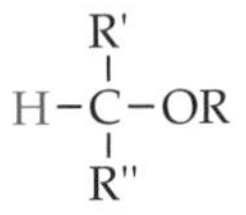 H—C(R')(R'')—OR
1.6–2.2	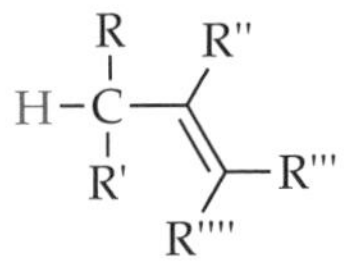 H—C(R)(R')—C(R'')=C(R''')(R'''')	3.2–3.6	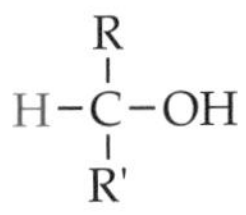 H—C(R)(R')—OH
1.8–2.6	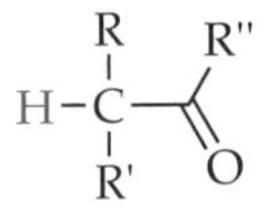 H—C(R)(R')—C(R'')(=O)	3.6–4.8	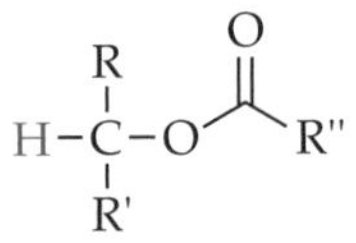 H—C(R)(R')—O—C(=O)R''
1.9–3.0	H—C≡CR	4.2–4.8	 H—C(R)(R')—F
2.0–2.8	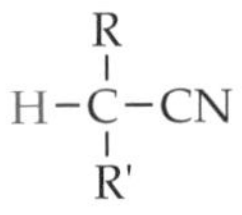 H—C(R)(R')—CN	4.6–5.7	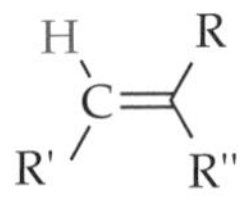 (H)(R)C=C(R')(R'')
2.1–3.1	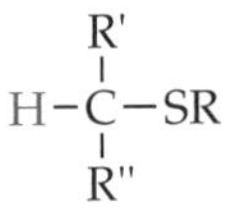 H—C(R')(R'')—SR	5.5–6.0	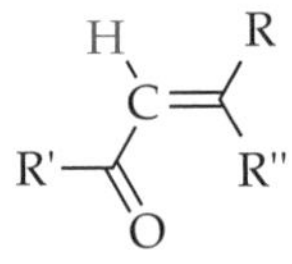 (H)(R)C=C(R')(R''); R'C=O
2.2–2.9	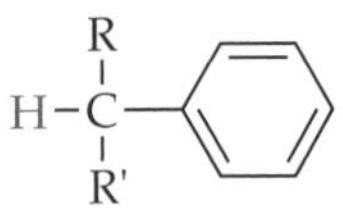 H—C(R)(R')—C$_6$H$_5$	6.0–7.5	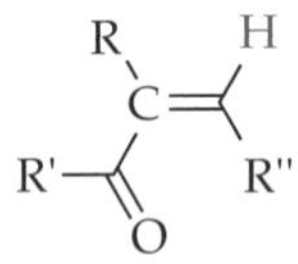 (R)(H)C=C(R')(R''); R'C=O
2.2–2.8	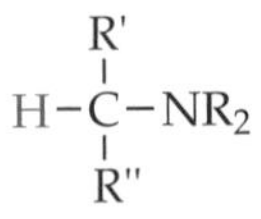 H—C(R')(R'')—NR$_2$	6.0–8.5	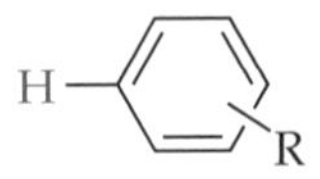 H—C$_6$H$_4$—R
2.2–4.2	 H—C(R)(R')—I	9.0–10.0	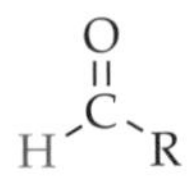 H—C(=O)R

Tetramethylsilane has been assigned a chemical shift value of 0.0 ppm and the resonance of all other hydrogen and carbon nuclei are measured in ppm relative to this value. Carbon is more electronegative than silicon so there is greater electron density around the methyl groups attached to silicon. Because of this, the carbons and hydrogens in TMS resonate at significantly lower frequencies than those found in most organic compounds lacking silicon. All NMR spectra are plotted with the frequency increasing from right to left with the signal for TMS at the far right of the spectrum. The resonances for the hydrogens or carbons in most organic compounds are therefore to the left of the signal for TMS. When a signal is to the left of another signal, it is sometimes referred to as **downfield** (less electron density). Similarly, if a signal is to the right of another signal, it is said to be **upfield**. If a resonance is located at a lower frequency than TMS (to the right of TMS), it is recorded as a negative chemical shift (minus($-$) ppm). Typical chemical shifts for common functional groups are shown in Tables 8.3 and 8.4.

Table 8.4 ^{13}C NMR Chemical Shifts of Common Functional Groups

Chemical Shift Range δ (ppm)	Type of Carbon	Chemical Shift Range δ (ppm)	Type of Carbon
0–5	 R'RC◁	50–80	C–OR alcohols and ethers
5–30	H–C–R (with H, H)	75–95	R–C≡C–R'
25–45	H–C–R (with H, R')	115–130	R–C≡N
35–60	H–C–R (with R', R")	105–145	R,C=C,R''' / R',R"
30–50	R–C–R" (with R''', R')	115–160	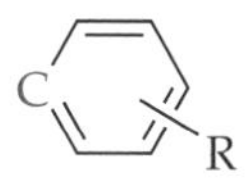 C⬡R
70–80	C–F	150–185	R₂N–C(=O)–R'
25–50	C–Cl	150–185	RO–C(=O)–R' carboxylic acids and esters
10–40	C–Br	185–220	R–C(=O)–R' aldehydes and ketones
0–30	C–I		
27–60	C–NR₂ amines		

Figure 8.7   ¹H and ¹³C NMR Chemical Shifts in Ppm for Dimethoxymethane

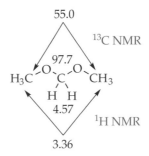

Chemical shift tables provide valuable information about the structure of a molecule. Electronegative elements such as the halogens, oxygen, and nitrogen have a tendency to shift resonances to higher frequencies (larger ppm) relative to alkyl groups. This effect can be additive as demonstrated for dimethoxymethane (Figure 8.7). The hydrogens of the two equivalent methyl groups have a chemical shift of 3.36 ppm that is within the expected range for ethers (3.3–3.9 ppm). The methylene group has two oxygen atoms attached to it and the hydrogens resonate at 4.57 ppm, a frequency higher than expected because of the additive effect of the two oxygens. The same trends are seen for the carbon chemical shifts.

When hydrogen is attached to an sp^2 hybridized carbon, the resonance is generally at a higher frequency than that for the hydrogens in most other functional groups, that is, the chemical shift (δ) in ppm is greater. The same is true for the ¹³C chemical shift of sp^2 hybridized carbons.

Some care must be used when drawing conclusions about structures based solely on chemical shift data. Many functional groups have overlapping chemical shift ranges. For example, the ranges for hydrogen attached to a carbon that is bonded to either a chlorine atom, bromine atom, or alkoxy group (an ether) overlap. In addition, no single table of chemical shift data can account for all possible arrangements of atoms and one must assume that there will be molecules for which the effect of a given functional group will lead to a resonance falling outside of the reported range.

8.1.3 Obtaining Structural Information from ¹H NMR Spectra

8.1.3.a Spin-Spin Coupling in ¹H NMR Spectra

As seen in Section 8.1.2, chemical shift data from NMR can provide us with information about the types of functional groups present in molecules. However, this information does not definitively reveal the connectivity of the atoms and the structure of a molecule.

Let's examine the ¹H NMR spectrum of 1,1,2-trichloroethane, a molecule that we predicted (Figure 8.6) would give rise to two unique hydrogen resonances. Figure 8.8a is a sketch of the predicted spectrum using the known chemical shifts of these two types of hydrogens. Figure 8.8b is an artist's rendition of the actual ¹H NMR spectrum of 1,1,2-trichloroethane showing five peaks for the three hydrogens! This is impossible if every peak represents a unique hydrogen so another explanation is needed.

Figure 8.8   The ^{1}H NMR Spectrum of 1,1,2-Trichlorethane

(a) Predicted ^{1}H NMR spectrum

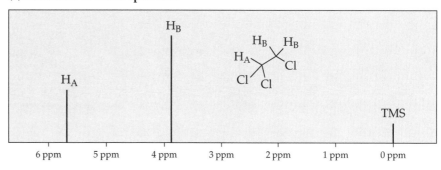

(b) Observed ^{1}H NMR spectrum

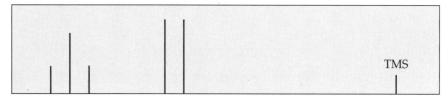

Figure 8.9a   Spin-Spin Coupling Leading to a Doublet

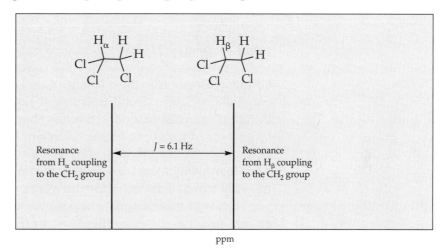

Notice that the two groupings of peaks are centered about the single peaks in Figure 8.8a. These groupings are known as **multiplets** and are the result of a phenomenon known as **spin-spin coupling**. That is, the spin state of H$_A$ (page 143) has an effect on the two equivalent methylene hydrogens (H$_B$) next to it and vice versa. The two resonances close to 4.0 ppm arise from H$_A$ coupling to the CH$_2$ group. This lone hydrogen has two possible spin states, α and β, each of which couple to the two equivalent hydrogens of the methylene group (H$_B$). Because there are approximately an equal number of hydrogens in the lower energy α spin state and in the higher energy β spin state, their effect on the CH$_2$ group will be equal in magnitude and opposite in direction. The CH$_2$ resonance is "split" into two peaks of equal height and equal distance from where the resonance would be if there were no coupling (Figure 8.9a). This multiplet is known as a **doublet (d)** and the chemical shift is reported as the midpoint of the two peaks.

Figure 8.9b   Spin-Spin Coupling Leading to a Triplet

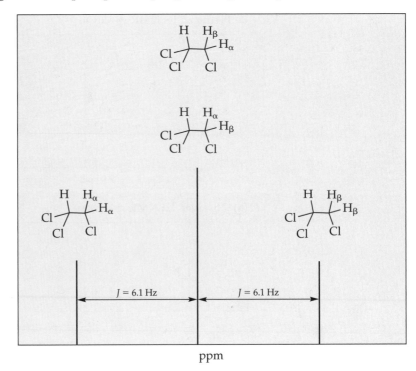

ppm

The three peaks centered at 5.8 ppm arise from the two equivalent H_B atoms coupling to H_A. Two hydrogens have four possible combinations of α and β spin states (Figure 8.9b). The two combinations for which one neighboring hydrogen is in the α spin state and one is in the β spin state have an equal and opposite effect so that they do not alter the resonance frequency. In the other two combinations, both neighboring hydrogens are either in the α spin state and result in a higher resonance frequency or in the β spin state to give a lower resonance frequency. The result is a three-line multiplet called a **triplet (t)**, where the intensities of the lines are 1:2:1. The chemical shift is reported as the midpoint of the three lines, the tallest peak of the triplet.

The distance between the peaks of a multiplet is measured in hertz (Hz) and is known as the **coupling constant (J)**. *Coupling constants are measured in Hz rather than ppm because, unlike chemical shifts, they are independent of the magnetic field strength used to record the spectrum.*

It is a common practice, when assigning the resonances in an NMR spectrum, to label all nonequivalent hydrogen and carbon atoms with a letter. We have used the notation CH_A and CH_BH_B for 1,1,2-dichlorethane. The coupling constant for H_A coupling to H_B is 6.1 Hz and is expressed as $J_{AB} =$ 6.1 Hz. Because H_A and H_B are coupled to one another, the distance between each peak of the triplet must be 6.1 Hz. That is, $J_{AB} = J_{BA} = 6.1$ Hz.

The ^{1}H NMR of methyl propanoate is shown in Figure 8.10. Based on chemical shifts alone, we can assign each of the three different types of hydrogens to a given multiplet.

The hydrogens of the methoxy group (3.62 ppm) show up as a single resonance known as a **singlet (s)**. The closest hydrogens are five bonds away, which is normally too far to cause coupling.

We have already discussed that two equivalent hydrogens will couple to the hydrogens on a neighboring carbon. This coupling leads to a triplet (t) as seen for the methyl group hydrogens at 1.09 ppm.

Figure 8.10   ¹H NMR Spectrum of Methyl Propanoate

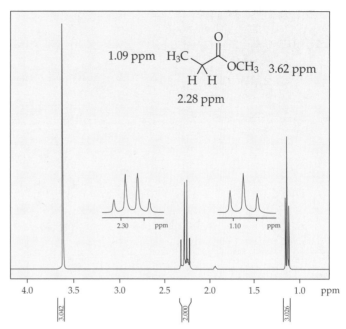

Note: The numbers in the brackets will be explained in Section 8.1.3.b.

The signal at 2.28 ppm is from the CH_2 group that is coupled to three hydrogens on the neighboring carbon atom. For these three hydrogens, there are a total of eight possible combinations of α and β spin states (α,α,α; α,α,β; α,β,α; β,α,α; α,β,β; β,α,β; β,β,α; β,β,β) and their effect on a neighboring hydrogen or group of equivalent hydrogens is shown in Figure 8.11. The result is known as a **quartet (q)** with peak intensities of 1:3:3:1. The chemical shift of a quartet is reported as the midpoint of the two inner resonances.

From the above analyses, it appears that there are predictable outcomes associated with spin-spin coupling. The general rule for the multiplicity of a signal caused from N equivalent neighboring nuclei is 2NI + 1, where I is the nuclear spin of the neighboring nucleus (Table 8.1). The nuclear spin of hydrogen is 1/2 so that coupling caused by N equivalent hydrogens can be summarized by the **N + 1 rule** that states: *The coupling of N equivalent hydrogens with one or more different hydrogens will lead to a multiplet of N + 1 lines* (Table 8.5).

Table 8.5 Application of the N + 1 Rule

N Equivalent Hydrogens	Multiplet Due to Coupling With N Equivalent Hydrogens	Idealized Peak Intensities of Multiplet
0	singlet (s)	1
1	doublet (d)	1:1
2	triplet (t)	1:2:1
3	quartet (q)	1:3:3:1
4	quintet (quin)	1:4:6:4:1
6	septet (sep)	1:6:15:20:15:6:1

Note: N = 5 is not shown because a CH_2 and CH_3 group cannot be equivalent.

Figure 8.11 Spin-Spin Coupling Leading to a Quartet

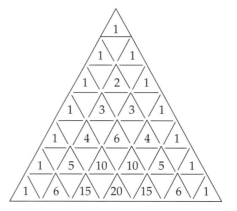

ppm

Figure 8.12 Pascal's triangle

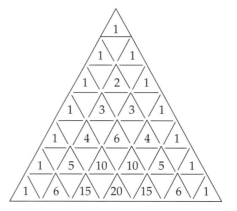

The relative intensity of these lines (Table 8.5) is predicted by Pascal's triangle (Figure 8.12).

According to the analysis of the ¹H NMR spectrum of methyl propanoate (Figure 8.10), the triplet and quartet should have peak ratios of 1:2:1 and 1:3:3:1 respectively. Careful inspection indicates that the two multiplets are not perfectly symmetrical. That is, the peaks closest to the other multiplet are taller and the peaks furthest from the other multiplet are shorter than predicted. This skewing of resonances occurs when two groups of hydrogens that couple with each other have similar chemical shifts. The closer the chemical shifts are, the more skewed the multiplets become. This point is illustrated in Figure 8.13 that shows a pair of nonequivalent geminal hydrogens (CH_AH_B).

Figure 8.13   The Effect of the Relative Chemical Shift on Coupled Resonances

(a) The chemical shifts are well separated

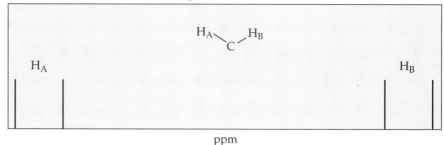

ppm

(b) The chemical shifts are more similar

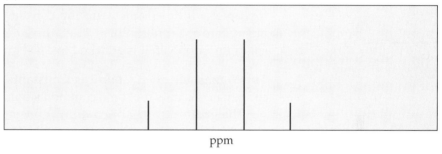

ppm

(c) The chemical shifts are almost equal

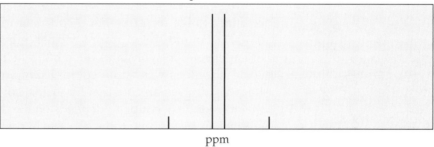

ppm

Note: The smaller resonances are often hidden in the baseline of the spectrum.

In Figure 8.13a, the chemical shifts are well separated and each signal is split into a doublet by coupling with the other geminal hydrogen.

As the chemical shift (δ) difference between the two doublets decreases, the height of the inner peaks increases and the outer peaks become smaller. There is a point where the two doublets resemble a quartet (Figure 8.13b). This pattern is often referred to as an **AB quartet**, but it does not represent a true quartet as defined by the N + 1 rule. There are three methods that can be used to distinguish the type of quartet being observed.

- Measure the distance between each of the four peaks. If these distances are not the same, it cannot be an N + 1 quartet.

- If the distances between the peaks are equal or close to being equal (±0.1 Hz), it is still possible that the multiplet is an AB quartet. Look for another multiplet with the same distance between its peaks. Multiplets from H–H coupling, J_{AB}, with a neighboring group always come in pairs so that there should be another signal with an identical coupling constant, J_{BA}, if the quartet is due to a neighboring methyl group.

- If all else fails, retake the spectrum on an instrument with a different magnetic field strength. Recall that chemical shifts are influenced by the strength of the external field, but coupling constants, *J*, are not. If it is a quartet that is due to the N + 1 rule, the distance between the peaks will not change.

Where the chemical shifts are almost equal (Figure 8.13c), the outer peaks become smaller as the inner peaks increase in size. Sometimes the small outer peaks are not distinguishable from the baseline and the signal resembles a doublet.

When the chemical shifts for H_A and H_B are identical, a singlet is observed. This is the same as one observes if H_A and H_B are equivalent.

Spin-spin coupling data is helpful in the assignment of atom connectivity. For example, if there is a quartet in a ^1H NMR spectrum, we know there is a methyl group next to the group in question. Deciphering the structure of the molecule relies on finding the adjacent methyl resonance, that is, a multiplet with the same coupling constant. (Remember $J_{AB} = J_{BA}$.) If the methyl group is split into a triplet, there is a CH_2 group attached to it. If it split into a doublet, it is attached to a CH group.

The Magnitude of ^1H Coupling Constants   Table 8.6 gives examples of ^1H coupling constants for some common situations.

The greatest range of coupling constants is 2–30 Hz and arises from nonequivalent geminal hydrogens (CH_AH_B) as previously discussed. This type of coupling is referred to as **two-bond** or **geminal coupling**.

Three-bond coupling, also known as **vicinal coupling**, occurs between hydrogens on adjacent carbon atoms. As shown in Table 8.6, the coupling constant range for three-bond coupling in saturated systems is between 0–10 Hz. For most alkyl groups, vicinal coupling ranges from 6–8 Hz. Smaller or larger coupling constants often indicate that the conformation of the molecule is somehow constrained, for example by being part of a ring. Martin Karplus[5] derived an equation for vicinal coupling constants that shows that the magnitude of coupling depends on the dihedral angle between two nonequivalent hydrogen atoms. For freely rotating bonds, this is not an issue because the data obtained from the NMR experiment is the average of all conformations. In rings, however, the number of accessible conformations is constrained. Plotting the Karplus equation leads to what is referred to as the **Karplus curve** (Figure 8.14). The curve shown in Figure 8.14 is for vicinal coupling between hydrogens attached to two neighboring sp^3 hybridized carbons. Notice that, if dihedral angles are in the range of 80–100 degrees, the coupling constant can be less than 1 Hz. Such data not only helps to establish connectivity in a molecule, but can also provide important information about bond angles and stereochemistry.

Four-bond coupling is relatively rare and is most often observed for coupling between an allylic hydrogen and an alkene hydrogen (allylic coupling), or between *meta* hydrogens (1,3 relationship) on a phenyl ring.

Five-bond coupling is important only in substituted benzenes where two nonequivalent hydrogens with a 1,4 relationship (*para*) couple with one another.

Obtaining Coupling Constants from NMR Spectra   The NMR spectrum records the signals in ppm from a standard (Section 8.1.2). Therefore, the

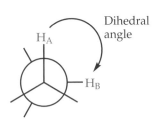

Dihedral angle

H_A

H_B

---

[5] Karplus, M., J. Am. Chem. Soc., *85*, 2870 (1963).

Table 8.6 ¹H Coupling Constants

Type of Coupling	Coupling Constant (Hz)	Type of Coupling	Coupling Constant (Hz)
H$_A$–C–H$_A$	0	R,R'C=CH$_A$H$_B$	0–3
H$_A$–C–H$_B$	2–30 (geminal)	R,R'C=C(H$_A$)(CH$_{3(B)}$)	4–10
H$_A$–C–C–H$_B$	0–10 (vicinal)	ortho	6–10 (1,2 or *ortho*)
H$_A$–C–C–C–H$_B$	0–1 (rare)	meta	1–3 (1,3 or *meta*)
H$_A$–C–C=C–H$_B$	2–3 (allylic)	para	0–1 (1,4 or *para*)
H$_A$,R C=C H$_B$,R'	6–12 (*cis*)	R–C(H$_A$)–C(=O)(H$_B$), R'	1–3
H$_A$,R C=C R',H$_B$	12–18 (*trans*)		

Figure 8.14   The Karplus Curve for Estimating Dihedral Angles from Vicinal Coupling in Rigid Systems

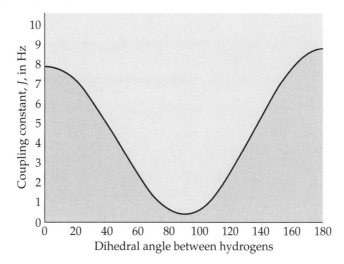

chemical shift data, δ, that is reported is independent of the strength of the magnetic field of the instrument. The coupling constant, *J*, is measured in Hz (page 152). *It is therefore necessary to multiply the distance between the signals in ppm by the frequency of the NMR spectrometer (MHz) in order to obtain the coupling constant in Hz (Equation 8.2).*

$$\text{Coupling Constant } J \text{ (Hz)} = \text{distance between the signals (ppm)} \times \text{frequency of the NMR spectrometer (MHz)} \quad (8.2)$$

Fortunately, the NMR spectrometer generates a **peak-print-out table** showing the location of all peaks in ppm and in Hz therefore simplifying the calculation of coupling constants to the subtraction of the Hz values.

8.1.3.b Integration of ^{1}H NMR Spectra

The multiplicity of a ^{1}H NMR resonance provides some information about the number of nearby hydrogens. On the other hand, a singlet does not provide any information about the group giving rise to such a resonance. For example, the ^{1}H NMR spectrum of *tert*-butyl methyl ether (Figure 8.5a) would not be very informative in assessing how many hydrogen atoms are present in the molecule if we did not have the elemental analysis or know the structure. Fortunately, the peaks in the spectrum are not straight lines emerging from the baseline. Each of the two peaks has a defined shape with an area underneath the peak. *The area under the peak is representative of the number of hydrogen atoms in the molecule that are associated with that resonance relative to all of the other resonances. This area is determined by integration of the peak and is shown as a bracketed number below the plot of the chemical shift.* Note that the actual values of 3 and 9 hydrogen atoms are not indicated in Figure 8.5a, but rather the lowest common denominator values of 1 and 3 are shown. The computer program that integrates the peaks has no way of knowing the actual number of hydrogens associated with any given resonance. Therefore, it integrates all resonances and assigns the peak with the smallest area the value of 1, specifying the area of the remaining peaks relative to it. The analyst does have the option of assigning a resonance a specific integral value, in which case all other integrals are reported relative to this value.

Integrals are useful in providing additional insight into the structure of a molecule; however, caution should be exercised. The values shown are the sum of all hydrogens responsible for a given resonance. If there are overlapping resonances, the values that are reported may be misleading.

Integrals are not accurate in ^{13}C NMR spectroscopy and are therefore never reported.

8.1.4 Interpreting NMR Spectra: More Challenges

8.1.4.a Complex Coupling in ^{1}H NMR Spectra

The compound 1,1,2-trichloro-3,3-dimethoxypropane has not yet been reported in the chemical literature, but we can predict what the NMR spectrum of a molecule should look like before taking the spectrum (or in this case, making the molecule). There are five types of hydrogen atoms in the molecule (Figure 8.15). The relative chemical shifts used are estimated from Table 8.3 and other information presented in Figures 8.7 and 8.8. The relative coupling constants have been estimated from data in Table 8.6. Note that the methoxy groups are different because there is no element of symmetry in the molecule. The hydrogens of each methoxy group (H_D, H_E) appear upfield (to the right) as singlets in the spectrum. The signals for H_A and H_C are doublets as predicted by the N + 1 rule because they are coupled to H_B.

Even though H_B has two hydrogens on neighboring carbon atoms, it is not a triplet. The reason is that the N + 1 rule states: *The coupling of N equivalent hydrogens with one or more different hydrogens will lead to a multiplet of N + 1 lines.* The key phrase in this rule is *N equivalent hydrogens.* Because H_A and H_C are not equivalent, the N + 1 rule must be applied in a step-wise fashion. That is, we first couple H_C to H_B to give a doublet for H_B. Each of the signals in this doublet is then split into a doublet from coupling to H_A. The result is two doublets, referred to as a **doublet of doublets (dd)**. It is important to note that these two doublets are considered one resonance because they are associated with only H_B. The diagram resulting from the step-wise application of the N + 1 rule in Figure 8.15 is known as a **splitting tree**. The tree starts at the top as a single peak representing an uncoupled hydrogen or group of equivalent hydrogens. The chemical shift of this peak is used as the chemical shift of the resulting multiplet. The angled lines that connect the vertical lines help in keeping track of the growth of the multiplet. In Figure 8.15, the tree was started by coupling H_C to H_B. The splitting from H_A was then added. The order was an arbitrary choice and does not affect the outcome of the final splitting pattern.

The appearance of the doublet of doublets changes as the two coupling constants are varied. Figure 8.16 shows what happens when J_{AB} is equal to J_{CB}. The result is the same as if we used the N + 1 rule where N = 2. Because the two middle lines common to J_{AB} overlap, the intensity of the resulting peak must be twice that of the others. Keep in mind, however, that even if a triplet is observed for H_B, H_A and H_C are still nonequivalent.

Figure 8.15   The Predicted ^{1}H NMR of 1,1,2-Trichloro-3,3-Dimethoxypropane

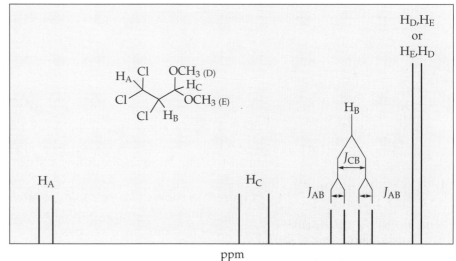

Figure 8.16 **The Step-Wise Application of the N + 1 Rule for Two Nonequivalent Neighboring Hydrogens where** $J_{AB} = J_{CB}$

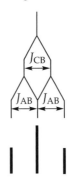

Figure 8.17 ¹H NMR Spectrum of 1-Chloro-2,4-Dinitrobenzene in CDCl₃

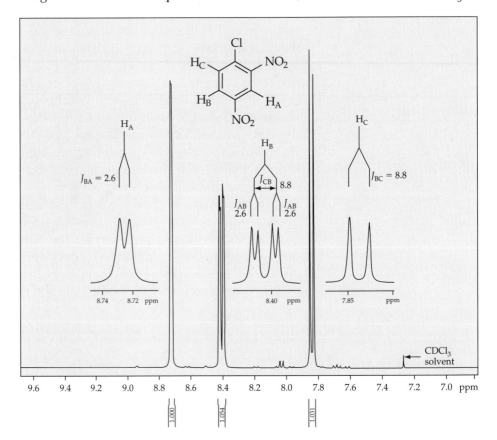

The ¹H NMR spectrum of 1-chloro-2,4-dinitrobenzene is shown in Figure 8.17 to highlight how step-wise application of the N + 1 rule can be used to determine the coupling patterns found in some substituted benzenes.

Figure 8.18 shows two examples of how the appearance of the multiplets changes for a CH₂ group coupled to a CH and a CH₃ group depending on the size of the coupling constants *J*.

Figure 8.19 shows the ¹H NMR of methyl butanoate. The two triplets are expected according to the N + 1 rule and are assigned based on their

Figure 8.18 Splitting Trees for a Doublet of Quartets

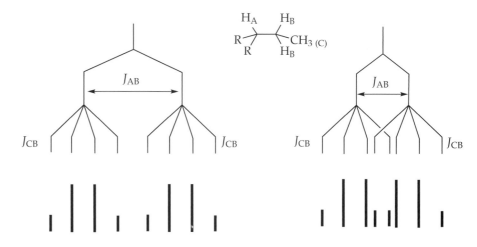

Figure 8.19 ¹H NMR Spectrum of Methyl Butanoate

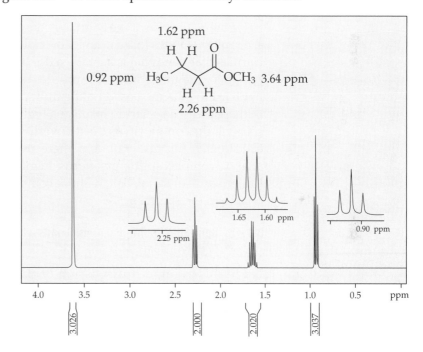

chemical shifts and integrals. The six-line pattern centered at 1.62 ppm is called a **sextet (sext)** and has peak intensities predicted by Pascal's triangle if N = 5. However, this is not possible because these hydrogens are coupled to a methyl group and a methylene group. The hydrogens on these groups are not equivalent (footnote for Table 8.5). As shown in Figure 8.16, if the coupling constants of two systems coupling to a third are equal or close to being equal, the resultant multiplet appears to follow the N + 1 rule. This is not uncommon for CH₂ groups found in the middle of alkyl chains, but should not be interpreted as meaning that the neighboring hydrogens are equivalent.

It is also possible to have three different groups of hydrogens couple to a fourth hydrogen. Figure 8.20 shows two different examples. Note that

Figure 8.20   Splitting Tree Analysis of Three Different Hydrogens Coupling to a Fourth Hydrogen

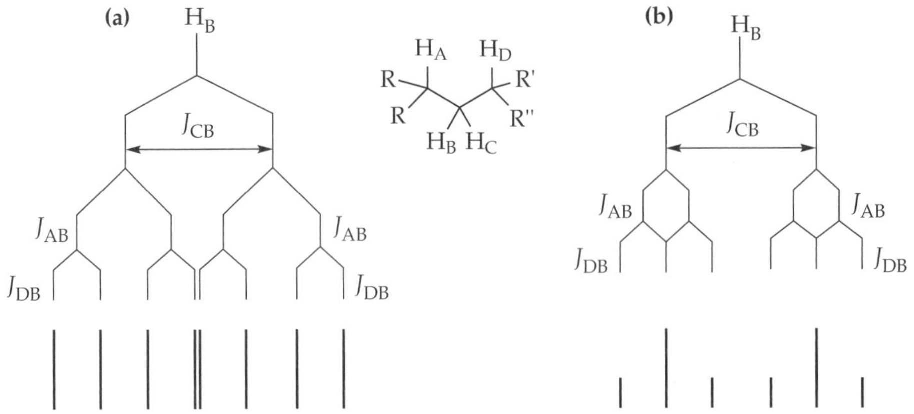

there is not a plane of symmetry in the molecule so that H_B and H_C are not equivalent. In Figure 8.20a, the coupling constants are significantly different and, as the tree shows, the result is a doublet of doublets of doublets for H_B. In Figure 8.20b, two of the coupling constants are the same ($J_{AB} = J_{DB}$) so that the multiplet looks like a doublet of triplets. Analysis using the splitting tree clarifies that the signal is actually a doublet of doublets of doublets.

There are times when a multiplet cannot be easily interpreted. This is not hard to imagine based on the multitude of possible coupling scenarios. In these cases, a spectrum taken at a higher magnetic field will determine or clarify if the complexity is due to overlapping mulitplets with similar chemical shifts. If this does not help, there are computer programs that can help analyze complex splitting patterns. In many cases, it is possible to determine the structure of a molecule without completely deciphering every single resonance in the spectrum.

8.1.4.b Spin-Spin Coupling in Compounds with Hydrogen Atoms Attached to Heteroatoms

Up to this point, our discussion of ^{1}H NMR spectroscopy has focused on hydrogens attached to carbon atoms. There are many functional groups that have one or more hydrogen atoms attached to a heteroatom (Figure 8.21).

To cause spin-spin splitting, nuclei must reside in their respective locations long enough to accommodate the time frame of the NMR experiment. Hydrogen atoms attached to electronegative elements such as oxygen and

Figure 8.21   Some Functional Groups with Exchangeable Hydrogen Atoms

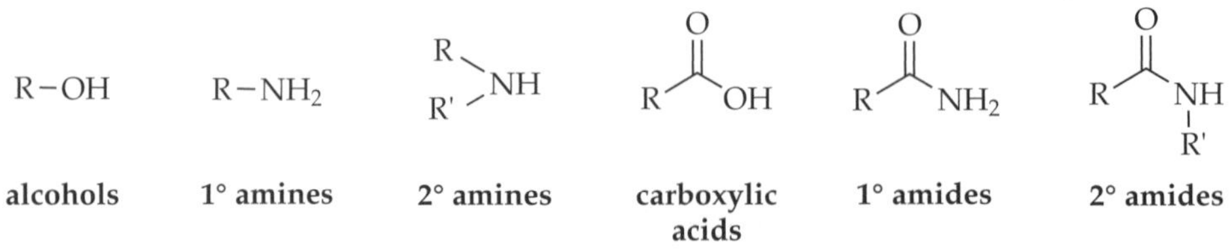

nitrogen have significant bond dipoles and undergo rapid proton exchange via hydrogen bonding (Section 5.3.2.a). Generally, these hydrogen atoms are not on one heteroatom long enough relative to the NMR time scale for coupling to take place. Such hydrogens appear as broadened singlets and often the signal is so broad that it is not even detectable.

A common method used to determine if a given resonance is due to an exchangeable hydrogen is to first record the ^{1}H NMR spectrum of the sample, then add a drop of deuterium oxide (D$_2$O) and retake the spectrum. The deuterons in the D$_2$O replace any exchangeable protons. Any resonances associated with these hydrogen atoms disappear because the deuterium atom that replaces the hydrogen does not have the same resonant frequency.

8.1.4.c Spin-Spin Coupling Between Other Nuclei

Just as nonequivalent hydrogens can couple to one another, other nuclei can couple to hydrogen, but the list of such nuclei is small. Of the nuclei that routinely appear in organic molecules only ^{13}C, ^{19}F, and ^{31}P exhibit coupling to hydrogens. All of these have a nuclear spin of 1/2 (Table 8.1) and follow the N + 1 rule of coupling (page 153).

One-bond coupling to ^{13}C is quite large as seen in Table 8.7 and two-bond coupling is smaller, but still significant. However this coupling is not visible in ^{1}H NMR spectra because the amount of ^{13}C present in naturally occurring materials is only approximately 1.1% relative to ^{12}C. The coupling of ^{13}C with hydrogen is observed in samples enriched in ^{13}C.

Because ^{13}C couples to hydrogen, hydrogen to ^{13}C coupling must also be observed in ^{13}C NMR spectra. However, the most important uses of ^{13}C NMR spectroscopy are the ability to reliably count the number of unique carbon atoms in a molecule and to easily identify functional groups by chemical shift so an instrument-based "trick" that decouples the hydrogens from the carbons is used to acquire the NMR data. This is called a **hydrogen (proton)-decoupled spectrum** and is designated as a ^{13}C[^{1}H] **NMR spectrum.** Without proton decoupling, it would be very difficult to decipher a ^{13}C NMR spectrum because of the extensive overlap of mulitplets caused by ^{1}H–^{13}C coupling.

Even though ^{13}C–^{13}C coupling also takes place, there are very few ^{13}C–^{13}C bonds in a molecule isolated or made from natural sources. Such multiplets are therefore not typically visible in the spectrum.

Naturally occurring organofluorine compounds are very rare compared with organochlorine and organobromine natural products. However, because of its unique electronic properties, fluorine is often incorporated into research chemicals and pharmaceuticals. Because ^{19}F has a natural abundance of 100%, it exhibits spin-spin coupling with both hydrogen and carbon. Typical coupling constants are shown in Table 8.7. The ^{1}H and ^{13}C[^{1}H] NMR spectra of 1-fluoro-2,4-dinitrobenzene (Sanger's reagent) are shown in Figure 8.22. To see the effects of ^{19}F–H coupling, compare the ^{1}H NMR spectrum of this molecule with its chlorine analog shown in Figure 8.17. Note that coupling between chlorine and carbon or hydrogen is generally not observed in solution NMR spectroscopy for reasons beyond the scope of this discussion.

Table 8.7 Coupling Constants Between Other Nuclei with a Spin 1/2

Type of Coupling	Coupling Constant (Hz)	Type of Coupling	Coupling Constant (Hz)
^{13}C-H (1-bond)	110–280	^{19}F-^{13}C	240–300
^{13}C-C-H (2-bond)	0–60	^{19}F-C-^{13}C	10–45
^{13}C-^{13}C (rare)	35–125	^{31}P-H	630–700
^{19}F-C-H	40–85	^{31}P-C-H	2–14
^{19}F-C-C-H	3–25	^{31}P-C-C-H	10–18
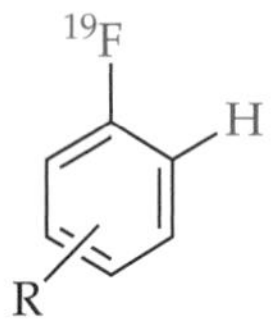	6–10 (1,2 or *ortho*)	^{31}P-O-C-H	10–12
		^{31}P-N-C-H	8–10
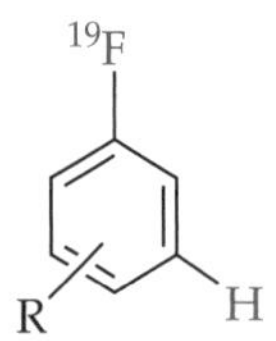	5–6 (1,3 or *meta*)	^{31}P-^{13}C	5–145
		^{31}P-C-^{13}C	4–15
		^{31}P-O-^{13}C	6–8
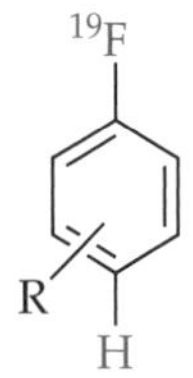	0–2 (1,2 or *para*)		

Phosphorus, in the form of phosphate and phosphate esters, is common in living systems. There are also many phosphorus-containing reagents used in organic synthesis. Like ^{19}F, ^{31}P has a natural abundance of 100% and a nuclear spin of 1/2. Typical coupling constants to both hydrogen and carbon are listed in Table 8.7. An example of phosphorous coupling is provided in the ^1H and $^{13}C[^1H]$ NMR spectra of triisopropylphosphine (Figure 8.23). The multiplet at 1.62 to 1.75 ppm is a doublet of septets. The signal at 1.03 to 1.09 is a doublet of doublets.

8.1.5 Preparing a Sample for NMR Analysis

A sample used for NMR analysis must be devoid of most impurities including any solvents that might have been used in the reaction or work-up phase of the experiment. The typical amount of sample used for routine NMR analyses is 10–50 mg. Much smaller amounts (< 1 mg) can be analyzed if

Figure 8.22   NMR Spectra of 1-Fluoro-2,4-Dintrobenzene in CDCl₃

(a) ¹H NMR

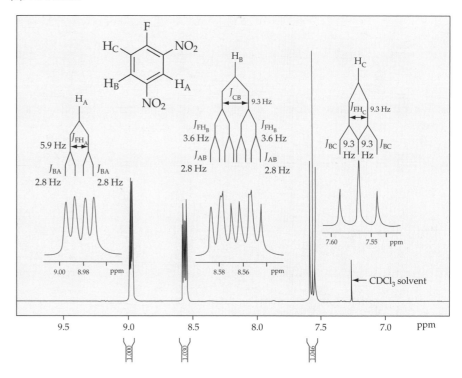

(b) ¹³C[¹H] NMR

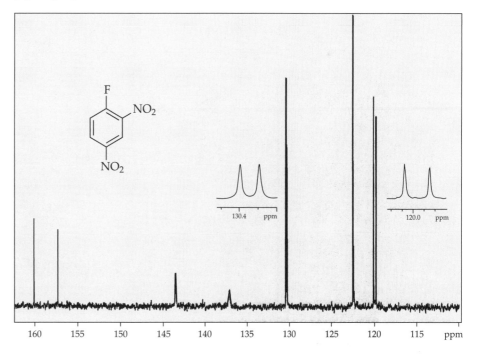

Figure 8.23   NMR Spectra of Triisopropylphosphine

(a) ¹H NMR

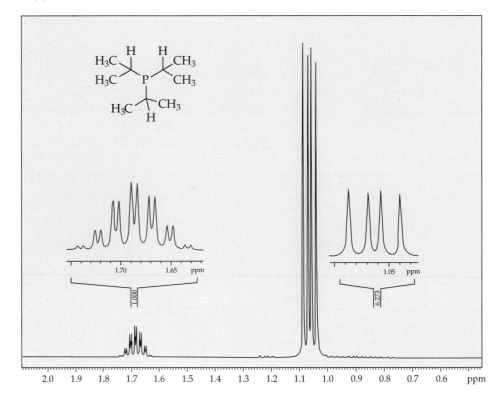

(b) ¹³C[¹H] NMR

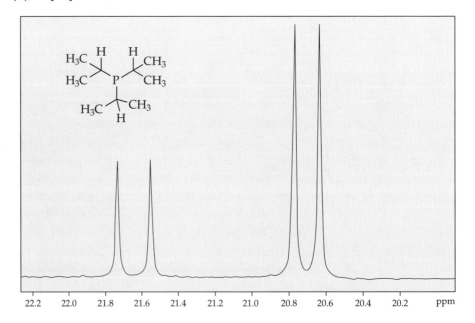

time is not a factor because the time needed to acquire the spectrum depends on the concentration of the sample. The compound must be soluble in 0.5–0.7 mL of a solvent that has no hydrogen atoms and as few carbon atoms as possible. Remember that the solvent is present in much larger concentrations than the analyte and, if the solvent has hydrogen atoms in it, the major resonance(s) observed will be those associated with the solvent. There are a variety of commercially available NMR solvents that meet these criteria. In all cases, any hydrogen atoms are replaced by deuterium that is not detected in ^{1}H NMR or ^{13}C NMR spectroscopy. However, all NMR solvents have a small amount of residual hydrogen present simply because the cost associated with making the sample completely hydrogen-free is prohibitive.

The most common NMR solvent is deuterochloroform, $CDCl_3$. The ^{1}H chemical shift for the residual amount of $CHCl_3$ in $CDCl_3$ is 7.27 ppm. In ^{13}C NMR spectra, the carbon of $CDCl_3$ shows up as a 1:1:1 triplet centered at 77.0 ppm. The triplet arises from the coupling of ^{13}C to a deuterium that has a nuclear spin of 1 (Table 8.1), so that the $2NI + 1$ rule (page 153) becomes $2N + 1$ where N is the number of deuteriums.

Other commonly used solvents are listed in Table 8.8. This table includes the chemical shifts of any residual hydrogens and the expected multiplets caused by the coupling with deuterium.

Table 8.8 Some Common NMR Solvents

Structure Name	^{1}H Chemical Shift(s) of Residual Hydrogen(s) (Number of Lines of the Multiplet Associated with Coupling to Deuterium)	^{13}C Chemical Shift(s) (Number of Lines of the Multiplet Associated with Coupling to Deuterium)
$CDCl_3$ deuterochloroform	7.27 (1)	77.0 (3)
CD_3OD tetradeuteromethanol	4.87 (1) and 3.31 (5)	49.2 (7)
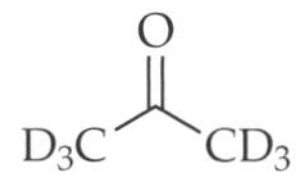 D_3C—C(=O)—CD_3 hexadeuteroacetone	2.05 (5)	206.7(13) and 29.7 (7)
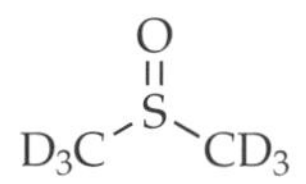 D_3C—S(=O)—CD_3 hexadeuterodimethylsulfoxide	2.50 (5)	39.5 (7)
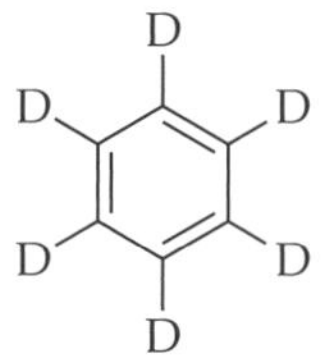 hexadeuterobenzene	7.16 (1)	128.4 (3)

Once the sample is dissolved in the NMR solvent, the solution is placed in an NMR tube. These tubes are designed to exacting standards for modern NMR instruments. Because ^{1}H and ^{13}C NMR chemical shifts are measured relative to tetramethylsilane (TMS), which is given a value of 0 ppm, a small amount of this material can be added to the sample. Alternately, $CDCl_3$ containing a small percentage of TMS is available from chemical suppliers. It is also common practice to use the residual hydrogen resonance of the NMR solvent as the reference point instead of adding TMS.

8.1.6 Reporting NMR Data

NMR spectral data for new compounds is reported in a standardized, logical, and abbreviated fashion. General guidelines have been established and the most commonly used format is summarized below. To avoid the task of labeling all of the hydrogen atoms in a molecule when hydrogen coupling constants are reported, the terminology J_{AB}, etc. is generally not used unless coupling to other nuclei such as ^{19}F or ^{31}P is observed. In addition, the complete assignment of all multiplets for more complicated molecules is not always possible and, more importantly, not always necessary to determine the structure of the molecule. In cases where a multiplet is not interpreted, it is reported as (m) and the chemical shift range is given.

For ^{1}H NMR, the convention for reporting data is: (Resonant frequency of instrument, NMR solvent) δ (ppm) (multiplicity if relevant, coupling constant(s) reported as J = X (Hz), number of hydrogen atoms from integration).

The data obtained for 1-chloro-2,4-dinitrobenzene (Figure 8.17) is reported as: (400 MHz, $CDCl_3$) δ 7.84 (d, J = 8.8, 1H), 8.41 (dd, J = 8.8, 2.6, 1H), 8.73 (d, J = 2.6, 1H).

For ^{13}C[^{1}H] NMR the convention for reporting data is: (Resonant frequency of instrument, NMR solvent) δ (ppm). When coupling to ^{19}F or ^{31}P is observed, it is reported after the relevant chemical shift using the notation (mulitiplicity, J = X (Hz)).

The data obtained for 1-fluoro-2,4-dinitrobenzene (Figure 8.22b) is reported as: (100 MHz, $CDCl_3$) δ 120.0 (d, J = 23.0), 122.5, 130.4 (d, J = 10.4), 137.2, 143.6, 158.9 (d, J = 275).

8.1.7 Problems

1. Sketch the ^{1}H NMR and ^{13}C[^{1}H] NMR spectra of the following molecules. For all spectra, chemical shifts should be reasonable. In the ^{1}H NMR spectra, draw any expected multiplets accurately. If there are any resonances in the same chemical shift region, be sure to separate them enough so that they are distinguishable. Assign all spectra by indicating which resonances belong to the different types of hydrogens and carbons in each compound.

2a. Using the symbols α and β for the two spin states, show how four equivalent hydrogens can couple to a CH_2 group leading to a quintet with peak ratios of 1:4:6:4:1.

2b. Provide an example of a molecule whose ^1H NMR spectrum would contain a quintet similar to what you have drawn in part a.

3a. Reconstruct a ^1H NMR spectrum from the following data. Refer to Table 8.5 for multiplet abbreviations.

δ 5.00 (sep, J = 6.2, 1H), 2.28 (q, J = 7.6, 2H), 1.23 (d, J = 6.2, 6H), 1.12 (t, J = 7.6, 3H).

3b. The molecular weight of this compound is 116. Using the ^1H NMR data given in part a, and the $^{13}C[^1H]$ NMR data shown below, determine the structure of this molecule.

δ 174.0, 67.4, 28.0, 21.9, 9.2.

4a. Based on the following coupling constants for the molecule shown, use a splitting tree to draw the multiplet expected for each hydrogen.

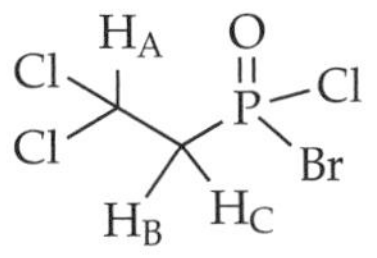

J_{AB} = 8 Hz, J_{P-H_A} = 16 Hz
J_{AC} = 6 Hz, J_{P-H_B} = 12 Hz
J_{BC} = 12 Hz, J_{P-H_C} = 10 Hz

4b. Draw the multiplet you would expect to see in the ^{31}P NMR spectrum of this molecule that would result from ^1H-^{31}P coupling. In practice, ^{31}P NMR spectra are usually run in a proton-decoupled mode ($^{31}P[^1H]$) using a technique similar to that used for $^{13}C[^1H]$ NMR spectra.

4c. How many peaks would you expect to see in the $^{13}C[^1H]$ NMR spectrum of this molecule? Refer to Table 8.7.

5. The ^1H and $^{13}C[^1H]$ NMR spectra for three of the six compounds shown below are provided. First, predict what you might expect to see in the spectra of each compound. For example, how many different carbons are there in each compound? What type of multiplet do you expect to see for any unique hydrogens? Don't forget that 4-bond coupling is common in allylic systems (Table 8.6). Then, match each set of spectra to a given compound. Justify your answers by assigning all resonances in the NMR spectra. In the ^1H NMR spectrum, indicate which hydrogens are coupled by calculating the coupling constants from the peak-print-out table (page 158) that is shown on each spectrum. Write them in the form of J_{AB} =, J_{AC} =.

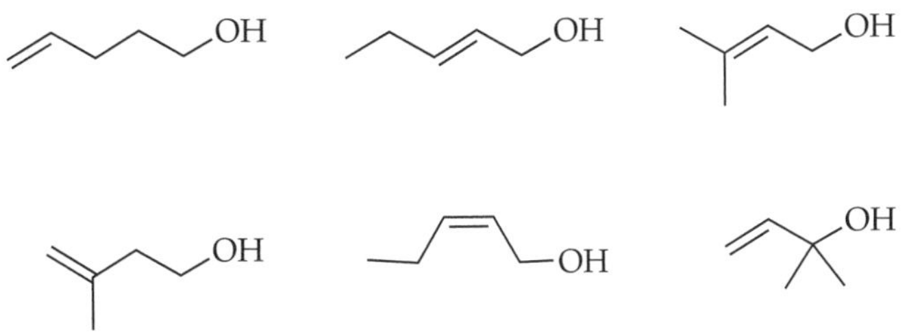

Compound A

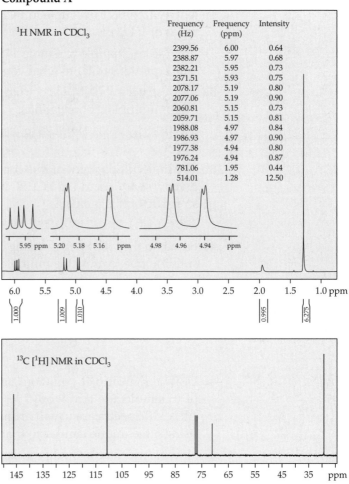

Frequency (Hz)	Frequency (ppm)	Intensity
2399.56	6.00	0.64
2388.87	5.97	0.68
2382.21	5.95	0.73
2371.51	5.93	0.75
2078.17	5.19	0.80
2077.06	5.19	0.90
2060.81	5.15	0.73
2059.71	5.15	0.81
1988.08	4.97	0.84
1986.93	4.97	0.90
1977.38	4.94	0.80
1976.24	4.94	0.87
781.06	1.95	0.44
514.01	1.28	12.50

Compound B

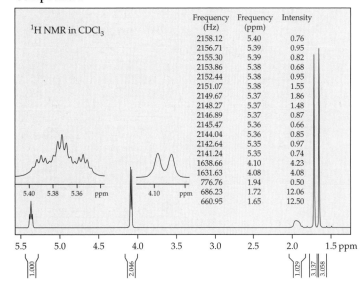

Frequency (Hz)	Frequency (ppm)	Intensity
2158.12	5.40	0.76
2156.71	5.39	0.95
2155.30	5.39	0.82
2153.86	5.38	0.68
2152.44	5.38	0.95
2151.07	5.38	1.55
2149.67	5.37	1.86
2148.27	5.37	1.48
2146.89	5.37	0.87
2145.47	5.36	0.66
2144.04	5.36	0.85
2142.64	5.35	0.97
2141.24	5.35	0.74
1638.66	4.10	4.23
1631.63	4.08	4.08
776.76	1.94	0.50
686.23	1.72	12.06
660.95	1.65	12.50

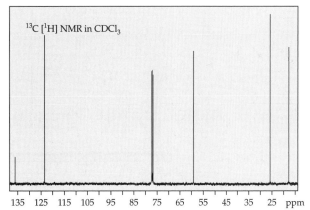

Compound C

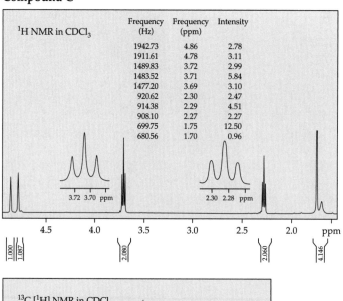

Frequency (Hz)	Frequency (ppm)	Intensity
1942.73	4.86	2.78
1911.61	4.78	3.11
1489.83	3.72	2.99
1483.52	3.71	5.84
1477.20	3.69	3.10
920.62	2.30	2.47
914.38	2.29	4.51
908.10	2.27	2.27
699.75	1.75	12.50
680.56	1.70	0.96

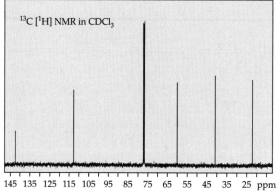

6. The ^{1}H and ^{13}C[^{1}H] NMR spectra of three compounds are shown. Identify the structure of each compound using these spectra along with the other information provided. Assign all resonances in the NMR spectra. In the ^{1}H NMR spectrum, if there are multiplets, indicate which hydrogens are coupled by calculating the coupling constants from the peak-print-out table (page 158) that is shown on each spectrum. Write them in the form of $J_{AB} =$, $J_{AC} =$.

Compound A
Molecular Formula C$_7$H$_{14}$O

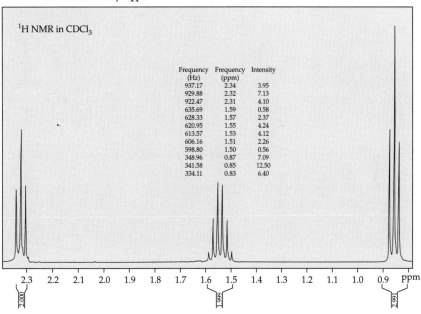

Frequency (Hz)	Frequency (ppm)	Intensity
937.17	2.34	3.95
929.88	2.32	7.13
922.47	2.31	4.10
635.69	1.59	0.58
628.33	1.57	2.37
620.95	1.55	4.24
613.57	1.53	4.12
606.16	1.51	2.26
598.80	1.50	0.56
348.96	0.87	7.09
341.58	0.85	12.50
334.11	0.83	6.40

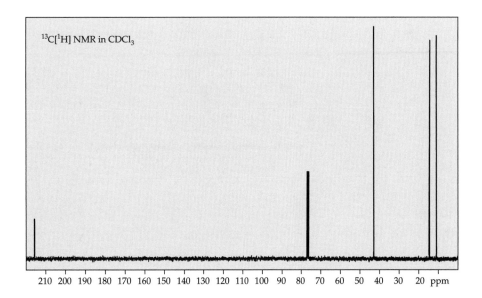

Compound B
Molecular Formula C₇H₁₂O₄

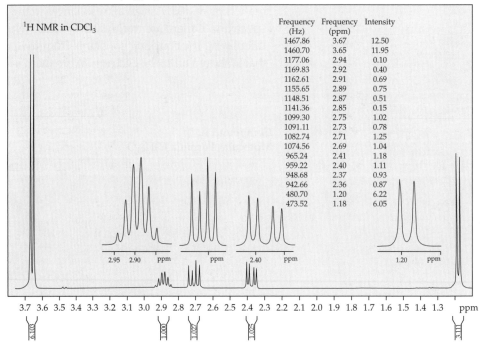

Frequency (Hz)	Frequency (ppm)	Intensity
1467.86	3.67	12.50
1460.70	3.65	11.95
1177.06	2.94	0.10
1169.83	2.92	0.40
1162.61	2.91	0.69
1155.65	2.89	0.75
1148.51	2.87	0.51
1141.36	2.85	0.15
1099.30	2.75	1.02
1091.11	2.73	0.78
1082.74	2.71	1.25
1074.56	2.69	1.04
965.24	2.41	1.18
959.22	2.40	1.11
948.68	2.37	0.93
942.66	2.36	0.87
480.70	1.20	6.22
473.52	1.18	6.05

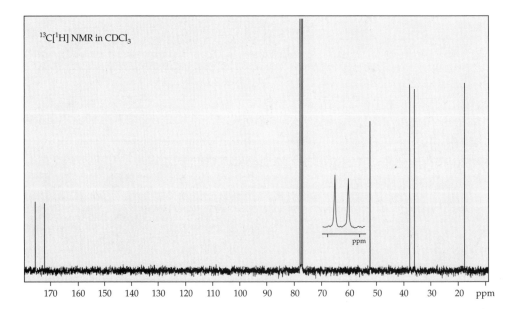

Compound C
Molecular Formula C₃H₅Br₃

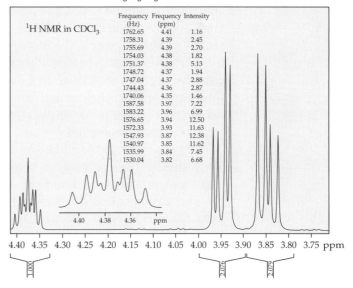

Frequency (Hz)	Frequency (ppm)	Intensity
1762.65	4.41	1.16
1758.31	4.39	2.45
1755.69	4.39	2.70
1754.03	4.38	1.82
1751.37	4.38	5.13
1748.72	4.37	1.94
1747.04	4.37	2.88
1744.43	4.36	2.87
1740.06	4.35	1.46
1587.58	3.97	7.22
1583.22	3.96	6.99
1576.65	3.94	12.50
1572.33	3.93	11.63
1547.93	3.87	12.38
1540.97	3.85	11.62
1535.99	3.84	7.45
1530.04	3.82	6.68

^{1}H NMR in CDCl₃

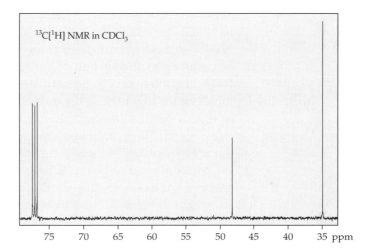

^{13}C[^{1}H] NMR in CDCl₃

8.2 MASS SPECTROMETRY

Mass spectrometry (MS) is an experimental method that aids in the determination of the mass of a molecule. Like NMR spectroscopy, mass spectrometry has its roots in physics. In fact, the 1906 Nobel Prize in Physics was awarded to Sir J.J. Thomson[5] for work that laid the foundations for modern mass spectrometry. In 1922, Francis Aston[6] received the Nobel Prize in Chemistry for his development of an improved mass spectrometer that allowed isotopes to be investigated. The advancement of mass spectrometric techniques since the early 1980's has been astonishing and there are no areas of chemistry that have not been influenced by these

[5] See http://nobelprize.org/nobel_prizes/physics/laureates/1906/index.html.
[6] See http://nobelprize.org/nobel_prizes/chemistry/laureates/1922/index.html.

developments. In fact, mass spectrometry is used in all scientific disciplines. Geology is one of the benefactors of mass spectrometry where it is used to analyze the isotope ratios of elements found in minerals. Astronomy, archaeology, engineering, and biology also use mass spectrometry. Many recent advancements in the biological sciences have been made possible by new techniques developed in mass spectrometric laboratories. In 2002, one half of the Nobel Prize in Chemistry was awarded to two chemists, John Fenn and Koichi Tanaka[7], for their work in developing methods that helped revolutionize the application of mass spectrometry to the study of large biological molecules.

In general, organic chemists use mass spectrometry to complement NMR studies. If a structure has been determined by NMR analysis, then the molecular mass can be determined from the molecular formula. Because mass spectrometry measures mass, it can be used to either verify the conclusions reached by NMR spectroscopy or indicate that something was missed in the structural assignment. The latter situation arises when a molecule has one or more functional groups that do not have carbons or hydrogens so they are not detectable in the NMR experiment. Structure determination by NMR can also be difficult if a molecule has one or more elements of symmetry. For example, if the molecule has a plane of symmetry, making its two halves equivalent, the analyst only sees one half of the molecule by NMR. Unless there are other clues from the NMR data, the structure might be assumed to be only half of the actual structure. These problems in structural determination can be resolved by examining the mass spectrum of the molecule. In Figure 8.24, the four compounds shown all exhibit one singlet in the ^{1}H NMR with very similar chemical shifts. Without support from both ^{13}C[^{1}H] NMR and mass spectrometry, it would be difficult to establish the identity of any of these compounds.

Molecules do not have to be symmetrical to provide misleading NMR spectra. For example, the NMR spectra of 1-fluoro-2,4-dinitrobenzene are shown in Figures 8.22 a and b. If we did not know the structure of this compound and had no evidence for the presence or absence of fluorine, the ^{13}C[^{1}H] and ^{1}H NMR spectra would be confusing. The ^{1}H NMR spectrum indicates that the hydrogens are attached to sp^2 hybridized carbons and the ^{13}C NMR spectrum confirms this. However, the ^{13}C NMR spectrum shows nine peaks suggesting the presence of nine different carbons, which is not correct. A mass spectrum would help clarify the mass of the compound and aid in correctly determining its structure.

Figure 8.24   Examples of Molecules Exhibiting One Singlet in their ^{1}H NMR Spectra

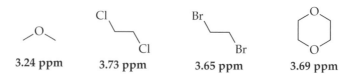

3.24 ppm	3.73 ppm	3.65 ppm	3.69 ppm

---

[7] See http://nobelprize.org/nobel_prizes/chemistry/laureates/2002/index.html.

There are many types of mass spectrometers and many ways to collect data from these instruments. For all of these techniques, the compounds being analyzed must be transformed into gas phase ions. The ions are then sorted according to their mass and charge by means of electric or magnetic fields. It is the mass to charge ratio, referred to as m/z, that is plotted on the *x*-axis of a mass spectrum. On the *y*-axis, the percent relative abundance of the ions found in the spectrum is recorded. Only ions are analyzed in mass spectrometry. Neutral molecules are present in a mass spectrometer, but are not detected and therefore not recorded.

Choosing the most suitable method to use for a given sample depends on the type of compound being analyzed. This section describes three techniques: electron ionization (EI), matrix assisted laser desorption ionization (MALDI), and electrospray ionization (ESI).

8.2.1 Electron Ionization Mass Spectrometry

EI mass spectrometry is the method of analysis most often used for small organic molecules. In EI mass spectrometry, a beam of high-energy electrons is generated inside a vacuum chamber. When a stream of gas-phase molecules is passed through this beam of electrons, collisions take place between the molecules and electrons. The typical energy of the electrons used in an EI MS experiment is 70 eV. To put this in perspective, the ionization potential for typical organic molecules is between 9 and 12 eV.

8.2.1.a The Molecular Ion

When one of these electrons collides with a molecule, enough energy is imparted to the molecule to ionize it to the **molecular ion** (Equation 8.3). The molecular ion differs from the original molecule by only one electron.

$$M + e^{\ominus} \longrightarrow \overset{\cdot}{M}^{\oplus} + 2e^{\ominus} \tag{8.3}$$

molecule   70 eV   molecular
electron   ion

The molecular ions generated under these conditions are cations and are referred to as radical cations because they have an odd number of electrons. These radical cations have a charge of +1 and therefore *the m/z ratio of the molecular ion represents the true mass of the compound.*

In order to analyze a molecule by EI MS, it must have a high enough vapor pressure to exist in the gas phase under the conditions of the experiment. Microgram to milligram quantities of liquids and solids are introduced into the vacuum chamber of the mass spectrometer by placing them on a specialized probe. If the molecule does not vaporize under these conditions, the probe tip can be heated while it is inside the vacuum chamber (Section 6.2). Most lower molecular weight (<500 daltons) organic molecules vaporize to some extent under these conditions. However, there are certain classes of compounds, such as sugars, amino acids and various organic salts, that are not volatile and decompose when heated. Such compounds can give rise to deceptive data because the decomposition products can be volatile and result in data that is unrelated to the starting compound.

Figure 8.25   The EI Mass Spectrum of Methane

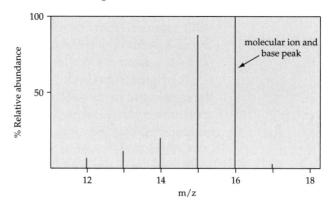

The EI mass spectrum of methane is shown in Figure 8.25. The molecular ion is located at m/z = 16, the mass of methane.

8.2.1.b The Base Peak

The tallest peak in the spectrum indicates the most abundant ion and is called the **base peak**. In the case of methane, the molecular ion is also the base peak. The base peak is set to a relative abundance of 100% and all other ions in a mass spectrum are reported as a percentage of the base peak. *It is important to note that the molecular ion is not always the base peak.* For example, in the EI mass spectrum of ethane (Figure 8.26), the base peak is at m/z = 28 while the molecular ion is at m/z = 30.

8.2.1.c Fragment Ions

Hydrocarbons The additional peaks in the spectra of methane and ethane are not due to impurities, but instead result from additional charged species known as fragment ions. As the name implies, they result from the fragmentation of other ions, most commonly the molecular ion. When the original uncharged molecule collides with a high-energy electron, the collision imparts more energy to the molecule than is necessary for ionization. Much of this excess energy would be dispersed by additional molecular collisions,

Figure 8.26   The EI Mass Spectrum of Ethane

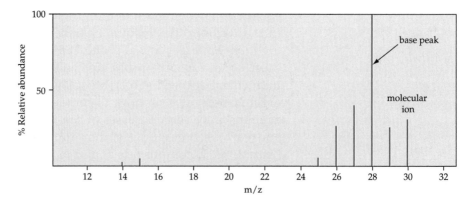

but such collisions are rare in the vacuum chamber. Therefore, the molecular ion dissipates the extra energy by fragmenting into smaller pieces.

Some of these bond-breaking reactions are completely predictable based on our knowledge of solution-phase reaction mechanisms. Fragmentation pathways can be characteristic of certain structural units and functional groups in organic molecules. Other fragmentations have no solution-phase analogies, but occur because of the high energy of the molecular ion. In many cases, we can draw useful conclusions about the connectivity of atoms in molecules based on an analysis of the fragmentation ions observed in an EI mass spectrum.

When writing rational mechanisms leading to fragmentation ions, it is important to indicate where the charge and odd electron are localized. This is very clear in some cases, but in others it can be somewhat arbitrary. In cases where there is more than one type of bond and/or lone pair of electrons, the molecular ion can be made up of several different species, all with the same mass but different electron distributions.

For example, in ethane there are two types of bonds and when the molecule collides with 70 eV electrons, the loss of an electron from either the carbon-carbon or carbon-hydrogen bonds can occur (Figure 8.27).

In Figure 8.27, the resulting radical cations are drawn with an electron between two atoms and the positive charge on the carbon. The distribution of the two molecular ions is not expected to be equal and would be predicted to favor radical cation **A** because it is formed by loss of an electron from the weaker of the two bonds. Some of the observed fragment ions can be predicted by starting with these radical cations and using rational arrow-pushing.

Loss of a single hydrogen atom can occur from either radical cation, leading to the ethyl cation at m/z = 29 (Figure 8.28). Loss of a methyl radical leading to the methyl cation (m/z = 15) is also observed and is most easily rationalized as coming from radical cation **A**. Note that in both of these cases, when a radical is lost from the even-mass molecular ion, a cation with an odd-numbered mass is observed. The loss of a hydrogen atom or methyl radical, as shown in Figure 8.28, are inferred from the ions observed because uncharged radicals are not observed in the mass spectrometry experiment.

Figure 8.27   Two Possible Molecular Ions of Ethane

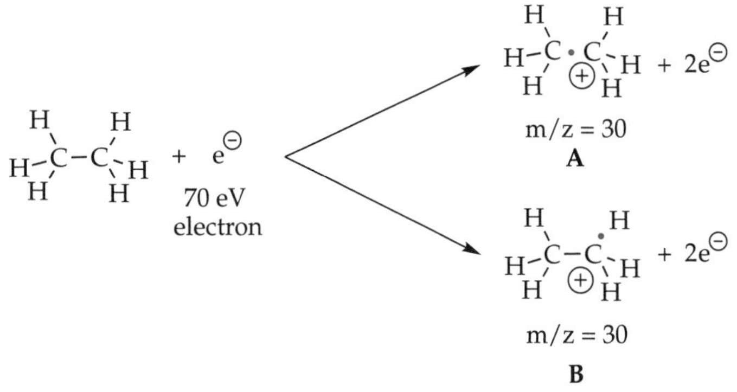

Figure 8.28   Fragmentation of Ethane Radical Cations

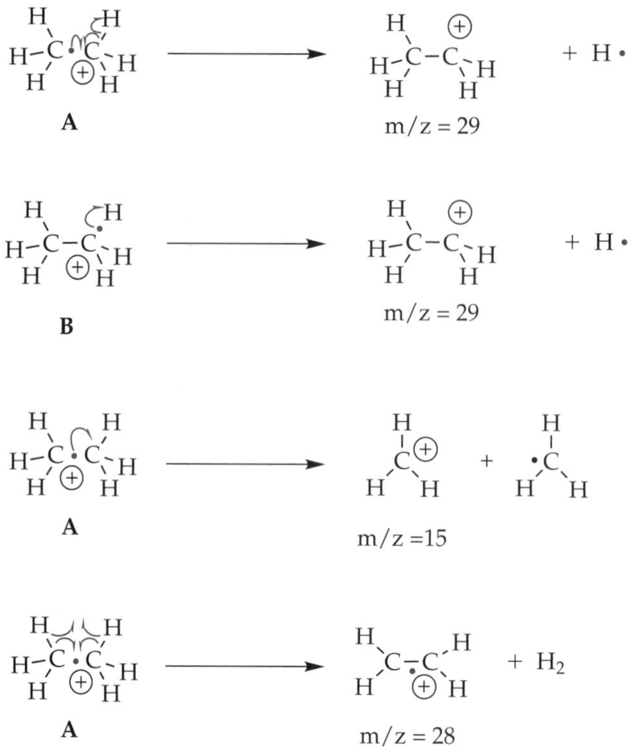

A $\longrightarrow$ m/z = 29   + H·

B $\longrightarrow$ m/z = 29   + H·

A $\longrightarrow$ m/z = 15   +

A $\longrightarrow$ m/z = 28   + H_2

The base peak at m/z = 28 is due to the loss of H_2 and a reasonable pathway based on experimental evidence is shown in Figure 8.28. The product is the ethylene radical cation. In this case, the loss of a neutral, even-electron molecule from an even-mass molecular ion leads to a new radical cation with an even mass.

The loss of H_2 from the same carbon also occurs as shown in the EI mass spectrum of methane (Figure 8.25). The peak at m/z = 14 results from loss of H_2 from a single carbon atom (Equation 8.4). When possible, the loss of H_2 from adjacent carbons is considered to be more favorable.

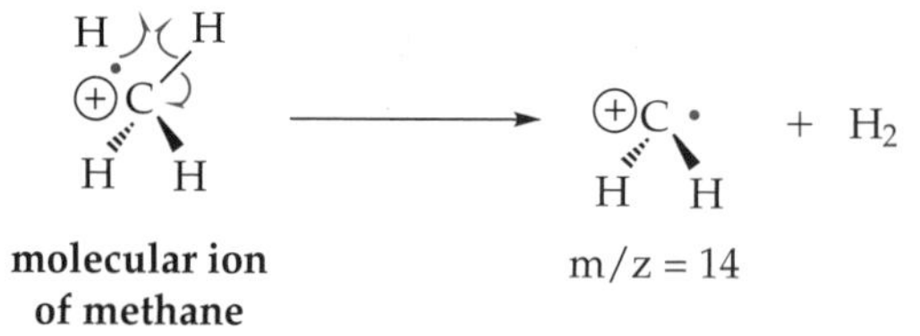

molecular ion
of methane $\longrightarrow$ m/z = 14   + H_2 (8.4)

Fragment ions can themselves break apart into other ions. Because fragment ions are less energetic than their precursor molecular ions, further fragmentation occurs through lower energy paths that are characterized by concerted losses of small, even-electron molecules. The fragment ions at m/z = 26 and 27 in the mass spectrum of ethane arise from the loss of H_2 from the ethylene radical cation and the ethyl cation respectively as shown in Figure 8.29.

Figure 8.29 Fragmentation of Fragment Ions in the EI Mass Spectrum of Ethane

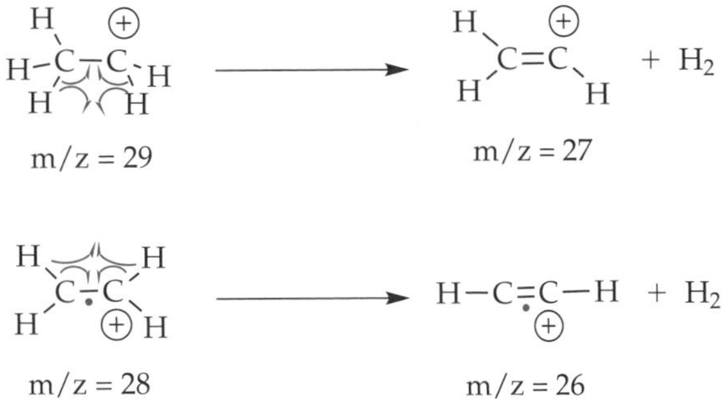

The intensity of the molecular ion in the mass spectrum is a direct reflection of the stability of the ion relative to fragmentation. That is, if there are few pathways by which the molecular ion can fragment, its relative abundance in the mass spectrum will be significant. However, if there are low energy bond-breaking mechanisms available by which the molecular ion can dissipate its excess energy, it will fall apart, resulting in a low relative abundance. In some cases, the molecular ion is so easily fragmented that it is not even detected.

The analysis of the mass spectrum of ethane is straightforward. However, as alkanes become larger, the interpretation of their fragmentation pathways becomes more complicated. This has to do, in part, with the fact that there are a greater number of carbon-carbon and carbon-hydrogen bonds in larger hydrocarbons, each of which can lose an electron in the ionization process. The greater number of bonds leads to a larger distribution of radical cations that make up the molecular ion and, therefore, a greater number of potential fragmentation pathways. To complicate matters further, it has been shown through deuterium labeling studies that a significant number of molecular rearrangements can take place in the molecular ions before they fragment. These molecular rearrangements can lead to fragmentation reactions that are not representative of the true structure of the starting molecule. For example, cyclohexane has no methyl groups in it, but it shows an M–15 peak at m/z = 69. This peak arises from ring opening of the molecular ion followed by intramolecular hydrogen abstraction and subsequent loss of a methyl group (Equation 8.5). The m/z = 69 ion might be either of the ions shown.

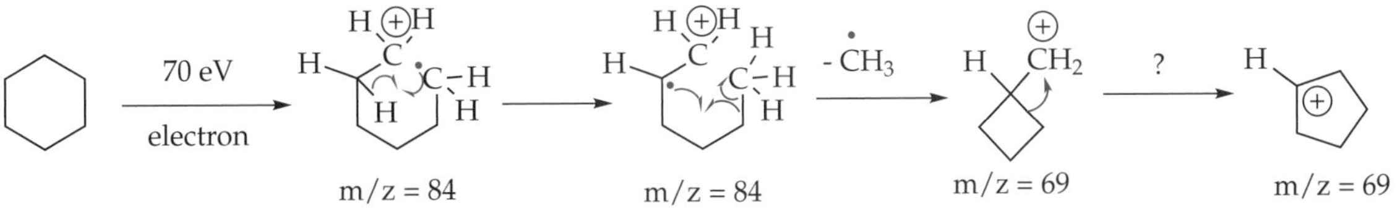

(8.5)

It is important to keep in mind that the arrow-pushing mechanisms used to rationalize the formation of certain fragment ions are just a method of keeping track of electrons and charges and may not be the exact pathway by which these fragmentations occur. However, it is still a useful exercise to write the mechanism if it allows conclusions to be drawn about the structural details of a molecule.

Molecules Containing Heteroatoms Molecules with functional groups containing one or more heteroatoms exhibit more predictable fragmentation behavior in EI MS than hydrocarbons because ionization is more likely to take place from a lone pair of electrons on the heteroatom than from a bonding pair of electrons within the molecule. The following discussion of the fragmentation patterns of some common functional groups that contain oxygen and/or nitrogen is an introduction to how to use rational arrow-pushing mechanisms to predict fragmentations in the EI mass spectra of molecules containing these heteroatoms.

Alcohols Two common fragmentation pathways seen in the EI mass spectra of alkyl alcohols are loss of water and α-cleavage.

Loss of Water   The relative abundance of the molecular ion for most alcohols is generally very small and is completely absent in some cases. *Most alkyl alcohols have a peak at M–18 from the loss of water from the molecular ion.* In solution-phase chemistry, loss of a hydroxyl group and a hydrogen on a carbon atom adjacent to the one bearing the hydroxyl group results in a 1,2-elimination of water to form an alkene. In contrast, in the gas phase, the EI mass spectra of labeled alcohols have shown that, if the alcohol has four or more carbons and a hydrogen on the fourth carbon from the hydroxyl function, a 1,4 elimination of water is observed. This fragmentation is favored by the six-membered transition state (Equation 8.6).

$$R' \text{...} + H_2O \quad (8.6)$$

α-Cleavage   Fragmentation by **α-cleavage** is found in many functional groups. In the case of alcohols, the α refers to the carbon attached to the hydroxyl group. Cleavage takes place between the α- and β-carbons leading to the loss of a neutral radical and the formation of an oxonium ion (Equation 8.7). The presence of an ion at m/z = 31 is often good evidence for the presence of a primary alcohol.

$$\quad (8.7)$$

primary alcohol   an oxonium ion
(m/z = 31)

Cleavage of the α C–H bond can also take place leading to an M–1 oxonium ion, but this is not a prominent ion in the spectrum of most alcohols. Another obvious source of the M-1 ion would be loss of hydrogen from the OH group; however, deuterium-labeling studies have shown that this does not occur.

In unsymmetrical secondary alcohols, α-cleavage leads to two different oxonium ions whereas for a symmetrical secondary alcohol only one oxonium ion is formed, making structural assignments possible. For example, the EI mass spectra of 3- and 4-heptanol are predicted to include the oxonium fragment ions shown in Figure 8.30. This analysis is confirmed by examining the actual spectra of these two compounds and noting the presence or absence of the expected ions as shown in Table 8.9. The data confirms that these two compounds can be distinguished from one another based solely on mass spectrometric analysis.

In Figure 8.30, the oxonium ion at m/z = 59 results from loss of a butyl radical from the 3-heptanol radical cation. Butyl radicals are more stable than ethyl radicals because of the greater inductive effect. Cleavage of the carbon-carbon bond that leads to the formation of the more stable radical is expected to be favored. Consequently the m/z = 59 ion is present in greater abundance than the ion at m/z = 87, which is due to loss of an ethyl radical. This is a general trend in fragmentation reactions. *When bond breaking occurs, two species are formed—the ion and the neutral fragment. Although it is tempting to compare the stabilities of the two ions that came from a similar fragmentation, the stabilities of the neutral fragments must also be considered in rationalizing relative ion abundances.*

The mass spectrum of 1-hepten-4-ol reinforces this point. α-Cleavage leads to the loss of either a propyl radical or an allyl radical (Figure 8.31).

Figure 8.30 Oxonium Ions from the α-Cleavage of the Molecular Ions of 3- and 4-Heptanol

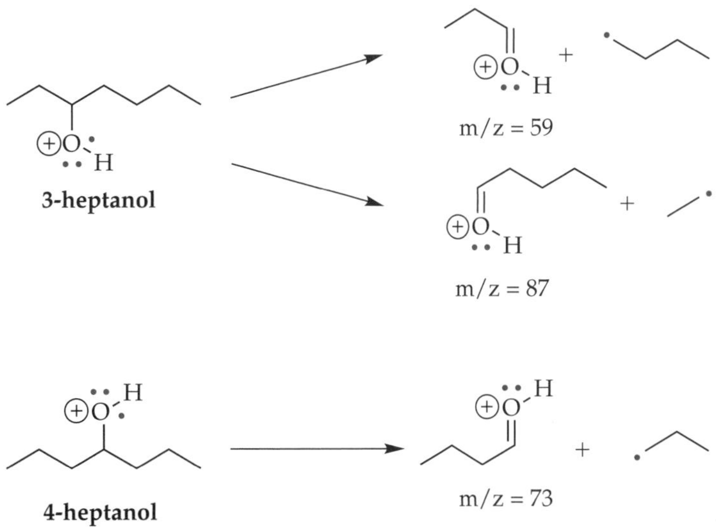

Table 8.9 Relative Abundances of the Oxonium Ions in the EI Mass Spectra of 3- and 4-Heptanol

	m/z = 59	m/z = 73	m/z = 87
3-heptanol	100%	<1%	30%
4-heptanol	<1%	70%	0%

Note: The values given are relative to the base peak in each mass spectrum

Figure 8.31   α-Cleavage of the Molecular Ion of 1-Hepten-4-ol

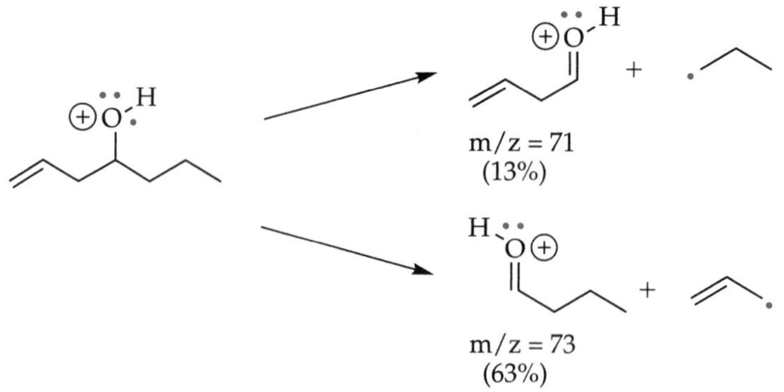

Note: Relative abundances are shown in parentheses.

The resulting oxonium ions are structurally similar and would not be expected to exhibit significant differences in stability. However, the ion at $m/z = 73$ is five times more intense than the one at $m/z = 71$. The resonance-stabilized allyl radical is significantly more stable than the propyl radical. It is this difference that leads to the skewed distribution of the oxonium ions.

Ethers   The molecular ions of ethers undergo α-cleavage. The reliability of this fragmentation pathway allows one to distinguish between the structures of many isomeric ethers and can help with the structural analysis of non-isomeric ethers. For example, the three isomeric six-carbon ethers shown in Figure 8.32 are distinguished from one another by analyzing their

Figure 8.32   α-Cleavage Reactions from the Molecular Ions of Three Isomeric Ethers

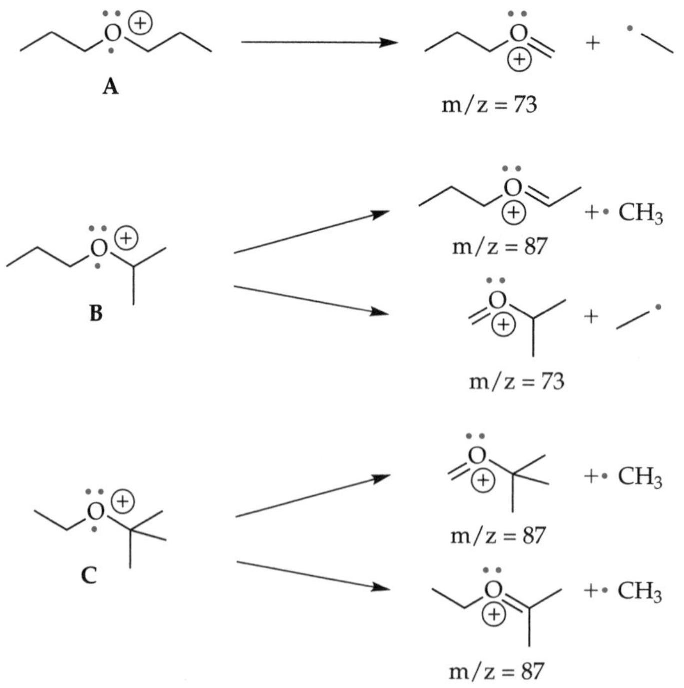

Table 8.10 Relative Abundances of the Oxonium
Ions in the EI Mass Spectra of Three Isomeric Ethers

Ether	m/z = 73	m/z = 87
A	20%	0%
B	11%	16%
C	<1%	58%

Note: The values given are relative to the base peak in each mass
spectrum.

expected oxonium ion fragments. Table 8.10 presents the actual fragmentation data and indicates that these ethers can be distinguished using their
fragmentation patterns.

Amines Amines also exhibit α-cleavage reactions as a primary form of fragmentation (Equation 8.8).

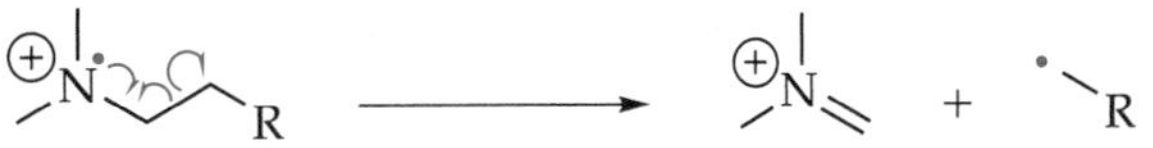

$$(8.8)$$

It should be noted that there is an important distinguishing feature in
the mass spectra of nitrogen-containing organic molecules. *Nitrogen is tri-
valent and is the only element of those commonly found in organic molecules that
can lead to an odd-mass molecular ion. If there are an odd number of nitrogen atoms in a molecule, the mass of the molecular ion is odd. If there are no nitrogens
or an even number of nitrogen atoms, the mass of the molecular ion is even.*

Aldehydes, Ketones, Esters and Amides

α-Cleavage Carbonyl-containing compounds such as aldehydes, ketones, esters, and amides all undergo α-cleavage reactions (Equation 8.9).
In these cases, the α-carbon is the one attached to the carbonyl carbon and
the resulting ions are referred to as acylium ions.

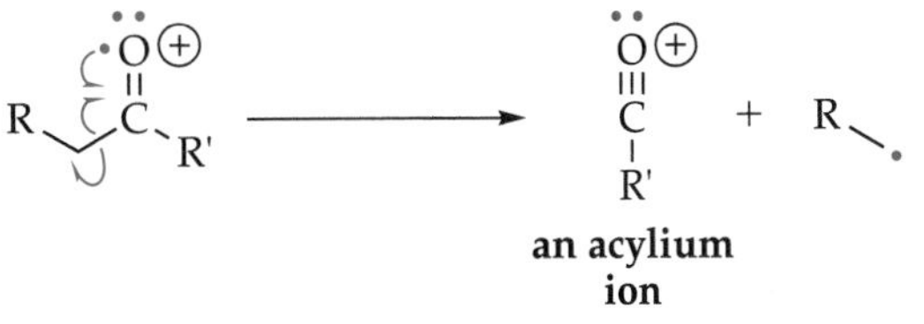

$$(8.9)$$

an acylium
ion

Further fragmentation of these ions by loss of carbon monoxide leads
to alkyl cations as shown in Equation 8.10.

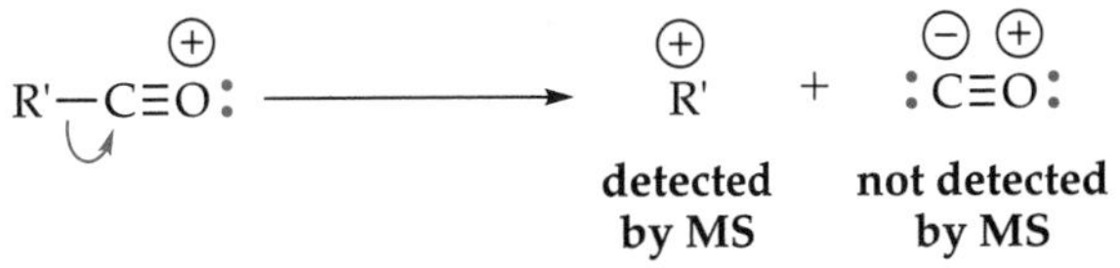

detected   not detected
by MS   by MS

$$(8.10)$$

This fragmentation pathway can be used to differentiate the structures of ketones, analogous to the analysis previously described for alcohols, ethers, and, by inference, amines.

For aldehydes, α-cleavage from the alkyl side of the carbonyl leads to formation of the HCO cation at $m/z = 29$ (Equation 8.11). The loss of hydrogen from the aldehyde molecular ion leading to an acylium ion is less common.

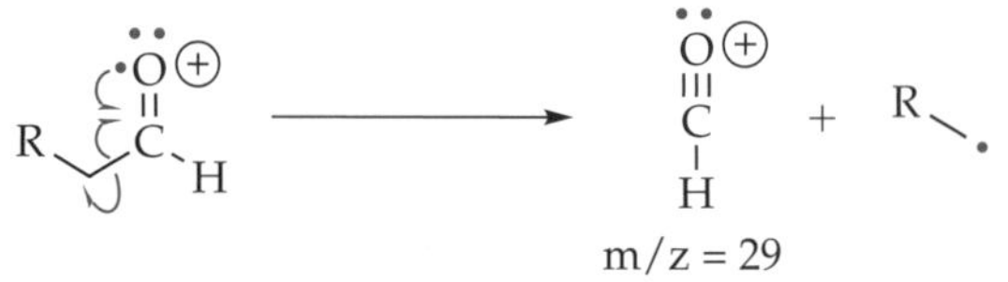

$$\text{(8.11)}$$

$m/z = 29$

Esters, especially methyl esters, can cleave on both sides of the carbonyl. The loss of the methoxy radical ($\bullet OCH_3$) from the molecular ion, giving rise to an M–31 ion, is very diagnostic of the presence of a methyl ester (Equation 8.12).

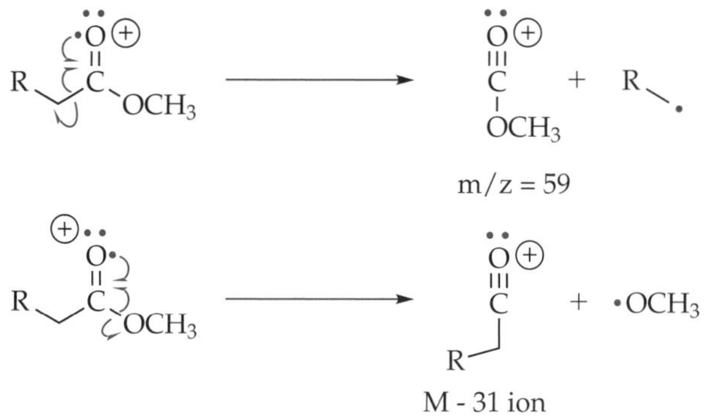

$$\text{(8.12)}$$

$m/z = 59$

M - 31 ion

The main mode of α-cleavage ions in the spectra of amides is cleavage of the bond between the α-carbon and the carbonyl (Equation 8.13).

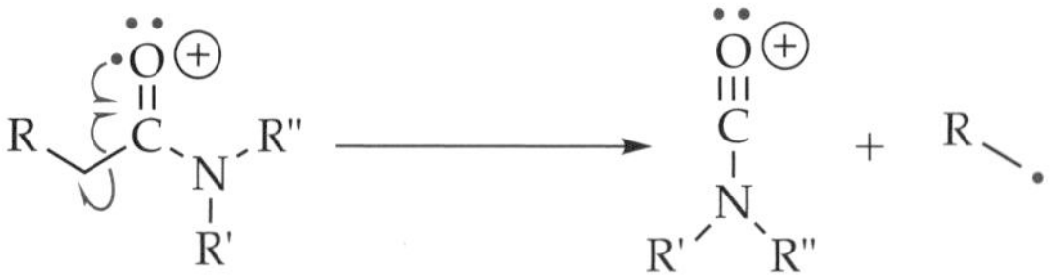

$$\text{(8.13)}$$

β-Cleavage or the McLafferty Rearrangement   Another common fragmentation pathway found in carbonyl-containing compounds is sometimes referred to as **β-cleavage**, but is more commonly called the **McLafferty rearrangement** after Professor Fred McLafferty of Cornell University. As shown in Equation 8.14, in order for this rearrangement to take place, the alkyl side chain of the carbonyl must have a γ-hydrogen, allowing the reaction to proceed through the optimal six-member transition state.

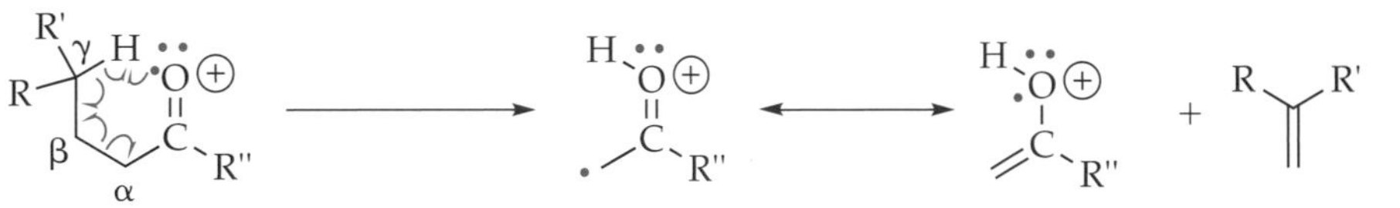

a McLafferty rearrangement

$$(8.14)$$

This reaction takes place in all molecular ions that have a carbonyl with a γ-hydrogen attached to an sp^3 hybridized carbon. These structural requirements of the McLafferty rearrangement, along with the consequences of α-cleavage, can provide a detailed picture of the structure of organic compounds containing the carbonyl function.

Alkylbenzenes The EI mass spectra of some alkylbenzenes have a prominent ion at m/z = 91. This corresponds to the benzyl cation, a very stable ion. However, the benzyl cation rapidly rearranges to the tropilium ion. Studies have shown that it is the tropilium ion that is the species actually observed in the mass spectrum (Equation 8.15).

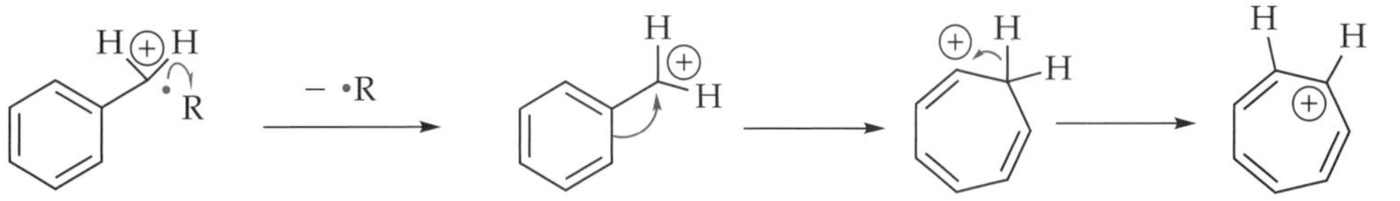

an alkylbenzene radical cation

benzyl cation

tropilium ion (m/z = 91)

$$(8.15)$$

8.2.1.d Gas Chromatography Mass Spectrometry (GC MS)

The theory and utility of gas chromatography was introduced in Section 7.8.6. This valuable analytical tool is capable of separating complex mixtures of organic molecules into gaseous bands of their constituent components. A simple gas chromatogram indicates the number and relative amounts of the components in a sample that have been successfully separated. The separated peaks may be single compounds or mixtures (Sections 7.8.2 and 7.8.6.f). Because the material leaving the GC column is already in the gas phase, the separated components can be introduced directly into the vacuum chamber of a mass spectrometer. Consequently, in addition to detecting the components, it is also possible to obtain structural information. The mass spectral data for each separated component is stored separately in a computer to allow analysis of each of the bands.

As we have seen, mass spectrometry can provide information about a molecule's mass and some insight into the compound's structure. However, identifying ten compounds separated by gas chromatography by interpreting ten different mass spectra is a formidable challenge. Fortunately, if certain experimental parameters are standardized, including the energy of the electrons used to ionize the sample, EI mass spectral data is quite reproducible from instrument to instrument. The m/z and relative abundance data from the EI mass spectra of many compounds has been digitized. The resulting library of mass spectral data can be searched using computer software. This means that, as compounds leave a gas chromatography column and are analyzed by mass spectrometry, their spectra can be compared with a library of spectra. If matches are found, the identity of all the compounds can be determined within minutes. In other cases, a list of several compounds with similar spectra is generated and it is necessary to do some spectral interpretation to narrow down the actual structure of the compound(s).

8.2.1.e Isotopes

Mass spectrometry measures mass and therefore isotopes are easily detected. The analyses of the EI mass spectra of organic compounds up until this point has focused on using nominal atomic masses, where carbon = 12, hydrogen = 1, nitrogen = 14 and oxygen = 16. All four of these elements have naturally occurring isotopes with higher atomic masses (Table 8.1), but these isotopes are found in relatively low abundance in small molecules.

In the mass spectrum of methane shown in Figure 8.25, there is a very small ion at m/z = 17, called the **M + 1 peak**, that is due primarily to $^{13}CH_4$. The intensity of this ion is approximately 1% that of the molecular ion, which is to be expected based on the 1.1% natural abundance of ^{13}C. For larger molecules, the chance of finding ^{13}C present in a sample becomes greater. The isotopes ^{13}C, ^{2}H, ^{15}N, ^{17}O and ^{33}S contribute to the M + 1 peak. The height of the M + 1 peak relative to the molecular ion is calculated as a percentage from Equation 8.16.

$$\frac{M + 1}{M} = (nC \times 1.1) + (nH \times 0.02) + (nN \times 0.36) + (nO \times 0.04) + (nS \times 0.8) \quad (8.16)$$

Because most organic molecules have more carbon than nitrogen or sulfur and the contribution from deuterium (^{2}H) is very small, the major influence on the M + 1 ion is from ^{13}C.

Equation 8.17 for decylamine ($C_{10}H_{23}N$) shows that the M + 1 ion is estimated to be approximately one-tenth the height of the molecular ion. The actual mass spectrum of this molecule confirms this prediction.

$$\frac{M + 1}{M} = (10C \times 1.1) + (23H \times 0.02) + (1N \times 0.36) = 11.82\% \quad (8.17)$$

From Equation 8.16, the intensity of the M + 1 peak for C_{60} (Buckminsterfullerene) is estimated to be 66% that of M. Figure 8.33, which shows the molecular ion region for the EI mass spectrum C_{60}, confirms this calculation for $^{12}C_{59}{}^{13}C$.

As can be seen from Equation 8.16, the M + 1 peak can be larger than the M peak for very large molecules.

Figure 8.33   **The Molecular Ion Region of the EI Mass Spectrum of C$_{60}$**

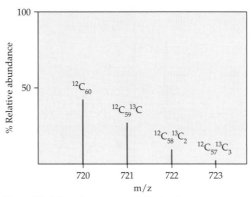

Note: All of these fragments are radical cations.

Figure 8.33 also illustrates that the chances of finding two or more ^{13}C atoms in one molecule increases for larger organic molecules. The contributions of two ^{13}C, one^{18}O and/or one ^{33}S to an M + 2 ion can be estimated using Equation 8.18. The contribution from any other isotopic combinations is negligible. The percent relative abundance is obtained by multiplying the ratio obtained from Equation 8.18 by 100.

$$\frac{M + 2}{M} \approx \left[\frac{n^2}{2} C \times (0.011)^2 \right] + (nO \times 0.0020) + (nS \times 0.044) \tag{8.18}$$

Isotopic patterns for molecules containing chlorine or bromine are diagnostic for the presence of these elements. Table 8.1 shows that for each of these elements there are significant quantities of each isotope separated by two mass units. *For molecules that have one chlorine atom, the ratio of the M to the M + 2 ion is approximately 3:1. Molecules with one bromine atom have an M: M + 2 ratio of approximately 1:1.* The 3:1 or 1:1 patterns persist in fragment ions if those ions still contain the halogen. The spectra of chloromethane and bromomethane demonstrate these characteristic patterns (Figure 8.34).

Figure 8.34a **EI Mass Spectrum of Chloromethane**

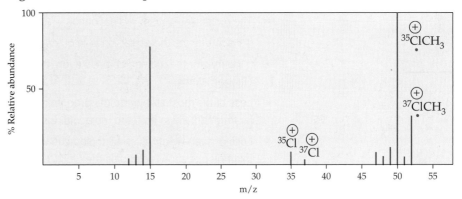

Figure 8.34b **EI Mass Spectrum of Bromomethane**

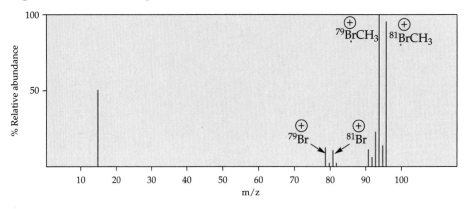

Figure 8.35 **EI Mass Spectrum of Dibromomethane**

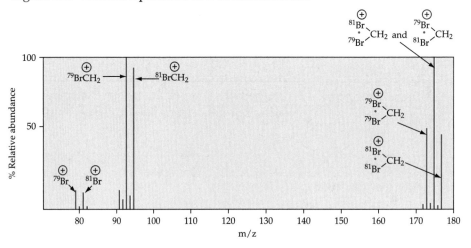

When two halogens are present M, M + 2, and M + 4 ions are observed. An example is shown in the EI mass spectrum of dibromomethane (Figure 8.35). The ratios of these peaks are predictable from a statistical analysis.

SUMMING IT UP!

The Important Features of EI Mass Spectra

- The molecular ion is a radical cation.
- If the m/z for the molecular ion is odd, the ion has an odd number of nitrogen atoms.
- If the m/z for the molecular ion is even, the ion has zero or an even number of nitrogen atoms.
- Loss of the most stable radical predominates in cases in which two or more similar fragmentations from the molecular ion can occur.
- Loss of even-electron, small molecules with strong bonds occurs when possible. Examples of such molecules are CO, CO_2, N_2, C_2H_4 and C_2H_2.
- Fragment ions can fragment further by the loss of even-electron species.

8.2.2 High Resolution EI
Mass Spectrometry

All of the spectra discussed up to this point have had integer values for m/z and are referred to as low-resolution mass spectra. With the exception of carbon-12, which is the basis of the atomic mass scale, the exact masses of elements are not integers as shown in Table 8.11. The differences between the exact masses and the nominal masses are very small, but can be of great utility in determining a molecular formula if sufficient accuracy is available to measure the mass. **High resolution EI mass spectrometry** is a technique that accurately detects differences in mass of 0.002 daltons. It is used to accurately measure the mass of one of the ions in a mass spectrum, usually the molecular ion.

Many different molecular formulas can give rise to molecular ions with the same nominal mass in a low-resolution mass spectrum. With prior knowledge of the sample's origin, many of the possible structures for a compound can be eliminated so that a low-resolution spectrum might be sufficient. However, in some cases, it is useful to know the exact mass of the molecular ion in order to establish the molecular formula. This information can provide a good starting point for further analysis of the mass spectrum and, potentially, the elucidation of the structure.

For example, if the molecular ion of an unknown compound is observed at m/z = 56 in the low-resolution mass spectrum, there are six possible compounds that could give rise to this molecular ion. Their formulas and exact masses are shown in Table 8.12. If analysis of the molecular

Table 8.11 **Exact Masses of Common Nuclei Found in Organic Molecules**

Nucleus	Exact Mass	Nucleus	Exact Mass
^{1}H	1.00783	^{31}P	30.9738
^{2}H	2.01410	^{32}S	31.9721
^{12}C	12.0000	^{33}S	32.9715
^{13}C	13.0034	^{34}S	33.9679
^{14}N	14.0031	^{35}Cl	34.9689
^{15}N	15.0001	^{37}Cl	36.9659
^{16}O	15.9949	^{79}Br	78.9183
^{17}O	16.9991	^{81}Br	80.9163
^{18}O	17.9992	^{127}I	126.9045
^{19}F	18.9984		

Table 8.12 **Molecules with the Nominal Mass of 56 and their Exact Masses**

Molecular Formula	Exact Mass
C_2O_2	55.9898
CN_2O	56.0011
N_4	56.0124
C_3H_4O	56.0262
$C_2H_4N_2$	56.0375
C_4H_8	56.0626

ion using high-resolution conditions indicates a mass of 56.0257, the best match is that for a compound with the molecular formula C_3H_4O.

The accuracy of high resolution EI mass spectrometry is usually on the order of ±0.002 daltons and this is often sufficient to tell the difference between various possibilities. As the mass of the molecule gets larger, there are more possible combinations of atoms that can give rise to the same nominal mass. Consequently, there are more molecules that have exact masses falling within the error limits of the high-resolution determination. Therefore, the high-resolution mass spectrum might lead to more than one molecular formula that matches the data. At this point, the fragmentation patterns must be analyzed to narrow down the list to one possibility.

8.2.2.a MALDI and ESI Mass Spectrometry

Electron ionization mass spectrometry is referred to as a hard-ionization technique because the gas phase molecules collide with high-energy electrons. The resultant radical cations have significant energy and undergo fragmentation. Larger organic molecules such as peptides, proteins, oligosaccharides, oligonucleotides and polymers are not sufficiently volatile to be analyzed under EI conditions. Attempting to heat these molecules in a vacuum to increase volatility leads to decomposition.

In the 1980's and 1990's, techniques were developed to overcome the volatility issues encountered with many molecules. **matrix assisted laser desorption ionization (MALDI) and electrospray ionization (ESI)** are the two methods most commonly used to generate gas-phase ions from non-volatile molecules. *MALDI and ESI are both considered soft-ionization techniques because very little excess energy is imparted to the analyte.* Consequently, the molecular ions generated from the ionization processes undergo little, if any, fragmentation. *In contrast to EI mass spectrometry where positive ions are generated by loss of electrons, these techniques produce ions from the addition of cations or the loss of protons.* Typical cations that can be added include protons (H^+) and, in some cases, metal ions such as sodium or potassium (Na^+, K^+). *The loss of protons generates negatively charged ions in contrast to electron ionization mass spectrometry, but signal detection relies only on the m/z of charged fragments, not on the sign of the charge.* These positive or negative ions are referred to as **quasi-molecular ions**.

MALDI In mass spectrometry, the ions need to be transferred into the gas phase in order to be analyzed. For MALDI, this is accomplished by mixing the analyte with a solid organic compound that is capable of absorbing laser light of a specific frequency. Three examples of these matrix compounds are shown in Figure 8.36.

A small amount of a strong organic acid such as trifluoroacetic acid is also added if protonated molecular ions are desired. The sample is then placed in the vacuum chamber of the mass spectrometer and a laser beam is focused onto it. The matrix absorbs some of the light, leading to excited state matrix molecules. These transfer some of their energy to the sample to give localized explosions that result in the sputtering of the sample into the gas phase. The resulting "debris" includes the charged ions of the analyte that can then be analyzed.

Figure 8.36   Three Common Matrix Compounds Used in MALDI Mass Spectrometry

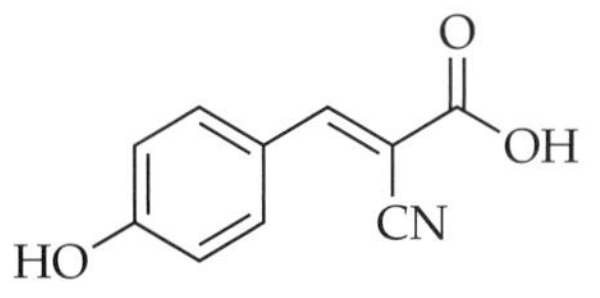

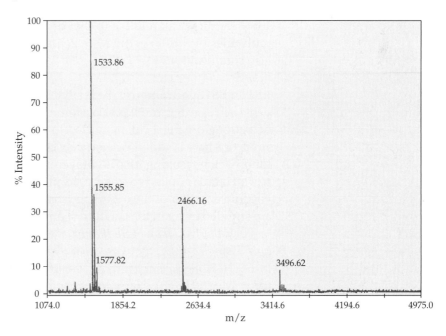

α-cyano-4-hydroxycinnamic acid 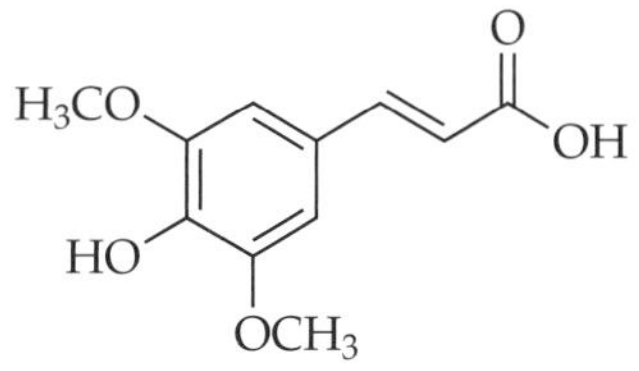   sinapic acid 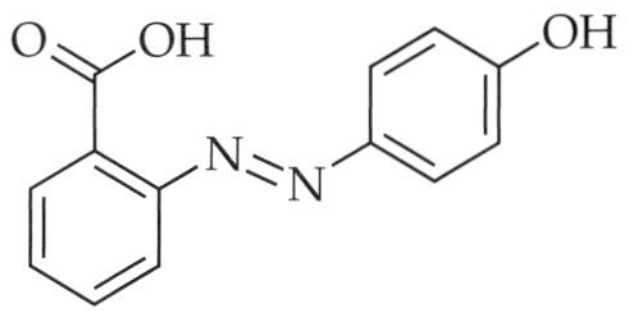   2-(4-hydroxyphenylazo)benzoic acid

A positive ion MALDI mass spectrum of a mixture of three peptides [Pro$_{14}$Arg (P14R), adrenocorticotropic hormone fragment 18–39 (ACTH 18–39), and insulin chain B oxidized] is shown in Figure 8.37a. Three ions are observed at m/z = 1533.86 (P14R), 2466.16 (ACTH 18–39), and 3496.62 (insulin chain B oxidized).

The ion at m/z = 1533.86 is the M + H ion or quasi-molecular ion, so the molecular mass of P14R is actually 1532.86. As can be seen in Figure 8.37a, there is an ion at m/z =1555.85 that is 22 mass units larger than the ion at m/z = 1533.86. This is a sodium adduct of P14R and is referred to as M + Na. Sodium has a mass of 23 confirming the m/z = 1555.85 assignment as the M + Na peak. The peak at m/z = 1577.82 is due to the M − H + 2Na ion.

Figure 8.37a   Positive Ion MALDI Mass Spectrum for a Mixture of Three Peptides[8]

[8] Spectrum courtesy of The University of California Berkeley Mass Spectrometry Facility.

Figure 8.37b Expansion of the m/z = 2466.16 Region for ACTH 18–39 Showing the Effect of Naturally Occurring Isotopes[9]

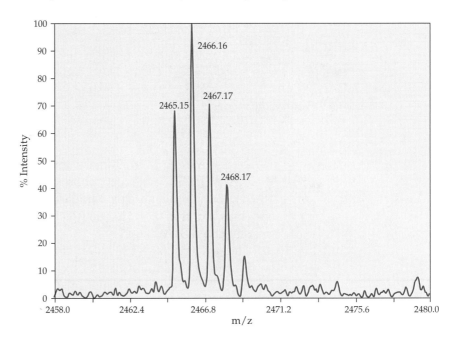

In Figure 8.37b, the region of the spectrum associated with m/z = 2466.16 is expanded to show that there are actually several peaks corresponding to the various isotopic contributions. The M + H peak, the quasi-molecular ion, is at m/z = 2465.15 and the largest peak in this pattern is due to the M + 1 + H ion. This result is consistent with a molecular formula of $C_{112}H_{166}N_{27}O_{36}$ for the M + H ion. Calculating the contributions of ^{13}C, ^2H, ^{15}N, ^{17}O to an M + H + 1 peak using Equation 8.16 gives an intensity of 138% for this peak relative to that of the quasi-molecular ion. Note that for molecules of this size, the M + 2 + H, and M + 3 + H isotopic peaks are also significant.

ESI In **ESI spectrometry**, the analyte is dissolved in a low boiling polar solvent such as water, methanol, acetonitrile, or mixtures of these. Small amounts of a strong organic acid are added to the solution if protonated ions are desired, or a base is added if negatively charged ions are being analyzed. The solution of the resulting ions is sprayed through a small diameter capillary that has a voltage applied to it. Tiny droplets are formed that are passed into the vacuum chamber of the mass spectrometer. The remaining solvent molecules are stripped away in the vacuum chamber, leaving the charged ions.

In both MALDI and ESI, the ions are formed by adding ions or taking away protons. Using MALDI, the quasi-molecular ions usually have a charge of + 1 or −1. In ESI, it is more common to see a distribution of differently charged ions (M + nH or M − nH). For example, in the negative ion ESI mass spectrum of the enzyme pepsin (Figure 8.38), the peak labeled −30 is the ion of pepsin that has lost 30 protons.

---

[9] Spectrum courtesy of The University of California Berkeley Mass Spectrometry Facility.

Figure 8.38 Negative Ion ESI Mass Spectrum of Pepsin (MW = 34,163)[10]

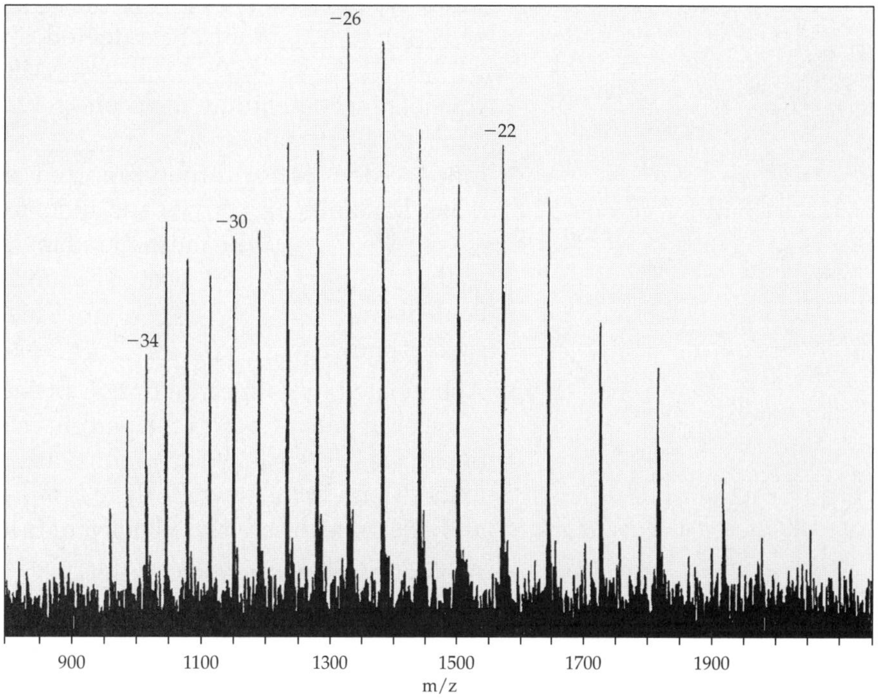

8.2.2.b LC MS

As discussed in Section 8.2.1.d, gas chromatography and EI mass spectrometry can be combined to provide a valuable means of separating and identifying the components of complex mixtures of organic molecules. High performance liquid chromatography, HPLC (Section 7.8.7), provides another method for separating organic molecules that are not sufficiently volatile to be analyzed by GC. In a method called **LC MS**, the compounds are dissolved in a volatile polar solvent such as methanol or acetonitrile and passed through a stationary phase. Because the separated fractions are in a suitable solvent, they can be used directly for ESI mass spectral analysis.

8.2.2.c MS MS

The fragmentation ions observed in EI mass spectra often help to determine the structure of a molecule. In MALDI and ESI, the quasi-molecular ions (M + H or M − H) can be separated in the mass spectrometer based on their mass to charge ratio and then passed into a separate chamber that has been charged with argon or xenon gas. The ions collide with the gas molecules, resulting in fragmentation. The mass spectrum of these fragment ions is analyzed. This type of experiment is known as **tandem mass spectrometry** or **MS MS**. The method has proven very successful in helping to determine the primary sequence of peptides and proteins.

---

[10] Spectrum courtesy of Julie A. Leary.

8.2.3 Problems

1. Fill in the blank spaces.

   In all types of mass _____, it is _____ phase _____ that are detected. In EI MS, otherwise known as _____ _____ MS, ions are generated in a vacuum chamber by bombarding molecules with high-energy _____. After a collision takes place, a _____ cation is formed. These ions are detected and then reported on a mass spectrum, where the *x*-axis has units of ___/___ and the *y*-axis is percent _____ _____. The most abundant ion in a mass spectrum is known as the _____ peak. The ion associated with the original molecule is the _____ ion. This ion can decompose into other species known as _____ ions.

2. Abbreviated mass spectral data is presented for three compounds. The number in parentheses is the percent relative abundance of the ion relative to the base peak. Write rational arrow-pushing mechanisms leading to these ions. In cases where there is more than one possible fragment ion with the same m/z, write all relevant mechanisms.

   A. Methyl pentanoate: m/z: 116(1), 85(52), 74(100), 57(25), 15(15)

   B. Triethylamine: m/z: 101(18), 100(6), 86(100), 58(27)

   C. 3,4,4-Trimethyl-3-pentanol: m/z: 115(4), 112(3), 101(35), 73(100)

3. The mass spectra of all possible structural isomers of methylhexanone are shown in the table below[11]. Using what you know about the common fragmentation pathways of carbonyls, assign each spectrum to a given ketone using the letter to the left of each structure in the table.

   • Begin by filling in the table below. The first row has been completed to help get you started.

   • Compare the data in the table to the spectra.

   • Keep in mind that all expected fragment ions might not be present in great abundance or even at all.

   • If there are any cases where two molecules give rise to the same m/z for all three common fragmentations, look at other ions in the spectra to help you decide between the two. Remember, strange things can happen in highly energized gas-phase molecules, so you might need to be creative in your analysis.

   • Explain briefly why you think your assignments are correct.

	Structure	IUPAC Name	m/z of α-cleavage	m/z of α-cleavage	m/z of McLafferty
A		5-methyl-3-hexanone	57	85	72
B		4-methyl-3-hexanone			

(continued)

---

[11] Spectra were taken from NIST Chemistry WebBook http//webbook.nist.gov/chemistry.

	Structure	IUPAC Name	m/z of α-cleavage	m/z of α-cleavage	m/z of McLafferty
C		2-methyl-3-hexanone			
D		5-methyl-2-hexanone			
E		4-methyl-2-hexanone			
F		3-methyl-2-hexanone			

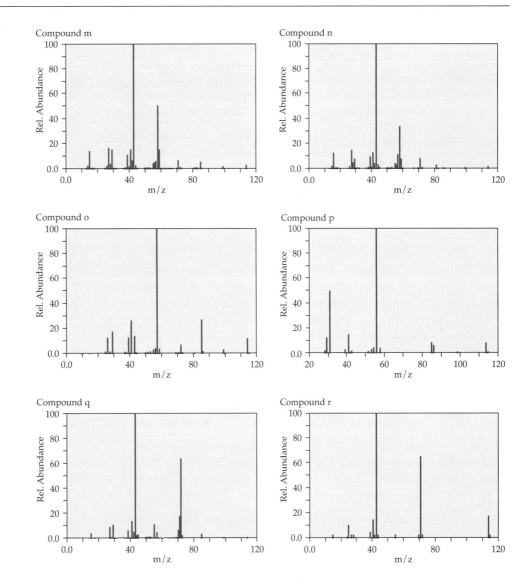

Compound m

Compound n

Compound o

Compound p

Compound q

Compound r

4. EI mass spectral data for 1,3-dibromopropane and 2,2-dibromopropane are provided below. Not all ions are included. Assign each data set to the correct compound and explain how you arrived at your conclusion. (The number in parenthesis is the percent relative abundance of the ion relative to the base peak.)

A   m/z: 204(15), 202(34), 200(19), 123(48), 121(51), 41(100)

B   m/z: 204(0.15), 202(0.35), 200(0.15), 189(0.65), 187(1.2), 185(0.8), 123(69), 121(72.5), 41(100)

5. A positive ion ESI mass spectrum of a protein showed a peak at m/z = 1001 resulting from the addition of 10 protons. What is the actual mass of the uncharged protein?

8.3 INFRARED SPECTROSCOPY

Infrared (IR) spectroscopy is an analytical technique that was commonly used by organic chemists. However, it has become less important as a tool for the determination of the structure of organic compounds because of the development of NMR spectroscopy and mass spectrometry. It is still valuable in the analysis of inorganic compounds because the ligands attached to metals often do not have carbon and/or hydrogen atoms and cannot be analyzed by ^{1}H or ^{13}C NMR spectroscopy. Furthermore, if the ligands do have carbon or hydrogen present and the metal is paramagnetic, analysis by NMR becomes difficult. Paramagnetism does not affect IR spectroscopic analysis.

IR radiation falls within the region of the electromagnetic spectrum that causes bond vibrations when it is absorbed by a molecule. When electromagnetic radiation of the right frequency is absorbed by a bond in a molecule, the bond stretches and contracts more than it does in the absence of electromagnetic radiation in the infrared range. The infrared spectrum records these stretching frequencies. An example of an IR spectrum is shown in Figure 8.39.

Figure 8.39 The Infrared Spectrum of Benzyl Cinnamate

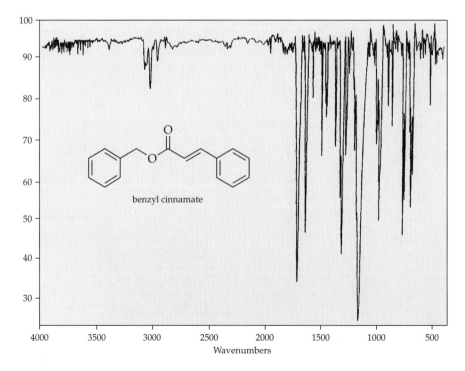

Table 8.13 IR Stretching Frequencies of Some Common Functional Groups

Bond	Stretching Frequency (cm^{-1})	Magnitude of the Band
C-H	2840–3000	Medium to strong
C=C	1640–1680	Weak to medium
C≡C	2100–2260	Weak to medium
O-H	3300–3650	Medium to strong
C=O	1650–1800	Medium to strong
C≡N	2220–2260	Weak to medium

The absorptions are reported as the wavelength of the electromagnetic radiation, λ, in µm or its reciprocal, $1/\lambda$, in cm^{-1}, which is called the **wavenumber**. The wavenumber is abbreviated as $\tilde{\nu}$ and is proportional to the frequency of the electromagnetic radiation as shown in Equation 8.19, where λ is the wavelength, c is the speed of light, and ν is the frequency.

$$\tilde{\nu} = 1/\lambda = \nu/c \tag{8.19}$$

Many functional groups have characteristic stretching frequencies that provide evidence of the absence or presence of these groups in a molecule. Two of the most easily recognized groups in an IR spectrum are the hydroxyl and carbonyl functions. Hydroxyl groups absorb in the 3300–3650 cm^{-1} region and are generally broad bands. The carbonyl groups, found in ketones, aldehydes, esters, and amides, have a strong absorption in the 1650–1800 cm^{-1} region of an IR spectrum. The stretching frequencies for some other common functional groups are shown in Table 8.13.

8.3.1 Recording an IR Spectrum

To record the IR spectrum of a high boiling liquid, a small drop is sandwiched between two discs made from a salt, such as KCl, KBr, or NaCl, to create a thin film of the compound. These salts are used because they do not absorb infrared radiation in regions of importance for characterizing organic functional groups. The discs are placed in an infrared spectrometer and any absorption of the infrared radiation is detected and recorded. If a thin film cannot be prepared, the compound is dissolved in a suitable solvent and the solution is placed in an IR cell consisting of two salt plates that have been sealed on their edges. The solvent chosen for this type of experiment should not strongly absorb in the same regions of the IR spectrum as the analyte. In general, the spectrum of the pure solvent must first be recorded and then stored in the spectrometer's database. The spectrometer subtracts the absorbances of the solvent to give the spectrum of the compound. If the compound is a solid, it can be analyzed in solution or it can be mixed with a salt such as KCl or KBr and ground to a fine powder. The resulting mixture is then pressed into a disc and infrared radiation is passed through the disc.

CHAPTER 9

Running a Synthetic Reaction

The theory section of this text is designed to give you a background in the techniques used in organic synthesis. The experiments provide you the opportunity to practice these techniques and prepare compounds according to the described procedures. With this knowledge and expertise, you will be able to synthesize compounds found in the literature and, eventually, compounds not previously known.

9.1 REPRODUCING A PUBLISHED SYNTHESIS

Before running a reaction, it is important to search the literature to determine if a compound has been made previously. Some of these resources are listed in Chapter 3.

As can be seen from the following examples however, even with a published literature preparation, replicating a synthesis is not straightforward and simple. Details of the procedure are usually very brief and considerable experience, knowledge, and understanding of the techniques and theory of organic chemistry are necessary for success.

9.1.1 Examples of Synthetic Experimental Data from the Literature[1]

1 **General Methods.**[2] Melting points are uncorrected. Air sensitive reactions were kept under N_2. Solvents used in
2 moisture sensitive reactions were dried using standard methods. Tetrahydrofuran (THF) and diethyl ether (Et_2O)
3 were distilled from sodium benzophenone ketyl. Dichloromethane (CH_2Cl_2) was distilled from CaH_2. Triethylamine
4 (Et_3N) was distilled and stored over molecular sieves prior to use. Dimethylsulfoxide (DMSO) and acetonitrile
5 (CH_3CN) were used directly from Aldrich "Sure Seal" bottles. The term "concentrated" refers to the removal of
6 solvent using a rotary evaporator (15 torr at 25°C) and then using a high vacuum line (<0.5 torr at 25°C) until a
7 constant weight was obtained. Thin layer chromatography (TLC) was performed using precoated Kieselgel 60
8 F-254 plates. Flash chromatography was performed using EM Science Silica Gel 60 (230–400 mesh).
9 NMR spectra were obtained using a Bruker AM-400 or AM-500 spectrometer. ^{1}H NMR chemical shifts
10 were reported in ppm relative to the solvent resonance: $CDCl_3$, δ 7.24; $(CD_3)_2SO$, δ 2.49; CD_3OD, δ 3.30.
11 Coupling constants (J) are reported in Hz. [13]C[^{1}H]NMR chemical shifts are reported in ppm relative to the
12 solvent resonance: $CDCl_3$, δ 77.0; $(CD_3)_2SO$, δ 39.5; CD_3OD, δ 49.0.
13 Fast atom bombardment mass spectra (FABMS) were performed using 3-nitrobenzyl alcohol (NBA) or
14 thioglycerol/glycerol (TG/G) as the matrix. In cases where a diol was not stable to the FABMS conditions, LiCl
15 was added to form stable adducts.[21] Infrared spectra were obtained using an ATI Mattson Gemini FTIR set at
16 2 cm^{-1} resolution. Optical rotation concentrations (c) are reported in g/100 mL. Elemental analyses were performed
17 by the Microanalytical Laboratory at the University of California, Berkeley.

[1] Kemp, S. Ph.D. Thesis, Dept of Chem. UC Berkeley **1995**; Park, J. Ph.D. Thesis, Dept of Chem. UC Berkeley **1992**.
[2] Note: The footnotes referenced in the following experimental procedures are from the original articles and are not included here.

18	**Procedure for the Formation of Cyclic Sulfates 4a-i from 1,2-Diols 3a-i.** (adapted from the procedure of
19	Sharpless, *et al.*[6a]). Thionyl chloride (220 μL, 3.0 mmol) in CH_2Cl_2 (280 μL) was added dropwise over 10
20	minutes to a stirring solution of diol (**3**)[1,2] (2.0 mmol) and Et_3N (1.1 mL, 8.0 mmol) in CH_2Cl_2 (10 mL) at 0°C.
21	Stirring was continued for 5 minutes at which time TLC (7:3 hexane:ethyl acetate) showed completion. Water
22	(1 mL) was added and stirring was continued for 5 min. Hexane (8 mL) and water (10 mL) were added and the
23	ice bath was removed. After stirring for an additional 5 minutes the aqueous layer was removed. The organic
24	layer was washed with water (3 x 5 mL), saturated $NaHCO_3$ (5 mL) and saturated NaCl (5 mL), dried ($MgSO_4$),
25	filtered, and concentrated to give the cyclic sulfite (pair of diastereomers) as an oil. After drying under
26	vacuum for 2 h, the oil was dissolved in CCl_4 (6 mL) and CH_3CN (6 mL). Water (9 mL), $RuCl_3 \cdot 3H_2O$ (6.0 mg),
27	and $NaIO_4$ (856 mg, 4 mmol) were added while stirring at 0°C. The reaction mixture was stirred at 0°C for about
28	1 hour while monitoring by TLC (7:3 hexane:ethyl acetate). Hexane (25 mL) and Et_2O (25 mL) were added, the
29	layers were separated, and the aqueous layer was extracted with hexane (2 x 20 mL). The combined organics
30	were washed with saturated NaCl (8 x 10 mL, or until the organic layer was nearly colorless), dried ($MgSO_4$),
31	filtered, and concentrated to a white solid. The cyclic sulfates could be further purified by flushing through a
32	plug of silica (7:3 hexane:ethyl acetate) to give cyclic sulfates **4a-i** as white solids or clear oils.

33	**(2S,3R,4R)-2-[N-(Benzyloxycarbonyl)amino]-1,6-diphenylhexane-3,4-cyclic sulfate (4b).** TLC of cyclic sulfite
34	intermediate: R_f 0.66, 0.70 (7:3 hexane:ethyl acetate). Yield 861 mg (92%) as an amorphous solid. An analytical
35	sample was obtained by chromatography (8:2 hexane:ethyl acetate) as a white solid: mp 67–68°C; R_f 0.67
36	(7:3 hexane:ethyl acetate); ^{1}H NMR (400 MHz, $CDCl_3$) δ 2.00–2.10 (m, 2H), 2.58–2.68 (m, 1H), 2.77–2.89 (m, 2H),
37	2.99 (dd, J = 7.5, 13.8, 1H), 4.19 (q, J = 8.6, 1H), 4.55 (d, J = 8.3, 1H), 4.80 (dd, J = 8.0, 12.8, 1H), 5.02
38	(d, J = 12.3, 1H), 5.13 (d, J = 12.3, 1H), 5.35 (d, J = 9.6, 1H), 7.16–7.43 (m, 15H); [13]C[^{1}H] NMR (100 MHz,
39	$CDCl_3$) δ 31.1, 33.2, 38.5, 51.4, 67.2, 74.6, 83.9, 86.5, 126.4, 127.2, 127.5, 127.7, 128.1, 128.3, 128.47, 128.54,
40	128.69, 128.90, 128.93, 129.42, 135.6, 135.9, 139.3, 156.2; FABMS (NBA w/LiCl) *m/z* 488 (MLi $^+$, 100), 438 (10),
41	408 (22), 397 (18), 390 (11), 318 (35); FAB HRMS *m/z* calcd for $C_{26}H_{27}LiNO_6S^+$ 488.1719. Found:488.1719.

42	**(2S, 3R, 4R)-2-[N-(Benzyloxycarbonyl)amino]-5-methyl-1-phenylhexane-3, 4-cyclic sulfate (4c).** TLC of cyclic
43	sulfite intermediate: R_f 0.54, 0.48 (5:1 hexane:ethyl acetate). Yield 1.72 g (97%) as a white solid. An analytical
44	sample was obtained by recrystallization from ether-hexane: mp 84–85°C; R_f 0.38 (5:1 hexane:ethyl acetate);
45	^{1}H NMR (400 MHz, $CDCl_3$) δ 0.86 (d, J = 6.9, 3H), 0.94 (d, J = 6.7, 3H), 1.97 (sext., J = 6.8, 1H), 2.84 (dd,
46	J = 9.5, 13.6, 1H), 3.04 (dd, J = 6.7, 13.7, 1H), 4.15 (m, 1H), 4.46 (dd, J = 7.2, 1H), 4.57 (d, J = 7.0, 1H), 5.06
47	(d, J = 12.3, 1H), 5.16 (d, J = 12.3, 1H), 5.23 (d, J = 9.9, 1H), 7.20–7.40 (m, 10H); [13]C[^{1}H]NMR (100 MHz,
48	$CDCl_3$) δ 17.4, 17.5, 30.7, 39.0, 52.9, 67.3, 84.2, 88.0, 127.4, 127.8, 128.2, 128.5, 129.0, 135.7, 135.9, 156.3;
49	FABMS (NBA w/LiCl) *m/z* 426 (MLi$^+$, 100), 420(MH$^+$, 35), 376 (46), 328 (46). Anal. Calcd for $C_{21}H_{25}NO_6S$: C,
50	60.13; H, 6.01; N, 3.34. Found: C, 60.20; H, 6.01; N, 3.28.

51	**Syntheses of Allylic Alcohols 3 from Diols 2.** Diol **2** (1.12 mmol) in tetrahydrofuran (50 mL) was added to
52	freshly washed sodium hydride (*ca.* 4.5 mmol from a 60% oil dispersion, washed with pentane, 2 × 20 mL). The
53	reaction solution was refluxed for 20–60 min, and then cooled to room temperature and quenched with water
54	(20 mL). Ether (50 mL) was added and the aqueous phase was separated and extracted with additional portions
55	of ether (3 × 30 mL). The combined organics were dried with $MgSO_4$, filtered, and concentrated to give a
56	colorless oil. The product was purified by bulb-to-bulb distillation or flash chromatography on a silica gel column
57	(2.5 × 10 cm) with an eluent of ethyl acetate in hexane (5% or 10%) to give the allylic alcohol.
58	In cases where the *Z*-allylic alcohol was obtained (i.e. the minor isomer), observable ^1H and $^{13}C[^1H]$ NMR
59	resonances are reported.

9.1.2 Reading and Interpreting a Published Synthesis[3]

Conventions for writing an experimental procedure are not uniform and you must become adept at interpreting what an author means. This ability comes with practice in running many reactions. In addition, it is helpful to know how to analyze an article that describes the synthesis of a new compound.

It is important to not simply go to the experimental section of an article and try to reproduce the work without first reading the entire article. Just as the introduction to the experiments in this laboratory text contain essential information, many important points are included in the result and discussion sections of an article that can be critical to the successful reproduction of an experiment. The main body of a paper may also be where acronyms for reagents are defined.

9.1.2.a The General Information Section

The general information section usually indicates where the starting materials and reagents came from, that is, whether they were purchased or synthesized. This section may also offer valuable information about the purity of the reagents and solvents used (lines 1–5). For example, "Tetrahydrofuran was used as received," means that you can use it straight out of the bottle. On the other hand, the statement, "Tetrahydrofuran was distilled from sodium benzophenone ketyl," means that the solvent must first be purified (lines 2–3). In this case, it might be necessary to find a literature procedure for purifying the solvent. Details associated with various spectroscopic methods are also outlined in the general information section (lines 9–17).

If a paper describes a synthetic procedure in which several derivatives of the same basic structure are prepared, a general procedure is often given followed by specific details for each derivative (lines 18–32, 33, 42).

9.1.2.b The Experimental Section

The experimental section usually begins with the amounts of reagents and solvent used. These are expressed in grams (milligrams) or moles (millimoles) or both. If the material is a liquid, the amount may be stated as a

---

[3] The line numbers refer to those indicated in Section 9.1.1.

volume that can be measured using a syringe, graduated cylinder, or, if the density is known, by converting the volume to grams and then weighing the material. The scale of the reaction must also be determined. If the published procedure is for 500 grams and only 1 gram is needed, the scale must be adjusted in order to avoid wasting reagents and increasing the cost associated with disposal. However, changing the scale of the reaction always requires thoughtful consideration of the mechanism and the thermodynamics of the reaction. You must be prepared for differences in reaction times, product yields, and other factors when changing the quantities from those reported in a published procedure. It is not always possible to successfully scale a procedure up or down by assuming a linear relationship between factors such as solvent volumes and reaction times. It is important to consider these issues ahead of time to ensure that the proper equipment is available.

Once the quantities of materials are determined, the reaction vessel size must be chosen. Sometimes the size of the container is stated in the experimental procedure, but usually it is not. Standard equipment can range from test tubes to round-bottom flasks and many factors affect the choice of the apparatus, including such simple considerations as availability. The single most important factor is to choose a large enough vessel so that the reaction mixture does not exceed a reasonable level in the container (lines 19–22).

Before any materials are added to the reaction vessel, some additional details should be considered.

- Is the order of addition of the starting materials, reagents, and solvents important? You must think about what is going to happen when the chemicals start reacting with one another. Should one of the reactants be diluted with solvent before other reactants are added? Should two of the reactants be allowed to react to form an intermediate product before the addition of a third reactant? Often these factors are not explicitly outlined in the experimental section and you must once again refer to the main body of the article.

- Should the reaction be stirred to maintain homogeneity? There are many methods used to stir reactions including overhead stirrers for large-scale reactions, stir bars in conjunction with stir plates, or some labs may be equipped with shakers.

- Is the reaction exothermic? In this case, slow addition or cooling the reaction might be necessary. Cooling is usually accomplished by placing the reaction vessel in an ice-water bath, an ice-salt-water bath, a dry ice-solvent bath, or a liquid nitrogen-solvent bath depending on the temperature necessary to control the reaction rate. On the other hand, some exothermic reactions, for example the Grignard reaction, have high heats of activation and must be initiated by first warming before they are then cooled to prevent the reaction from getting out of control.

- Is the activation energy of the reaction such that heating is necessary to obtain a reasonable reaction rate? Hot water, sand, or oil baths can be used. A heating mantle or block allows more control over a wider range of temperatures. When a reaction is heated, it is important to consider the volatility of the solvent and reactants and to add a reflux condenser to prevent evaporation during the reaction if necessary.

Preparing a "Slush" (Low-Temperature) Bath

For slush baths made from a solvent and liquid nitrogen, the solvent is placed in an insulated container and the liquid nitrogen is poured in *slowly* with stirring until the mixture has the consistency of slush.

To prepare a slush bath using dry ice (solid carbon dioxide), the solvent is put in an insulated container and small pieces of dry ice are added *slowly* while stirring the mixture. The temperature listed below will be achieved when there is excess dry ice visible in the solution. In both cases, the temperature can be checked with a low temperature thermometer. Maintenance of the temperature in both types of baths is achieved by adding liquid nitrogen or dry ice as needed.

Some examples of slush-bath temperatures:

ethylene glycol/dry ice:	$-15°C$
acetonitrile/dry ice:	$-42°C$
n-octane/liquid nitrogen:	$-56°C$
ethanol/dry ice:	$-72°C$
acetone/dry ice:	$-77°C$
ethyl acetate/liquid nitrogen:	$-84°C$

For a comprehensive list of slush-baths, see *The Chemists Companion*, Gordon, A.J. and Ford, R.A., Wiley Interscience, New York, 1972.

9.1.2.c Monitoring the Progress of the Reaction

Some procedures indicate exactly how long to allow the reaction to proceed (lines 21–23). However, if it is not specified or if you change the scale of the reaction, it is important to devise a method to follow the progress of the reaction. Visual changes such as a color change, formation of a precipitate, or pH change are convenient, but often not sufficiently precise. Thin layer chromatography (Section 7.8.3) is an excellent method because it is quick, inexpensive, and requires very simple equipment (lines 21, 27–28). Conditions must be found so that the starting material and product have different R_f values. In addition, it is necessary to find a visualization method(s) that allows both of them to be identified. Visualization techniques are rarely stated in published procedures (lines 7–8, 21, 33–34, 42–43). Alternately, HPLC (Section 7.8.7) or GC (Section 7.8.6) can be used to monitor the progress of a reaction.

9.1.2.d Isolating and Purifying the Product

Once the reaction is completed, the product must be isolated. This is called **working up a reaction** and is a crucial part of any synthesis. In general, it is the least detailed of all of the steps in a published procedure. The authors assume that anyone doing experimental chemistry is well versed in techniques such as extraction (lines 28–29, 54–55), washing (lines 24, 30), filtration (lines 31, 55), crystallization (lines 44), drying (lines 24, 25, 30, 55), distillation (line 56), and chromatography (lines 31–32, 34–35, 56–57). For these reasons, it is necessary to have a good understanding of and practice in the techniques introduced in Chapter 7.

After the product is isolated, it is usually necessary to purify it. Again, the methods of purification (lines 31–32, 35–36, 43–44, 56–57) are usually very vague for the same reason as those for the workup. The same knowledge of recrystallization, distillation, and chromatography are required to achieve adequate purification without significantly sacrificing the yield of the product.

9.1.2.e Cleanup

Procedures do not usually include details about how to properly dispose of reagents, solvents, and by-products of a reaction. It is important to know how these materials should be properly recycled or disposed of according to state law. A good general rule is to assume that no organic chemicals should be disposed of down a drain that leads to a municipal sewer or an open body of water.

9.1.2.f Characterizing the Product

Finally, it is necessary to characterize the product and document the analytical data. In a published procedure, the methods of obtaining this data are rarely detailed (lines 9–17). Sometimes information on the magnetic field strength of the NMR used is provided (line 9), but this is often not the case. If the material is a solid, a melting point (Section 6.3) is always included (lines 35, 44). Liquids are usually characterized by NMR spectroscopy (Section 8.1) and/or mass spectrometry (Section 8.2) and this data is also included for solids (lines 36–41, 45–50). Analytical data is typically presented in a standard format (lines 41, 50 and Section 8.1.6).

It is important to have sufficient knowledge of NMR spectroscopy and mass spectrometry in order to decide if the author's interpretation of the analytical data is in fact correct. In addition, if the synthesis for some reason did not result in the same product as the published procedure, it is crucial to be able to determine the structure of the product actually obtained.

Therefore, the first step in becoming a synthetic organic chemist is to master the theory and techniques and then to reproduce the results obtained by others by following documented experimentals. Only then are you ready to develop and run your own syntheses of new compounds.

9.2 RUNNING A NEW SYNTHETIC REACTION

If the synthesis of the compound you need has not been published or you wish to develop a more environmentally-friendly method of synthesis, it is still important to use information from a literature search as a guideline. The synthesis of a closely related compound may be available to provide a starting point. It is critical to apply the skills of retro-synthesis learned in the lecture course.

9.2.1 Determining the Reactants and Conditions

In choosing a method of synthesis, the availability and cost of all starting materials should be assessed. The toxicity, potential hazards, and proper disposal of all reagents, solvents, products, and side-products must be taken into consideration in the planning of a new synthetic procedure. The most environmentally-friendly method should be chosen. Keep in mind that a pathway with fewer steps might be preferable because at each step some material is usually lost and minimizing steps can sometimes maximize yield.

Once the starting materials have been determined, the reaction conditions must be assessed. For example, it is usually necessary to get materials into solution to interact and react so an appropriate solvent must be chosen. Experience and knowledge about the thermodynamics of similar reactions helps in establishing the conditions necessary to achieve a reasonable reaction rate. Also, the reagents used must be compatible with other functional groups in the starting materials, otherwise it is necessary to use protecting groups.

9.2.2 Planning, Executing, and Documenting the Laboratory Procedure

Once a potential pathway has been determined, a great deal of planning and preparation is necessary before working in the lab. Information corresponding to the prelab writeup for experiments from this text (Section 2.1) and journal procedures should be documented in the laboratory notebook. In addition to all of the issues addressed and the skills developed in running published literature procedures, the following points must be considered:

- Determine the starting materials and reagents and record their physical properties.
- Predict possible side-products of the reaction and determine their physical properties.
- Choose an appropriate scale for the reaction and secure the needed reactants. Do not purchase an excess of reagents.
- Determine the amounts of each material to be used. What is the limiting reagent? Should stoichiometric amounts be used or is an excess of one reagent necessary to drive the reaction to completion?
- Determine the method of dispensing materials. Note that the limiting reagent should always be weighed, but solvents and other liquids can be measured by volume.
- Determine the method and the order of addition of reagents.
- Determine a method to follow the progress of the reaction.
- Analyze the properties of the final reaction mixture and predict the properties of the desired material and by-products in order to plan an appropriate method for isolating the product from the reaction mixture and for purifying it.

The information in the laboratory notebook is essentially equivalent to a published procedure. However, it has not been tested! In the lab, it is important to have enough skill and experience to recognize and adjust to unexpected observations. A careful, complete record of each step as it is performed and the observed changes should be documented in the laboratory notebook. The synthesis may not necessarily proceed according to plan and an experienced chemist is able to be flexible in recovering from the unexpected and in rescuing the synthesis.

The structure of the product and its purity should be confirmed using spectroscopic methods and mass spectrometry. The spectral data and all relevant physical properties, including melting point, color, and crystalline form, should be recorded in the laboratory notebook.

9.2.3 The Result

Organic synthesis is fun and the rewards are many. However, just like many endeavors, making new compounds requires a commitment of time and effort to become an expert and to be successful. The new material you have synthesized might lead to a cure for a disease or might serve as an intermediate for new technologies.

PART 2

Experiments

Calibrating a Pasteur Pipette

In the organic chemistry laboratory, all of the solvents and many of the reactants and reagents you will be using are liquids. There are two methods for measuring liquids: by weight and by volume. The material should be weighed if it is a limiting reactant. If the liquid is a reactant that is involved in the equation for the reaction, but is not the limiting reactant, a syringe can be used to provide an accurate measurement of the material. On the other hand, if the liquid is being used, for example, in a recrystallization or as a solvent for the reaction where the exact amount is not critical, it is easiest to deliver liquids by volume. Therefore, it is useful to learn how to estimate liquid volumes using a Pasteur pipette.

Background Reading:

Before the laboratory period, read Chapter 1, *Safety in the Chemistry Lab*; Chapter 2, *The Laboratory Notebook and the Laboratory Period*; Chapter 3, *Where to Find It: Searching the Literature*; and Chapter 4, *Things You Need to Know Before You Begin*. See Section 7.1.1 and Figure 7.1 for how to hold a Pasteur pipette.

Prelab Checklist:

Record the densities of water, ethanol, and ethyl acetate in your laboratory notebook before the laboratory period.

Experimental Procedure

Wear your safety glasses at all times in the lab.

Pour approximately 6 mL of water into a beaker. Using a Pasteur pipette equipped with a pipette bulb, add water to the 0.5 mL mark of a graduated cylinder. Empty any water remaining in the pipette back into the beaker. Use the empty pipette to draw up the 0.5 mL of water from the graduated cylinder. Make a mental note of the level of water in the pipette. Expel this water back into the beaker. Then, draw up what you think is approximately 0.5 mL of water from the beaker and expel it into the graduated cylinder. Record the volume in your laboratory notebook. Repeat this until you hit the 0.5 mL mark three times in succession.

Repeat this procedure for 1 mL of water.

TECHNIQUE TIP   Steady the pipette with the middle and ring fingers while squeezing the bulb with the thumb and forefinger (Section 7.1.1 and Figure 7.1).

To get a feel for the maximum amount of liquid that can be drawn up into a Pasteur pipette, add approximately 6 mL of water, 95% ethanol, or ethyl acetate to separate clean beakers. Weigh an empty container and record the weight in your laboratory notebook. Expel as much air as possible from a clean Pasteur pipette fitted with a pipette bulb. Draw up as much water as possible into the pipette from the beaker containing water. Expel the water from the pipette into the tared, empty container. Weigh the container and record the weight in your laboratory notebook. Repeat the process again, drawing up as much liquid as possible. Reweigh the container and record the weight. Repeat the process once more.

Use clean pipettes to repeat the procedure three times with ethanol and three times with ethyl acetate.

Using the density, convert the weights of the liquids to mL and round to the nearest 0.1 mL.

Did you know?

ethyl acetate

Ethyl acetate is a common solvent used in the organic chemistry laboratory. It is also a natural product found in many fruits. Consequently, it is used in the flavor and fragrance industry as a flavoring and olfactory agent for foods and perfumes. You might have encountered it as the main ingredient in some fingernail polish removers. If you happen to like studying bugs (entomology), you probably have used ethyl acetate vapors in a "killing jar" to render specimens unconscious and eventually dead while keeping the body malleable for the purpose of mounting them.

Cleanup

The water can be poured down the drain. The ethanol and ethyl acetate should be placed in the appropriately labeled liquid waste containers. The Pasteur pipettes should be placed in the container for contaminated laboratory debris.

Discussion

- Are the volumes about the same for each of the three trials using 0.5 and 1.0 mL of water or are any of the measurements consistently different? Explain any differences.

- Do your results suggest that you can reliably use a full Pasteur pipette to deliver a given volume of solvent to a reaction or a recrystallization?

Questions

1. If an experiment calls for adding exactly 3.1 mL of a given solvent, which of these three methods would you recommend? Explain your choice.

   a. Measure the solvent using a graduated cylinder.

   b. Weigh the solvent on a balance.

   c. Estimate the volume using a Pasteur pipette.

2. If an experiment calls for adding approximately 1 mL of a given solvent, which of these three methods would you recommend? Explain your choice.

   a. Measure the solvent using a graduated cylinder.

   b. Weigh the solvent on a balance.

   c. Estimate the volume using a Pasteur pipette.

3. What is the molecular weight of Compound X given the following information?

   • The density of Compound X is 1.32 g/mL.

   • 0.52 mL of Compound X is equal to 1 mmol.

4. Why should you use a new Pasteur pipette each time you use a different solvent?

5. Fifteen drops of ethyl acetate from a Pasteur pipette weigh 0.16 g. What is the average volume of each drop?

Investigating Solubility and Acid-Base Reactions

An understanding of the solubility of compounds and how acid-base properties affect solubility is important in isolating and purifying the products of chemical reactions.

Background Reading:

Before the laboratory period, complete the following background reading.

> **Acids and Bases:** In addition to the discussion of acids and bases in Section 4.1, you may also wish to review this topic in your introductory chemistry text. Remember that carboxylic acids are organic acids. They can act as proton donors. Amines are organic bases and can act as proton acceptors.

> **Properties of Organic Molecules:** In order to understand solubility, it is necessary to understand the intermolecular forces between identical and dissimilar molecules discussed in Chapter 5.

> **Solubility:** Solubility and Inert vs. Reactive Solvents are discussed in Sections 6.1 and 6.1.1.

Prelab Checklist:

Enter the names, structures, molecular formulas, and molecular weights of all reactants and reagents in tabular form in your laboratory notebook before the laboratory period. Include the liquid densities and boiling points of all liquids and the melting points of all solids. See Chapter 3 for references that have detailed information about chemical compounds.

For each of the parts of the experiment, predict in the prelab write-up if the process involves a change in only the intermolecular forces or a chemical reaction. When chemical reactions are predicted, write a balanced equation for each reaction and identify the bonds being broken and those being formed.

Experimental Procedure

This experiment is to be done in groups of two. Collaboration is important and you are expected to discuss all results and observations with your partner. Be sure to record your partner's observations in your laboratory notebook.

Wear your safety glasses at all times in the lab.

CHEMICAL SAFETY NOTE   Organic solvents, acids, and bases can injure your skin and eyes. It is important to avoid contact with these materials. Should you accidently get some on your skin, rinse the area well with running water for several minutes. If the area of exposure is your eyes, it is very important that you use the eye wash for a minimum of five minutes. Because bases are particularly damaging to the eye, it is extremely important that your eyes be protected by safety glasses at all times. Notify your instructor or have a fellow student alert the instructor to any accident.

TECHNIQUE TIP   Use a sand bath to heat the solutions. In general, the rheostat for the sand bath should be set at a low setting. A digital thermometer is used to measure the temperature of the sand bath at the depth at which the reaction tubes/small test tubes are placed. *Never* put a mercury thermometer in a sand bath!

SAFETY NOTE   In this experiment, a boiling stick should be used to prevent bumping (superheating) when a solution is heated because it is more easily removed from the solution than a boiling chip. (See Section 7.1.1.)

Each test will be done in a reaction tube/small test tube. In Part I you will need clean, dry tubes, but for the remaining parts, the tubes should be clean but they may contain traces of water. In all of the parts, use a different Pasteur pipette to deliver each of the liquids to avoid contamination. Record all observations, both before and after mixing. To mix the contents, use the paddle end of a clean stainless steel spatula and spin it in the tube.

Part I: Solute Miscibility with a Non-polar Solvent

Add 1 mL of hexane to six *dry* reaction tubes/small test tubes and 5 drops of one of the following solutes: diethyl ether, ethyl acetate, acetone, ethanol, dichloromethane, or water.

Part II: Solute Miscibility with a Polar Solvent

Add 1 mL of water to six reaction tubes/small test tubes and 5 drops of one of the following solutes: diethyl ether, ethyl acetate, dichloromethane, acetone, ethanol, or toluene.

Part III: Water Solubility of Alcohols

Add 1 mL of water to four reaction tubes/small test tubes and 5 drops of one of the following alcohols: ethanol, 1-propanol, 1-butanol, or 1-pentanol. If the alcohols dissolve with stirring, add 5 more drops. Repeat to a total of 25 drops, or until the solubility limit is reached, whichever occurs first. Make sure to record your observations after each set of additions.

Part IV: Temperature Dependence of Solubility

Weigh approximately 40–50 mg of benzoic acid in a reaction tube/small test tube. Connect a clamp to a reaction tube/small test tube so you can hold the hot tube and secure it in a clamp holder. Add 1 mL of water to the solid in the tube. Add a boiling stick and stir vigorously. Record your observations and then heat the mixture on the sand bath until the water just boils. Record your observations. Remove the tube from the sand bath, remove the boiling stick, and allow the solution to cool slowly to room temperature undisturbed. Record your observations and save the resulting mixture for Part VII.

Did you know?

benzoic acid

Nostradamus, the famed prophet, has been credited with being the first person to document the isolation of benzoic acid by heating gum benzoin, the resin from trees in the genus Styrax, and collecting the distillate. Benzoic acid is a starting material for the synthesis of many compounds containing the phenyl group. It is also a food preservative, inhibiting the growth of mold, bacteria, and yeast in many products. It is commonly added to these products as the sodium or potassium salt (that is, sodium or potassium benzoate).

Part V: The Solubility of an Acid

Weigh 40–50 mg of benzoic acid (an organic acid) in each of five reaction tubes/small test tubes and add 1 mL of water to each tube. Add one of the following to these solutions: 10 drops of 6M NaOH, 10 drops of 6M NH_4OH, 10 drops of 10% $NaHCO_3$, 5 drops of 6M HCl, or 20 drops of diethyl ether. Stir each solution thoroughly with a stainless steel spatula and record your observations.

For the reaction tube/small test tube containing diethyl ether, use a Pasteur pipette to mix the layers by drawing the lower layer into the pipette and expelling it through the top layer (Figure 7.21). Repeat this procedure twice and then allow the layers to separate. Remove as much of the ether layer as possible with a Pasteur pipette and transfer it to a watch glass. Put the watch glass in the fume hood and allow the ether to evaporate. Record your observations.

Cool the reaction tube/small test tube containing the NaOH solution in an ice-water bath and carefully add 10 drops of 6M HCl. Record your observations.

Part VI: The Solubility of a Base

Put 5 drops of diisobutylamine (an organic base) in each of three reaction tubes/small test tubes and add 2.5 mL of water to each. Add, with vigorous stirring, one of the following: 5 drops of 6M NaOH, 10 drops of 6M HCl, or 25 drops of diethyl ether. Record your observations.

Part VII: The Solubility of an Acid-Base Mixture

Add 3 drops of diisobutylamine to the benzoic acid solution from Part IV. Stir the mixture vigorously and record the results. Add 3 more drops, again stirring vigorously and recording the results. Finally add 3 additional drops followed by vigorous stirring, making sure that you record the results.

Cleanup

All liquid solutions should be discarded in appropriate waste containers. Pasteur pipettes, weigh paper, and boiling sticks should be disposed of in the container for chemically contaminated laboratory debris.

Discussion

- For each section, analyze the results you obtained as well as those from your partner in terms of the intermolecular forces involved in the pure solute, the pure solvent, and the mixture.

- If a chemical reaction occurred, account for your observations by explaining the difference in the behavior of the product of the reaction when compared with the original solute in terms of the intermolecular forces.

- In all sciences, it is important to extrapolate from information acquired in an experiment. Based on the data you collected in this experiment, make predictions about the solubility of the compounds listed below in the corresponding solvent. Rationalize your predictions.

    1,2-dichloroethane in hexane

    1,2-dichloroethane in water

    dibromomethane in water

    methanol in water

    1,4-butanediol in water

Questions

1. For each term, provide the structure of two molecules used in this experiment.

    a. polar, protic

    b. polar, aprotic

2. Provide the structure of two organic molecules used in this experiment that, in their pure liquid states, interact mainly through induced dipole-induced dipole forces.

3a. When 157 mg of dipentylamine was added to 122 mg of benzoic acid and 2 mL of water, a completely homogenous solution was formed. Draw the structures of the species present in the water.

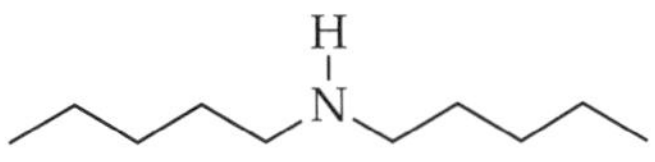

dipentylamine

3b. When 1.57 g of dipentylamine was added to 122 mg of benzoic acid and 2 mL of water, two distinct liquid layers were observed. Explain.

4. Liquid-liquid extraction (Section 7.6.1) is a technique used in organic chemistry that requires the use of two immiscible solvents. A compound distributes between two immiscible layers in an amount proportional to its relative solubility in each solvent.

4a. Referring to your experimental results, indicate which of the following solvent pairs could be used in a liquid-liquid extraction experiment. Where the layers are immiscible, use the densities of the liquids from your prelab write-up to predict which layer would be on the top.

     water-acetone

     water-ethanol

     water-diethyl ether

     water-dichloromethane

     water-ethyl acetate

     hexane-acetone

     hexane-toluene

4b. Did you perform a liquid-liquid extraction in any part of the experiment? Explain.

5a. How many mg of sodium bicarbonate are needed to completely react with 122 mg of benzoic acid? What are the products of this reaction?

5b. Explain why the following statement is incorrect: "Benzoic acid dissolved in water upon the addition of a saturated solution of sodium bicarbonate."

Mixed Melting Points

The use of a mixed melting point of an unknown and known compound can be a powerful technique for confirming the identity of the unknown.

Background Reading:

Before the laboratory period, read the discussion of melting point ranges in Section 6.3.

Prelab Checklist:

Provide the structures and melting points of the compounds for Part I by completing Table E3.1 in your laboratory notebook before the laboratory period. See Chapter 3 for references that have detailed information about chemical compounds.

Experimental Procedure

Wear your safety glasses at all times in the lab.

CHEMICAL SAFETY NOTE   Some of the unknown compounds can be mild irritants if left in contact with the skin for prolonged periods. Do not handle them with your bare hands. Wash your hands after using them and avoid contact with your mouth or eyes.

Part I: Finding an Identical Compound

Your instructor will assign you an unknown. Possible unknowns for the experiment are shown in Table E3.1.

Fill two melting-point capillaries with small amounts of your sample (Section 6.3.3). Use one to determine an approximate melting point range for your unknown by allowing the temperature to rise at about 6°C per minute. Then cool the melting point apparatus to at least 5°C below the temperature at which melting began and make a slow, careful determination of the melting point range on the second sample by allowing the temperature to rise at a rate of approximately 2°C per minute. It is important that you measure your melting point carefully because finding another student in the lab with the identical compound depends on it.

TECHNIQUE TIP   The same sample should never be used to determine a melting point a second time. The value obtained might not be accurate because some compounds decompose when heated. The temperature at which this occurs is referred to as the **decomposition point**.

Table E3.1 Mixed Melting Point Unknowns

Unknown	Structure	Melting Point Range
α-hydroxyisobutyric acid (2-hydroxyisobutyric acid) (2-hydroxy-2-methylpropanoic acid)		
α-methylcinnamic acid		
2-carboxybenzaldehyde		
glutaric acid		
DL-mandelic acid[1] (RS-mandelic acid[1])		
S-(+)-mandelic acid[1]		
benzoic acid		
DL-malic acid[1] (RS-malic acid[1])		
urea		
trans-cinnamic acid		
cholesterol		
4-hydroxyphenylacetic acid		
succinic acid		
dulcitol		
hippuric acid		

Record the melting point range in your laboratory notebook and also enter it on the data sheet provided by your instructor along with your name. From this data sheet, locate another student in the lab whose unknown has a melting point range that is within 4°C of yours. Use a small amount of that student's compound to prepare a sample for a mixed melting point. The error range of ±4°C accounts for experimental error and the fact that the thermometers are not calibrated so that the ranges from different instruments may vary slightly.

To prepare the mixed melting point sample, use one part of your unknown to one part of the other student's compound (Section 6.3.3). Mix the sample thoroughly by gently crushing the mixture with your glass stirring rod until it looks homogeneous.

Determine the melting point range of this sample. Record the result in your laboratory notebook.

Continue your search until you have identified a student in the lab with a compound identical to yours and a second student in the lab with an unknown with approximately the same melting point range, but whose compound is not the same as yours. In your laboratory notebook, clearly identify the students with the same and different unknowns with melting points within the specified range.

---

[1] See Section 6.3.2.a and the Did you know? on page 66.

Did you know?

hippuric acid

Hippuric acid is found in the urine and blood of people who have been exposed to toluene and/or benzoic acid. The body recognizes toluene as a toxin and it is oxidized by enzymes to benzoic acid. The benzoic acid couples with a molecule of glycine to give hippuric acid. Hippuric acid is readily excreted by passing through the kidneys and into the urine. This is one example of how detoxification occurs in the body.

Part II: Generating a Melting Point-Composition Diagram

Your instructor will provide you with an individual assignment to prepare a mixed melting point sample of two organic compounds. For example, if your assignment is to prepare an 80:20 mixture of Compounds A and B, this means that you will prepare a sample that contains 80% by weight Compound A and 20% by weight Compound B. The total weight of your sample should be 40 mg. Weigh the compounds on separate pieces of weighing paper, being as accurate as possible.

Transfer both compounds to a small test tube or vial, making sure that all of the solids are at the bottom of the container. Mix these together by rotating the broad end of the spatula through the mixture. When the sample is mixed, use a clean, dry glass stir rod to gently grind the mixture together, making sure to break up any large crystals. Mix the sample again with the spatula and then repeat the grinding with the glass stir rod. Mix the sample one more time with the spatula. At this point, you should have a free flowing, homogenous powder for melting point analysis.

Prepare a melting point capillary with a small amount of the sample and measure the melting point. If you find that the sample started to melt at or below the beginning temperature you chose, prepare another capillary and begin your measurement at a lower temperature. Record the melting range in your laboratory notebook. There will be a chart posted in the lab similar to Table E3.2. Write your name and the data you obtained for your mixture in the corresponding column on this chart. The melting points of the pure compounds (100:0 and 0:100) will be provided by your instructor.

Table E3.2 Melting Point-Composition Data

Name	Ratio of A:B										
	100:0	90:10	80:20	70:30	60:40	50:50	40:60	30:70	20:80	10:90	0:100

Continue.

Part III: Determining the Composition of an Unknown Mixture by Melting Point Analysis

Determine the melting point range of the mixture of A and B of unknown composition assigned by your instructor. As with Part II, if you find that the sample started to melt at or below the beginning temperature you chose, prepare another capillary and begin your measurement at a lower temperature. Record the melting point range in your laboratory notebook along with the unknown sample code.

Cleanup

Used melting point capillaries and weigh paper should be disposed of in the containers for chemically contaminated laboratory debris. Any unused compound mixtures should be disposed of in the appropriately labeled containers for solid chemical waste.

Discussion

Your discussion should be divided into three parts based on Parts I through III of this experiment. Be sure to include the following information for each part.

Part I: Briefly discuss how you used mixed melting points to find another student in the lab room with an unknown solid identical to yours.

Part II: Prepare a melting point-composition diagram from the class data posted in the lab for Part II. To ensure uniformity, follow the instructions below.

- For each ratio of A:B from the posted chart, calculate the average value of the final temperature of a given melting range. If one of these values seems dramatically different from the others in that column (that is, it is 5°C lower or higher), do not use it. If you leave a value out of your average calculation, you must state why you didn't use it and offer one or more reasons why this value might be in error.

- Plot the data as a graph. The *x*-axis should be the A:B composition and the *y*-axis should be the average value of the final temperature for each composition.

  Discuss your diagram briefly. Estimate where you think the eutectic point is. Be sure to justify your estimate, not only using the melting point-composition diagram, but also all of the raw data collected by the students in the lab room.

Part III: Estimate the composition of your unknown mixture of A and B based on the diagram generated in Part II. If you conclude there are two possible answers, then outline a quick melting point experiment that would allow you to rule out one or the other possibility.

Questions

1a. A researcher accidentally spilled some sand into his beaker containing crystalline compound X. Knowing what he did about melting points, he was not worried about obtaining the melting point of compound X with a few grains of sand present. Why?

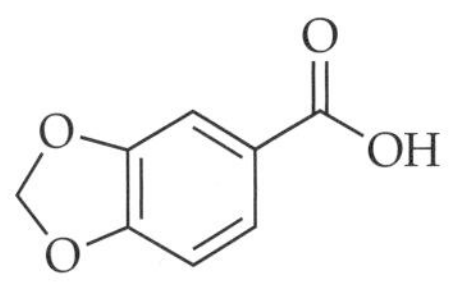

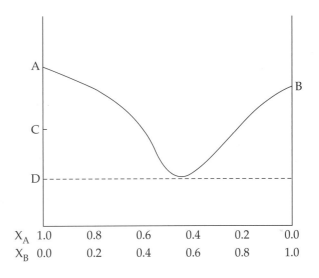

compound X

1b. Clumsy researcher that he was, later that day he spilled a little baking soda ($NaHCO_3$) into the beaker. He figured if a little sand didn't hurt, then a little baking soda shouldn't matter either when recording his melting point. Is this a valid assumption? Explain.

2. The melting point-composition diagram for Compounds A and B is shown below. If a mixture of A and B had a final melting range temperature of C, what are the approximate mole fractions of A and B in the mixture?

3. You and one of your peers might have observed the following situation in Part I of this experiment. Student A's sample melted at 89–92°C. Student B's sample melted at 96–98°C. Because Student A's and Student B's samples were within ±4°C of each other, they prepared a mixed melting point sample. Each student recorded the following results: each sample began to melt at 89°C and melting seemed to stop at around 92°C. The temperature increased with continued heating. The solid began to melt again at a temperature of 96°C. By 98°C, all of the solid in the melting point capillary was gone. Explain what happened.

4. A student was not sure that she accurately recorded the final temperature of her melting point sample. Therefore, she decided to rerun the experiment with the same sample. Was this a wise decision? Explain.

5. Why is a melting point range always observed when a melting point is measured using a conventional melting point apparatus?

Whittling Down the Possibilities: Identifying an Unknown Using Molecular Dipole Moment, Solubility, Density, and Boiling Point Data

Determining the identity of unknown compounds is one of the most challenging and rewarding tasks undertaken by organic chemists. The development of nuclear magnetic resonance spectroscopy and mass spectrometry (Chapter 8) has made this task much easier. Before these techniques were available, chemists relied on qualitative observations and quantitative measurements of a compound's physical properties to pinpoint its identity, assuming such properties had already been reported in the literature.

Whether determining the identity of an unknown compound by modern measurements or more time-honored methods, the information obtained from each technique is used to narrow down the possibilities. Eventually one last observation or measurement reveals the correct structure of the compound in question. In this experiment, you and a partner will make three observations and one physical measurement for four unknowns. The observations include the qualitative determination of the molecular dipole moment, the solubility in water, and an indirect measurement of the density. Finally, you will determine the boiling point of the unknowns. Although it is tempting to proceed to the boiling point measurement immediately in hopes of identifying the unknown from a list of boiling points, you will find that several of the unknown compounds have similar boiling points. Therefore, more sleuthing is needed to unmask the true identity of the mystery compounds.

Background Reading:

Before the laboratory period, review Sections 5.1 and 5.2. In Chapter 6, read Section 6.1, Section 6.2, and Section 6.4.

Prelab Checklist:

Include a flowchart of the procedures you will use to identify the unknown in your pre-lab write-up. In addition, generate Tables E4.1 and E4.2 in your laboratory notebook before the laboratory period. For each of the potential unknowns shown in Table E4.3, draw its structure and check one of the two columns under the Molecular Dipole Moment and the Soluble in Water categories in Table E4.1. Base your decisions on any prior experiences with these properties and the suggested background reading. Use the data provided for each unknown in Table E4.3 for the Density Relative to Water and the Boiling Point (BP) columns. Table E4.2 should be used during the laboratory period to record your observations for the four unknowns.

Table E4.1 Predictions for Knowns

Predictions for Knowns	Structure	Molecular Dipole Moment		Soluble in Water		Density Relative to Water		BP
		yes	no	yes	no	greater than	less than	
dichloromethane								
acetone								
methanol								
hexane								

Continue for all unknowns.

Table E4.2 Observations on Unknowns

Observations		Molecular Dipole Moment		Soluble in Water		Density Relative to Water		BP	Identity of Unknown
Student Name	Unknown Number	yes	no	yes	no	greater than	less than		

Table E4.3 List of Potential Unknowns

Solvent	Boiling Point°C	Density g/mL
dichloromethane	40	1.3
acetone	56	0.79
methanol	65	0.79
hexane	69	0.66
carbon tetrachloride	76–77	1.59
ethyl acetate	77	0.9
ethanol	78	0.79
cyclohexane	81	0.78
1,2-dichloroethane	83–84	1.26
1-propanol	97	0.80
heptane	98	0.68
α, α, α-trifluorotoluene (trifluoromethylbenzene)	101	1.20
2-bromopentane	116–117	1.22
octane	124–126	0.70
1-pentanol	136–138	0.81
para-xylene (1,4-dimethylbenzene)	138	0.87

Did you know?

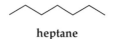

heptane

Aqueous solutions of bromine and iodine are both brown. Heptane is used to distinguish them because bromine is brown in heptane, but a solution of iodine is purple.

Experimental Procedure

Make sure you are wearing your safety glasses during the entire laboratory period.

In this experiment, you and a partner will be assigned four unknown liquid samples. The possible unknowns are listed in Table E4-3. Each student is responsible for performing the tasks outlined in Parts I to IV on two of the unknowns. You should record the results from your observations and those from your partner in your laboratory notebook. Be sure to discuss your observations with your partner. After each part, use the information obtained and your predictions from Table E4.1 in your prelab write-up to rule out possible structures for the unknown. Deciding on the identity of each unknown should be a joint effort.

CHEMICAL SAFETY NOTE   Because you are working with unknowns, it is prudent to assume that any of the compounds can act as skin irritants. Therefore, if any part of your skin comes in contact with any unknown liquid, you should thoroughly wash the affected area with soap and water and report the incident to your instructor. If any unknown liquid comes in contact with your eyes, you should immediately proceed to the eyewash and flush the affected eye(s) with water for at least five minutes. Again, make sure that your instructor is aware of the situation.

Part I: Dipole: Yes or No? A Qualitative Test for the Presence of a Molecular Dipole Moment

Molecules that have permanent molecular dipole moments are, by definition, attracted to an electrical charge. The extent of this attraction depends on the magnitude of the dipole. Although there are sophisticated instruments that measure the exact molecular dipole moment of a compound, you will use a charged plastic rod to determine whether or not a stream of the liquid is diverted toward the rod. If an extreme diversion is noted, then the compound has a molecular dipole moment. If no diversion is observed, you can assume the molecule has no molecular dipole moment or a very small molecular dipole moment as seen for some alkanes (Section 5.2.1).

In the fume hood, there will be a plastic rod attached to the laboratory rack or a ring stand along with a piece of fabric for charging the rod. Place a clean 150-mL beaker directly under the plastic rod to collect the unknown. Wrap the piece of fabric around the rod and rub it for a few seconds to generate an electrostatic charge. Draw up approximately 1 mL of one of the unknown liquids into a Pasteur pipette and hold it approximately 1.5 cm to the side and slightly above the top of the rod. Depress the pipette bulb to release a stream of the unknown and note whether or not the liquid is diverted toward the rod. If the compound has a significant molecular dipole moment, the diversion will be quite obvious. If only a slight diversion is observed, this is not indicative of the compound having a molecular dipole moment. Transfer the unknown from the beaker to a labeled reaction tube/small test tube. Cap the tube.

TECHNIQUE TIP   See Section 7.1.1 and Figure 7.1 for how to hold a Pasteur pipette.

EXPERIMENTAL NOTE   If some of the unknown touched the plastic rod, wipe the rod with a paper towel.

Repeat the molecular dipole moment determination on the remaining unknown.

Part II: Soluble in Water: Yes or No?

Place 0.5 mL of the unknown in a reaction tube/small test tube and add 0.5 mL of distilled water. In order to ensure complete mixing of both liquids, draw the mixture up into a Pasteur pipette and then forcefully expel the mixture back into the reaction tube/small test tube (Figure 7.21). Repeat this twice more and then wait a few moments to see if two layers form. If there is no clear separation of layers, the unknown is soluble in water and you should proceed to Part IV. If two layers are observed, the unknown is at least partially insoluble in water. Record your observations and go to Part III. Do not discard this mixture. You will use it in Part III.

Part III: Density of the Unknown Relative to Water: Greater Than or Less Than?

EXPERIMENTAL NOTE   Remember, if in Part II the unknown proved to be soluble in water, then you will not be performing this part of the experiment. If this is the case, go on to Part IV.

It is simple to accurately determine the density of the unknown. However, we are only interested in determining if the unknown is more or less dense than water. First, weigh 10 mg of sodium chloride and place it on a watch glass using the spatula to make a little mound of salt. With a Pasteur pipette, carefully remove one half of the top layer of liquid from the mixture obtained in Part II. Be sure not to get any of the bottom layer into the pipette. Add this liquid directly to the mound of salt and swirl the watch glass. Determine whether or not the sodium chloride dissolves (Section 6.1). If it does, then the top layer is water. If it does not dissolve, then the top layer is the unknown because none of the unknowns will dissolve sodium chloride. This observation gives you information about the density of the unknown relative to water.

Part IV: Boiling Point Determination

 To determine the boiling point of the unknown, add 0.3–0.5 mL of the unknown liquid and a boiling chip to a reaction tube/small test tube. Place the tube in the sand bath and bring the liquid to a boil. Measure the temperature of the vapors of the boiling liquid by placing the digital thermometer probe a few millimeters above the surface of the liquid. It is important to measure the temperature of the hot vapor rather than the temperature of the liquid. When the temperature reading is constant for approximately 10 seconds, record this measurement as the boiling point. The accuracy of this method is approximately ±4°C.

EXPERIMENTAL NOTE   After you have taken the boiling point measurement, be sure to thoroughly wipe off the probe tip so that the unknown samples will not become cross-contaminated.

The boiling point measurement along with the information determined in Parts I to III should enable you to identify each of the unknowns.

Cleanup

Under no circumstances should you dispose of any of the unknowns or aqueous mixtures of unknowns down the sink drains. Place all liquid samples in the appropriate waste containers. Rinse any tube that contained such materials with two 1-mL portions of acetone and discard these rinses in the appropriate waste containers. At this point, you may wash out the tube with soap and water in the sink. Dispose of all contaminated Pasteur pipettes, paper towels, and weigh paper in the container for contaminated laboratory debris.

Discussion

- Summarize how you deduced the identity of the unknowns from the methods used in this experiment. Be sure to discuss how you kept narrowing down the list of possibilities.

- Refer back to your predictions in Table E4.1 to determine if the results you obtained are the same as those you predicted for the unknowns. If there are any discrepancies, discuss your reasoning at the time you made your predictions and how the actual experimental results have made you rethink these predictions.

Questions

1. Explain how you would accurately determine the density of an unknown liquid.

2. Why is the thermometer not placed directly in the liquid when measuring the boiling point?

3. Assume you find four bottles in an empty laboratory, each containing a liquid. The labels that were on these bottles have fallen off and are strewn about the bench top. Therefore, you know the contents of the four bottles, but do not know which bottle belonged to which compound. The names on the labels are: *trans*-1,2-dichloroethylene, *cis*-1,2-dichloroethylene, cyclooctane, and acetic acid. Without determining the boiling point, would you be able to identify the contents of each bottle using information obtained from Parts I to III of this experiment? Explain your reasoning.

4. What effect would you expect glucose to have on the boiling point and freezing point of water? Explain your answer in some detail.

Recrystallization and Melting Points: Recrystallization of Adipic and Salicylic Acids

In this experiment, you will learn how to recrystallize a compound to remove impurities. You will use recrystallization to separate mixtures of salicylic acid and adipic acid based on their solubility differences. In addition, you will assess the effectiveness of recrystallization under different conditions. Solubility data for the two acids are given in Table E5.1 to enable you to calculate the theoretical percent recovery.

Melting points will be used to assess the purity of the separated components. The identities of the organic compounds will be confirmed by measuring the mixed melting points with authentic samples.

Did you know?

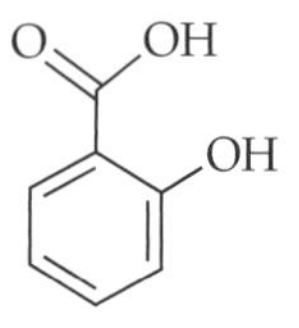

salicylic acid

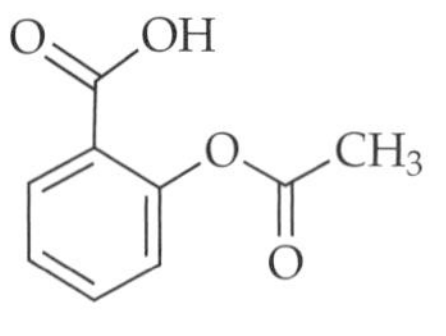

acetylsalicylic acid

Salicylic acid, a plant hormone biosynthesized from phenylalanine, is found in many plants. Ancient Egyptians treated pain with dried myrtle and American Indians used a tea obtained from the bark of willow trees to treat pain and fever. It is thought that Hippocrates of Kos (460 BC–370 BC) also used willow bark tea to relieve pain. Europeans became aware of the advantages of a component in willow bark in 1758, but it wasn't until the 1850's that salicylic acid, the compound in willow bark responsible for the medicinal properties, was synthesized in the laboratory, allowing mass production. It was the first drug that was mass-marketed. Unfortunately, it caused upset stomachs and irritated the mouth and throat. In 1897, Hoffmann, who was working at the German company Bayer, synthesized acetylsalicylic acid (aspirin). This compound did not have the severe side effects and, in 1899, aspirin replaced salicylic acid for the treatment of pain, inflammation, and fever. Today salicylic acid is used in skin-care products, wart removal solutions, and dandruff shampoos because it causes the epidermal cells to shed, allowing for new cell growth.

Background Reading:

Before the laboratory period, it is important to understand the discussion of solubility in Section 6.1 and recrystallization in Section 7.1,

Table E5.1 Solubility Data for Adipic and Salicylic Acids

| Temperature°C | Solubility in g/100mL water | |
	Adipic Acid	Salicylic Acid
100	>95	8.3
25	2.5	0.2
0	0.3	0.1

including all sub-headings within the section. You should also review the information on changes in physical state in Section 6.2, and melting point ranges in Section 6.3 and all of the sub-sections. You will use the filtration method described in Section 7.2.1.

Prelab Checklist:

Include the structure, molecular weight, solubility data, and melting points for the two organic acids in your prelab write-up.

The procedure for these recrystallizations should be summarized in flowchart form in your laboratory notebook, including all of the steps to be followed in recrystallizing a compound.

Construct Tables E5.2 and E5.3 in your laboratory notebook for use in recording data collected during the experiment.

Experimental Procedure

Make sure you are wearing your safety glasses during the entire laboratory period.

CHEMICAL SAFETY NOTE   The compounds used in this experiment are weak acids, but they will cause irritation if left in contact with the skin for prolonged periods. Do not handle them with your bare hands. Wash your hands after using them and avoid contact with your mouth or eyes. Do not inhale the vapors coming from the boiling solutions because the compounds can be volatilized by boiling water.

SAFETY NOTE   Always add a boiling stick or chip to any solution before you begin heating it to prevent the solution from bumping (Section 7.1.1). In this experiment, a boiling stick should be used because it is easily removed from the solution.

SAFETY NOTE   Use the digital thermometer to measure the temperature of the sand bath. *Never* put a mercury thermometer in a sand bath.

TECHNIQUE TIP   A clean piece of weighing paper should always be used for each different compound to avoid cross-contamination.

Part I: Separation of a Mixture of Salicylic Acid and Adipic Acid Based on their Solubility Differences

Each student should do Part I of the experiment.

Weigh approximately 75 mg salicylic acid (2-hydroxybenzoic acid) and approximately 75 mg adipic acid (hexanedioic acid). Use a clean piece of weighing paper for each compound. Make sure you record the exact weights in your laboratory notebook. Transfer both compounds to the same reaction/small test tube. Add a boiling stick and 3 mL of distilled water. Use a clamp to hold the reaction tube/small test tube.

Heat the mixture to boiling on the sand bath. All of the white solid should dissolve in boiling water. If some white solid remains, add hot distilled water dropwise and continue boiling until all of the material dissolves.

Put the reaction tube/small test tube in a beaker and insulate it with cotton. Allow the solution to cool to room temperature without disturbing it.

At this point, you should go on to Part II, checking the solution occasionally for crystal formation.

When the solution has cooled to room temperature, remove the solvent from the crystals using the square-tip Pasteur pipette method (Section 7.2.1 and Figure 7.6). Make sure that you use an unchipped pipette and that you seat it firmly on the bottom of the reaction tube/small test tube. The filtrate (solvent) should be expelled into a clean 10-mL beaker. The crystals should be washed with a small amount of ice-cold distilled water, again using the square-tip Pasteur pipette method to remove the wash. Combine these washes with the filtrate.

Using the stainless steel spatula, remove the crystals from the reaction tube/small test tube and blot them dry on a piece of filter paper. Transfer the crystals to a preweighed watch glass and spread them out. Weigh the watch glass with the crystals and then reweigh it five minutes later. If the weight has decreased, continue drying until a constant weight is obtained. Record the weight of the crystals in your laboratory notebook. These crystals are referred to as **Crop A**. Transfer the dry crystals to a sealable plastic bag or container and label it Crop A.

Add a new boiling stick to the solvent (filtrate) in the 10-mL beaker and heat the solution on the sand bath until the volume has been reduced to about 1.5 mL. Check the volume with a Pasteur pipette (Experiment 1). After the approximate volume has been achieved, transfer the solution to a reaction tube/small test tube and allow the concentrated solution to cool to room temperature. In your laboratory notebook, record whether or not crystals form after cooling to room temperature and then immerse the tube in an ice-water bath and cool the solution to 5–10°C. It may be necessary to scratch the inside of the reaction tube/small test tube *gently* with a glass stirring rod to initiate crystal formation (Section 7.1.6).

Remove the filtrate from the crystals using the square-tip Pasteur pipette method. The filtrate should be expelled into a clean reaction tube/small test tube. After all of the solvent is removed, rinse the crystals with a small amount of ice-cold distilled water. Stir the mixture to wash the crystals and remove the wash using the square-tip Pasteur pipette method. Using the spatula, transfer the crystals from the reaction tube/small test tube to a piece of filter paper. Follow the procedure described for Crop A to dry the crystals and record the weight of the crystals in your laboratory notebook. These crystals are referred to as **Crop B**. Transfer the dry crystals to a sealable plastic bag or container and label it Crop B.

Part II: Assessing the Effectiveness of Recrystallization

Part II is to be done in groups of three, with each of the members of the group completing one of the sections and sharing the data obtained with the other two group members. The purpose of these sections is to assess the effects of the method of cooling and the amount of contamination on the purity of the material obtained from recrystallization.

Student 1: Weigh 50 mg salicylic acid and 50 mg adipic acid, and transfer both compounds to a reaction tube/small test tube. Add 1 mL of distilled water and a boiling stick, and then heat to boiling on the sand bath according to the procedure used for Part I. If the solution contains white solid, add hot distilled water dropwise and reheat to boiling until all of the material dissolves. Allow the solution to cool to room temperature slowly and undisturbed. Do *not* cool the solution in an ice-water bath. Remove the solvent using the square-tip Pasteur pipette method and wash the crystals with a small amount of ice-cold water. Use the procedure described for Crop A to dry the crystals and record the weight of the crystals in your laboratory notebook. These crystals are referred to as **Crop C**. Transfer them to a sealable plastic bag or container and label it Crop C.

Student 2: Weigh 50 mg salicylic acid, and 20 mg adipic acid, and transfer both compounds to a reaction tube/small test tube. Add a boiling stick and recrystallize the mixture from 1 mL of distilled water. Cool the solution slowly to room temperature. Do *not* cool the solution in an ice-water bath. Remove the solvent using the square-tip Pasteur pipette method and wash the crystals with a small amount of ice-cold water. Use the procedure described for Crop A to dry the crystals and record the weight of the crystals in your laboratory notebook. These crystals are referred to as **Crop D**. Transfer them to a sealable plastic bag or container and label it Crop D.

Student 3: Weigh 50 mg salicylic acid and 20 mg adipic acid, and transfer both compounds to a reaction tube/small test tube. Add a boiling stick and dissolve the mixture in 1 mL of distilled water by heating on the sand bath. After all of the solid is in solution, plunge the hot solution into an ice-water bath immediately. Collect the crystals by removing the solvent using the square-tip Pasteur pipette method and wash the crystals with a small amount of ice-cold water. Use the procedure described for Crop A to dry the crystals and record the weight of the crystals in your laboratory notebook. These crystals are referred to as **Crop E**. Transfer them to a sealable plastic bag or container and label it Crop E.

Part III: Melting Points

You should identify each of the crops of crystals obtained based on melting points, mixed melting points with the authentic compounds, and solubilities. Authentic samples of the acids will be available in the lab.

Determine the melting point of Crops A and B and of these crystals (A and B) mixed with authentic samples of salicylic acid and with adipic acid. Also measure the melting point of the crystals you isolated in Part II (Crop C, D, or E) and these crystals mixed with either salicylic acid or adipic acid, not both. You should include the data for the other two crops isolated by your partners in Part II in your laboratory notebook. This section can be completed during the next laboratory period if necessary.

Cleanup

All aqueous waste should be discarded in the appropriate waste container. All solid products should be placed in the container for solid chemical waste. Dispose of all contaminated Pasteur pipettes, used weigh paper, and melting point capillaries in the container for contaminated laboratory debris.

Discussion

- Calculate the percent recovery from the weight of the material originally used in the mixture. Present your experimental data in tabular form as shown in Table E5.2.
- From the solubility data given at the beginning of the experiment, calculate the maximum amount of salicylic and adipic acid that is recoverable in each of the crops. Summarize the information in Table E5.3.

Table E5.2 Summary of Experimental Data

Crop	Weight of Salicylic Acid Used	Weight of Adipic Acid Used	Weight of Recrystallized Product	% Recovery	MP Crop	Mixed MP with Salicylic Acid	Mixed MP with Adipic Acid	Identity of Crop
A								
B								
C								
D								
E								

Table E5.3 Comparison of the Amount Recovered with the Maximum Recoverable Amount

Crop	Weight of Salicylic Acid Used	Maximum Recoverable Amount of Salicylic Acid	Actual Amount Recovered	Weight of Adipic Used	Maximum Recoverable Amount of Adipic Acid	Actual Amount Recovered
A						
B						
C						
D						
E						

Table E5.4 Mixed Melting Point Data for Adipic Acid (A) and Salicylic Acid (B)

A:B	Initial Melting Point°C	Final Melting Point°C
100 A: 0 B	151.2	154.0
90 A: 10 B	137.2	151.0
80 A: 20 B	124.8	148.6
70 A: 30 B	124.5	145.8
60 A: 40 B	124.0	142.1
55 A: 45 B	123.7	140.7
50 A: 50 B	123.7	139.2
45 A: 55 B	123.6	139.7
40 A: 60 B	124.0	140.4
30 A: 70 B	124.8	146.3
20 A: 80 B	130.6	153.6
10 A: 90 B	154.1	157.7
0 A: 100 B	159.3	161.7

- Compare your experimental results with the maximum amount of compound recoverable at the temperature and volume used, and explain the melting point and mixed melting point behavior of all crops. Include an explanation for any variation in the purity of the crops and discuss the limitations of using recrystallization to separate mixtures of salicylic and adipic acids. Suggest changes in the procedures to improve the results.

- Plot the mixed melting point data for adipic and salicylic acids given in Table E5.4. Using the sample you collected in Part II, determine the percent composition of the product you obtained (Experiment 3).

Questions

1. A student uses 1 mL of water to wash the crystals from Crop A. Briefly discuss whether this is a good or faulty technique and explain why.

2. A sample consists of, at most, 100 mg of impure salicylic acid. A student adds 1 mL of hot water and a boiling stick and heats the resulting mixture on the sand bath. Not all of the solid dissolves so the student continues to add hot water dropwise. After another 0.5 mL of water has been added, for a total of 1.5 mL, there is still some white solid remaining. Provide an explanation and advise the student as to how to proceed.

3. The solubility of benzoic acid in water is 5.9 g/100 mL at 100°C; 0.30 g/100 mL at 25°C; and 0.17 g/100 mL at 0°C. Can a mixture of approximately equal amounts of benzoic acid and salicylic acid be separated by recrystallization? Explain.

4. The solubility of salicylic acid in diethyl ether is 1 g/3 mL at 25 °C. Is diethyl ether an appropriate solvent to use for recrystallizing salicylic acid? Explain.

5. Explain why it is not possible to recover 100% of the adipic acid from a recrystallization using water.

Recrystallization and Melting Points: Recrystallization of an Unknown Solid and Decolorization of Brown Sugar

When you need to use recrystallization to purify a compound that you have prepared in the lab, solubility data is often not available. It is necessary to find a solvent in which impurities are either insoluble so they can be removed by filtration, or very soluble that so they remain in solution (Section 7.1.2 and Figures 7.2 and 7.3). The desired compound must be much more soluble in the hot solvent than in the ice-cold solvent so that it can be recovered. This experiment is designed to give you experience in choosing a solvent for recrystallization and in purifying and identifying an unknown.

Often organic compounds are contaminated with highly colored impurities. One traditional way of eliminating these contaminants is to adsorb them onto activated charcoal. In order to demonstrate an actual commercial application of the decolorization process, in Part II you will use activated charcoal to remove the colored substances found in brown sugar (Sections 7.1.4 and 7.1.5).

Background Reading:

Before the laboratory period, review the steps of recrystallization found in Section 7.1, including all of the subheadings. These steps are summarized on page 98. In addition, read the discussions of solubility (Section 6.1) and melting point ranges (Section 6.3, including all of its subheadings).

Prelab Checklist:

In your prelab write-up, draw the structures and record the boiling points of any solvents you are using for recrystallization. Copy Table E6.1 into your laboratory notebook and fill in the structural formulas and the melting point ranges for all of the unknowns.

Explain how authentic samples of the unknowns can be useful in verifying the identity of the unknown.

Experimental Procedure

Make sure you are wearing your safety glasses during the entire laboratory period.

Part I: Recrystallization of an Unknown

CHEMICAL SAFETY NOTE Some of the compounds used as unknowns can be mild irritants if left in contact with the skin for prolonged periods. Do not handle them with your bare hands. Wash your hands after using them and avoid contact with your mouth or eyes. Do not inhale the vapors coming from the boiling solutions because some of the compounds can be volatilized by boiling water.

SAFETY NOTE In this experiment, a boiling stick should be used rather than a boiling chip to prevent superheating (bumping) because it is easily removed from the solution before crystallization takes place (Section 7.1.1).

SAFETY NOTE Use the digital thermometer to measure the temperature of the sand bath. *Never* put a mercury thermometer in a sand bath.

Your instructor will divide the laboratory section into groups of four. You should record the names of the members in your group in your laboratory notebook. Each group will be supplied with approximately 500 mg of a crude unknown solid. Possible unknowns are shown in Table E6.1.

Table E6.1 Unknowns for Recrystallization

Compound	Structure	Melting Point	Solubility
meso-erythritol			1.5 g/1 mL water insoluble in ether
4-acetylbiphenyl			insoluble in water
nicotinamide			1g/1 mL water 1g/1 mL ethanol
benzamide			1g/74 mL water 1g/6 mL ethanol
benzoin			1g/3335 mL water slightly soluble in ether
urea			1g/1 mL water insoluble in ether
camphoric acid			1g/125 mL water freely soluble in ether
succinic acid			1g/13 mL cold water 1g/113 mL ether
hippuric acid			1g/250 mL cold water 1g/400 mL ether

Each group member should take approximately 100 mg of the unknown. Divide the responsibility for examining the solubility of the unknown in the following solvents: 95% ethanol, reagent grade acetone, hexane, and water. One member of the group should also determine the solubility of the unknown in diethyl ether at room temperature. Although diethyl ether will not be used as a recrystallization solvent, the information you obtain may help in establishing the identity of the unknown.

Put a few crystals of the compound in a reaction tube/small test tube and add a few drops of the solvent. If the crystals dissolve immediately, the solvent will not be useful for recrystallization because the compound will not be recoverable. If the crystals do not dissolve, add a boiling stick and heat the mixture gently on the sand bath to determine if the crystals are soluble in hot solvent. A few drops more of solvent can be added if necessary. Cool the solution to room temperature and then in an ice-water bath to determine if the crystals can be recovered from the solution (Section 7.1.2, 7.1.2.a, Summing It Up on page 92, and Figure 7.3).

EXPERIMENTAL NOTE   The solubility tests should be done in dry reaction/small test tubes. If the tube is wet with water, rinse it with a very small amount of the solvent being tested. In the case of hexane, rinse with a small amount of acetone first, followed by hexane.

Record all of the results from your group's solubility tests in your laboratory notebook and decide on a suitable solvent for the recrystallization of the unknown. Before proceeding, check with your instructor on your choice of the solvent for recrystallization.

EXPERIMENTAL NOTE   Because of its low boiling point, extreme flammability, tendency to form explosive peroxides, and anesthetic properties, diethyl ether is not an appropriate solvent for recrystallization (Section 7.1.2).

Each member of the group should purify the remainder of his or her sample by recrystallization in the group's choice of solvent. Weigh the material before recrystallizing it. Dry the recrystallized product and determine the weight. Calculate the percent recovery. If the recovery is low, consider obtaining a second crop of crystals. This is accomplished by evaporating some of the solvent from the mother liquor and again cooling the solution (Section 7.1.10).

Measure the melting point of the sample. Identify the unknown from the possibilities listed in Table E6.1 using the melting point and solubility data.

Record the results obtained by all members of the group in your laboratory notebook.

Did you know?

diethyl ether October 16th is referred to as Ether Day. On this date in 1846, Willliam T. G. Morton successfully anesthetized a patient at Massachusetts General Hospital using diethyl ether vapors. The surgeon, John C. Warren, then removed a lump from underneath the jaw of the patient. The patient awoke after the surgery claiming that he had no recollection of the procedure and had experienced no pain.

Part II: Decolorization of Brown Sugar

Weigh two 120-mg quantities of brown sugar and place them in separate reaction tubes/small test tubes. Dissolve each sample in approximately 0.4 mL of distilled water. Stir the solution with a boiling stick to help dissolve all of the material in each tube. (Spinning the boiling stick between two fingers works well.) Retain one of the solutions for comparison.

Add approximately 8–10 mg of activated charcoal to the other solution. (For this experiment, weigh this amount to get an idea of what 8–10 mg of charcoal looks like.) Place a boiling stick in the tube and boil the solution for 5–7 minutes.

Filter the solution using the glass-fiber Pasteur pipette method described in Section 7.2.2 and Figure 7.7. Collect the filtrate in a reaction tube/small test tube. It will be necessary to exert pressure from a pipette bulb on the filtration pipette to force the solution through the microfilter.

Rinse the reaction tube/small test tube with approximately 0.1–0.2 mL of water and pass this through the filter into the collection tube. Compare the difference in color of the solution that was treated with activated charcoal with the solution that was not. Record your observations. To achieve the best comparison, it is useful to hold the two tubes against a white background, for example, a sheet of paper or your laboratory notebook.

Cleanup

Part I: Put the recrystallized product in an appropriate container. All liquids should be discarded in the appropriate liquid waste containers.

Part II: Any aqueous waste containing brown sugar or its decolorized variant may be discarded down the drain. Contaminated Pasteur pipettes and weigh paper should be placed in the container for contaminated laboratory debris.

Discussion

• How did your group decide on the solvent to use for the recrystallization of the unknown?

• How did your group decide on the identity of the unknown after recrystallization?

- Was the solubility information helpful in narrowing down the possibilities?
- Provide a short discussion about the decolorization of brown sugar using activated charcoal.

Questions

1. One explanation for the fact that the color of the brown sugar solution has diminished or disappeared is that all of the material (including the sugar) is adsorbed onto the activated carbon. With the equipment available in the lab, how could you determine if this did or did not happen?

2. Why is it not a good idea to use a large excess of activated charcoal to remove colored impurities?

3. A student started with 100 mg of an unknown solid and recrystallized it from water. The weight of the crystals isolated was 120 mg and the melting point range was very broad and lower than the literature value. Explain the problem that the student encountered and how the student can solve it.

4. A student started with 100 mg of an unknown solid and recrystallized this material from water. The weight of the crystals isolated was 27 mg. Explain why the student might have obtained this result and what the student can do to try to correct the problem.

Thin Layer Chromatography (TLC)

The production of prostaglandins in the cells of the body is catalyzed by two cyclooxygenase enzymes, COX-1 and COX-2. Although both of the enzymes produce protaglandins that promote inflammation, pain, and fever, only the COX-1 enzyme produces compounds that protect the lining of the stomach from acid and help in the clotting of blood. Non-steroidal anti-inflammatory drugs (NSAIDs) reduce prostaglandin synthesis by interfering with these two enzymes. These drugs, therefore, provide relief from pain, inflammation, and fever. Aspirin interferes more with the COX-1 enzyme and promotes bleeding for a longer period (4 to 7 days) compared with other NSAIDs. This property has prompted the use of aspirin for preventing blood clots that cause heart attacks and strokes. Newer NSAIDs are being developed that selectively interfere with the COX-2 enzyme, therefore avoiding the problems associated with loss of the prostaglandins that protect the lining of the stomach. Technically, acetaminophen is not considered an NSAID because it has no effect on inflammation.

In Part I of this experiment, you will determine the TLC behavior and the R_f values for the components of some over-the-counter analgesics (pain killers) in four different solvent systems. In Part II you will use this data to identify an unknown pain killer.

Background Reading:

Before the laboratory period, read about the technique of chromatography discussed in Section 7.8. In addition, read Sections 7.8.1, 7.8.2, and the section on thin layer chromatography (Section 7.8.3) before doing this experiment.

Prelab Checklist:

The prelab write-up should include a flowchart of the procedure. Include the structures and names of all the compounds and solvents that you will use in this experiment. Predict the relative R_f values of the four components of the analgesics in 100% ethyl acetate.

Did you know?

In 1911, the U.S. Government sued the Coca-Cola company in an attempt to force them to remove the caffeine from the formulation. The suit was based on a claim that excessive use of Coca-Cola led to inappropriate and wild behavior at a girls' school. Even though the suit failed, in 1912, the Pure Food and Drug Act was amended to add caffeine to the list of "habit-forming" and "harmful" substances and the caffeine content must be listed on a product's label.

Experimental Procedure

Wear your safety glasses at all times in the lab.

Part I: TLC of Analgesics

The compounds you will examine are aspirin, acetaminophen, ibuprofen, and caffeine.

aspirin   acetaminophen   ibuprofen   caffeine

Your instructor will divide the lab into groups of four students. Be sure to record the names of the members of your group in your laboratory notebook. Each student should spot solutions of acetylsalicylic acid (aspirin) (see the experimental note below), 4-acetamidophenol (acetaminophen), caffeine, and 2-(4-isobutylphenyl)propanoic acid (ibuprofen) on a TLC plate that can accommodate four lanes.

Each of the students in the group is responsible for evaluating the TLC behavior of the four compounds using one of the four solvent systems: 100% ethyl acetate; 92:8 ethyl acetate:acetic acid (by volume); 60:40 hexane:ethyl acetate; or 10:10:1 toluene:ethyl acetate:acetic acid.

Deliver 10 mL of your assigned solvent system into a 150-mL beaker that has been fitted with a filter paper as described in Section 7.8.3.g and Figure 7.24 and then cover the beaker with a piece of aluminum foil. It is important to keep the aluminum foil on the beaker so the chamber remains saturated with the solvent vapor and the composition of the solvent does not change due to the evaporation of the more volatile components. Make sure the filter paper is saturated with solvent before introducing the TLC plate. Retain the chamber for use in the analysis of your unknown in Part II.

Develop your plate by allowing the solvent to flow to within 0.5 cm of the top of the plate. Visualize the plate under the UV lamp. *Remember the warning: Never look into a UV lamp* (Sections 7.8.3.d and 7.8.3.g). Use a pencil to circle the spots that are visible under the UV lamp. Compare your results with the others in your group and then place your plate in an iodine chamber. Circle any new spots that were not visible under the UV lamp. Note any differences in how the spots are stained by the iodine. These differences can often be very indicative of a given compound, and can be used as a means of identification when compounds have similar R_f values.

WARNING

Make full-size sketches of your plate and your coworkers' plates in your laboratory notebook. Clearly indicate on or near each of the plates the following information:

- The name of the person who developed the plate
- The solvent system used to develop the plate
- The compound spotted in each lane on the plate
- The spots that were visualized under the UV lamp, those that were stained by iodine, and those that were visualized by both methods
- The R_f value of each of the spots as determined from Equation 7.4.

EXPERIMENTAL NOTE   Acetylsalicylic acid sometimes hydrolyzes to a small extent to form acetic acid and salicylic acid. Salicylic acid will be available from your instructor if needed for examining the TLC behavior and R_f of this compound.

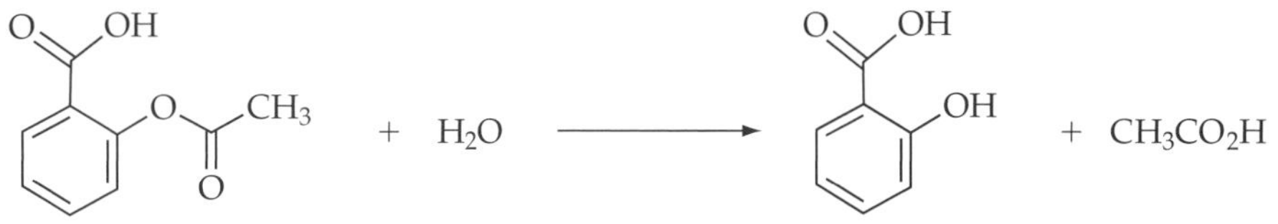

acetylsalicylic acid   salicylic acid   acetic acid

Part II: Identification of an Unknown 'Pain Killer'

Now you are ready to identify your unknown pain killer. Before you begin this part of the experiment, be sure that all four solvent systems have been evaluated and that the results are clear. Plates containing large broad spots, indicating that too much material was spotted, or spots that do not line up with their origin should be redone.

Each group of four students will be given an unknown pain killer that has been crushed to a powder. Place the entire sample in a reaction tube/small test tube and add approximately 0.5 mL of absolute ethanol. Use a glass rod to mix the solution, using a gentle grinding motion, and then allow the solid to settle. Not all of the material will dissolve because the binder used in the tablet is insoluble in ethanol.

Choose the two solvent systems that gave the best separation for the known compounds. In order to determine the R_f values of the unknown and the four knowns in the two different solvent systems, divide the tasks as shown in Table E7.1.

Table E7.1 Tasks for Determining the R_f Values

Group Member	Compounds to be Spotted on the TLC Plate			Developing Solvent
1	aspirin	unknown	acetaminophen	1
2	aspirin	unknown	acetaminophen	2
3	caffeine	unknown	ibuprofen	1
4	caffeine	unknown	ibuprofen	2

Table E7.2 Components of Analgesics

	Active Ingredient(s)	Common Name
Tylenol	4-acetamidophenol	acetaminophen
aspirin	acetylsalicylic acid	aspirin
Excedrin	acetylsalicylic acid (44%) 4-acetamidophenol (44%) caffeine (12%)	aspirin acetaminophen
ibuprofen	2-(4-isobutylphenyl)propanoic acid	ibuprofen

Record the results from all four of the TLC plates in your laboratory notebook. Clearly identify the person who developed each TLC plate. It is important to draw full-size reproductions. Develop additional plates if necessary to clarify the results. Use Equation 7.4 to determine the R_f values of the spots. Based on the possibilities shown in Table E7.2, decide which pain killer you have.

Cleanup

Dispose of the solvent in your TLC chambers and the solutions of the unknown analgesics in the appropriate liquid waste containers. After all the members of the group have drawn the full-sized TLC plates in their laboratory notebooks, the TLC plates and the used micropipettes should be put in the container for contaminated laboratory debris.

Discussion

- What criteria did your group use to narrow down the choice of TLC solvents used in Part II?

- Discuss how your group identified your unknown analgesic.

- Compare the relative R_f values you predicted in the prelab write-up with those obtained experimentally for the four components of the analgesics. In cases where the relative order you predicted differed from the experimental results, explain how the results helped you rethink your predictions.

Questions

1. Why is it a problem to have the spotting line on a TLC plate below the solvent level in the developing chamber?

2. A mixture of compounds is spotted on a TLC plate and developed in 90:10 hexane:ethyl acetate. Visualization with both UV and iodine shows a single spot at the origin. What mixture of hexane and ethyl acetate should you try next? Explain.

3. One spot with an R_f of 0.62 is observed after you develop the TLC of a reaction mixture. Does this prove that you have only one product present? Explain.

4. Which two of the following compounds would show up as a dark spot under a UV lamp when spotted on a TLC plate that contains a fluorescent indicator?

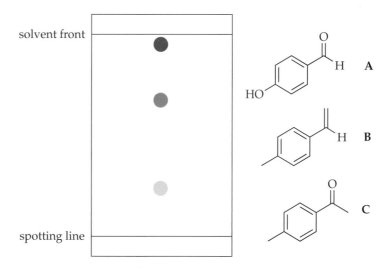

5. On the developed TLC plate, match each spot to one of the compounds shown. The developing solvent was 90:10 ethyl acetate:hexane.

Identification of an Adulterated Herb or Spice by Thin Layer Chromatography (TLC)

The use of herbs and spices in both healing and cooking is documented throughout the course of written history. For example, the Ebers Papyrus, a famous document that lists more than 800 medical prescriptions used by the ancient Egyptians, makes mention of many herbs. In the 16th and 17th centuries, explorers sought new lands to find sources of exotic spices. The Portuguese occupied the island of Ceylon in 1536 in order to secure a continuous supply of cinnamon bark. Two hundred years later, the Dutch began cultivating cinnamon trees on the same island only to learn a difficult lesson in the theory of supply and demand. The amount of bark produced far outweighed the demand, leading to many a fragrant fire on this beautiful island. Despite such setbacks, the herb/ spice industry has continued to grow. As is the case with many successful enterprises, there are always those who attempt to maximize their profit by sacrificing quality. For example, it was once a common practice to adulterate or contaminate shipments of valuable herbs and spices with less valuable substitutes. With the development of chromatography, it became possible to readily discern if a batch of herbs or spices had in fact been adulterated. This was accomplished by examining one portion of the chemical makeup of herbs and spices, namely the volatile oils. Such oils are referred to as volatile because they are readily separated from other chemicals by steam distillation (Sections 6.4.1.d and 7.3.4). Another method of obtaining these oils, along with some other non-volatile substances, is to extract the herbs or spices with a non-polar solvent (Section 7.6.2). Although this method does not result in a pure sample of volatile oil, it does provide the opportunity for a relatively quick analysis of the major constituents of the oil by TLC.

The herbs and spices being investigated in this experiment are: anise, caraway, cinnamon, cloves, coriander, and cumin. Your task is to analyze the volatile oils of some of these botanical products and then identify the adulterant in a sample.

Did you know?

Between 1652 and 1672, the Dutch and English fought three wars, called the Spice Wars, over claims to the Spice Islands. In the settlement of the second of these, the English gave up their claim to one of the Spice Islands in return for the island of Manhattan, a part of the Dutch settlement in America.

Background Reading:

The technique of chromatography is discussed in Section 7.8. Before the laboratory period, read Sections 7.8.1, 7.8.2, and 7.8.3. Pay particular attention to the discussion regarding the use of dinitrophenylhydrazine in Section 7.8.3.d.

Prelab Checklist:

In your prelab write-up, include the structures and the physical properties of the TLC solvents and dinitrophenylhydrazine. A flowchart of the procedure you will use should be included in your prelab write-up.

Experimental Procedure

Wear your safety glasses at all times in the lab.

Your instructor will divide your laboratory section into groups of four or five students. The division of tasks within each group for TLC analysis of both known and unknown samples is shown in Table E8.1. Students 1, 2, and 3 will each use a different solvent system; however, they will all visualize their TLC plates with ultraviolet light (UV) and iodine (I_2) (Sections 7.8.3.d and 7.8.3.g). Student(s) 4 (and 5) will use the same solvent systems as student(s) 2 (and 3) respectively, but will visualize the TLC plates by UV and then dip the plate into an acidic solution of dinitrophenylhydrazine (DNPH). Review the information about this reagent in Section 7.8.3.d. The procedure is described in detail in Part I. Each student will develop a minimum of five TLC plates during the course of this experiment.

Part I: Preparation of the Herb/Spice Samples for Analysis

Each group will receive four sample vials that are labeled with a sample code. Three of these vials contain a pure herb or spice (H/S) that has been ground to a powder. The fourth vial, the one with the highest code number, contains a pure herb or spice from one of the first three vials plus a pure herb or spice sample from another group in the laboratory section. This is the adulterated sample.

Table E8.1 Division of Tasks for TLC Analysis

Student	Solvent System	Visualization Method
1	95% hexane:5% ethyl acetate	UV, I_2
2	90% hexane:10% ethyl acetate	UV, I_2
3	80% hexane:20% ethyl acetate	UV, I_2
4	90% hexane:10% ethyl acetate	UV, DNPH
(5)	80% hexane:20% ethyl acetate	UV, DNPH

Each vial contains approximately 0.5 g of material. By visual inspection, remove one half of the sample from each vial and place it on a watch glass labeled with the appropriate sample code. Reserve this material to compare its appearance and odor with other samples later in the laboratory period.

Add 1.5 mL of hexane to each sample vial. Cap and shake the vial for 20–30 seconds. Allow the mixture to settle while you proceed with the first set of analyses.

Part II: **Thin Layer Chromatography of Known Compounds Found in Some Herbs and Spices: TLC Plates 1 and 2**

The chemical makeup of herbs and spices is often complex. The structures of some of the compounds that could be present in a hexane extract of your herb or spice are shown in Figure E8.1. Analyze the six known compounds using two TLC plates, spotting the stock solutions of three of the compounds on each plate. Develop and visualize the plates using your assigned solvent system and visualization technique.

EXPERIMENTAL NOTE   Some of the stock solutions may contain minor impurities that are not necessarily related to herb/spice chemistry; therefore, assume that only the major spot is the desired compound.

EXPERIMENTAL NOTE   Full-sized sketches of all TLC plates are required in your laboratory notebook.

TECHNIQUE TIP   Do not draw conclusions about the presence or absence of compounds based on only one visualization technique. Use all of the techniques assigned before drawing any conclusions.

Figure E8.1   Structures of Known Compounds

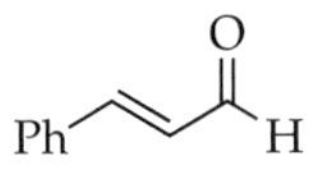

3-phenylpropenal

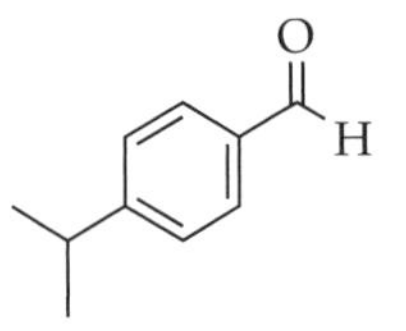

4-(1-methylethyl)benzaldehyde

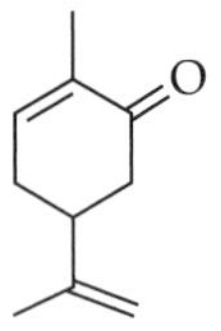

2-methyl-5-(1-methylethenyl)-2-cyclohexene-1-one

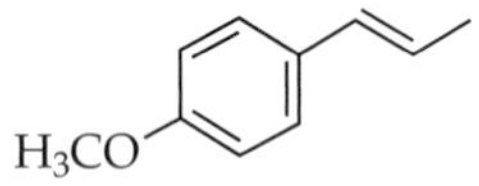

1-methoxy-4-(1-propenyl)benzene

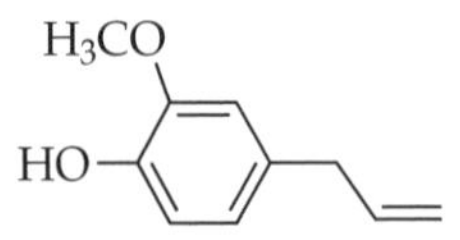

2-methoxy-4-(2-propenyl)phenol

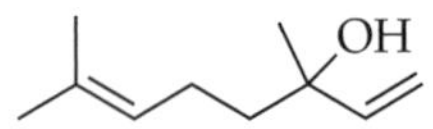

3,7-dimethyl-1,6-octadien-3-ol

Procedure for Students Using DNPH: Before dipping the plate into the DNPH solution, examine it with the UV lamp and outline any spots that are visible. Do not visualize the plate with iodine. Make sure that you sketch an outline of the plate in your notebook and record the UV visible spots before continuing. The DNPH solution will be in the fume hood. Use tweezers to grasp the plate below the spotting line. Dip the plate into the solution with the solvent front down, making sure that the solution covers the entire plate, including the spotting line. Pull the plate out and let any residual solution drip back into the beaker. Place the plate face up on a paper towel in the fume hood. Do not touch the plate. Sketch any new spots that were not UV active in your laboratory notebook as best you can without touching the plate. In addition mark those spots that were visualized by both UV and DNPH in your laboratory notebook. Dispose of the plate in the specially labeled container in the fume hood.

> **CHEMICAL SAFETY NOTE**   The reagent solution is made with 85% phosphoric acid, an extremely corrosive material. Do not take the plates to the lab bench or place them on your laboratory notebook.

Part III: **Thin Layer Chromatography of Pure Herb/Spice Samples: TLC Plate 3**

Spot the previously prepared hexane solutions of the three pure H/S samples on TLC plate 3, develop the plate in your assigned solvent, and visualize the spots using your assigned method. Remember, these H/S samples are likely to have more than one compound present in the hexane extracts. You should focus on circling the major spots for each sample. However, note any minor spots that are clearly separated from other spots or behave uniquely when visualized using one of the three techniques. Sometimes minor spots can help you establish the difference between two samples that have a similar, but not identical chemical makeup.

> **EXPERIMENTAL NOTE**   The hexane extracts of the H/S samples are dilute relative to the solutions of the compounds used to spot Plates 1 and 2. Therefore, on Plate 3, spot each H/S extract 8 to 10 times, allowing the hexane to evaporate before spotting again in the same place.

> **TECHNIQUE TIP**   If your TLC capillaries become clogged with particulate matter from the H/S sample, break the TLC capillaries off close to where the constriction begins.

As a group, discuss your combined results and complete the following tasks.

- For each solvent system, provide the following information on the data sheets posted in the laboratory room: group number; sample code; odor of H/S; appearance of H/S; R_f data for the UV, I_2, and DNPH spots. This information will be used by other groups in the lab to identify their adulterant.

- Compare the TLC data of your group's H/S samples with that of the known compounds analyzed on Plates 1 and 2. Identify at least one of these known compounds in each of the three pure H/S samples. If you wish to confirm the presence of a known compound in one of the pure H/S samples, cospot both on a new TLC plate and develop it (Section 7.8.3.f). Be sure that all of your results are consistent.

- Guess the identity of each pure H/S sample based on its odor. If the odor from the sample on the watch glass is not prevalent, take some of the H/S and rub it between your thumb and index fingers. The heat generated in this process will help volatilize some of the oils in the herb or spice.

- Present your results to your instructor who will verify your conclusions.

Part IV: Identification of the Adulterant in the Adulterated Herb/Spice Sample: TLC Plates 4 and 5

Each person in a group should spot TLC Plate 4 with the previously prepared hexane extract of the adulterated H/S sample, the one with the highest code number. On the same plate, spot all three of the pure H/S samples and then develop the plate and visualize the spots.

As a group, compare the results to determine which one of the pure H/S samples is mixed with the adulterant. Then identify all important aspects of the adulterant from the TLC of the mixed sample, for example, a positive DNPH test, I_2 active spots, etc.

EXPERIMENTAL NOTE   It is acceptable to make an educated guess about the components of the mixture based on odor and appearance. However, you must confirm your guess by TLC.

From the class' results on the data sheet in the laboratory room, identify another group of students that has a pure H/S sample with characteristics similar to your adulterant.

Each student in the group needs to prepare TLC Plate 5 with the following samples:

- The adulterated sample
- The hexane extracts from another group's pure H/S sample that is believed to be the adulterant
- The pure H/S sample from your group (identified on Plate 4) that is mixed with the adulterant

Develop and visualize Plate 5 and determine if your adulterant matches the other group's pure H/S sample. If so, ask the other group for the identity of the herb or spice. If the two are not the same, repeat the procedure with a sample from a different group until you have located a match.

With other members of your group, take your results to your instructor and present your case. Your instructor will inform you of your adulterant-detective abilities. If your conclusions were slightly off, you should go back and examine the data more carefully and run additional TLC analyses. Don't despair, even the best detectives can be led astray.

Cleanup

Dispose of the solvent in the TLC chambers and the sample vials in the appropriate liquid waste containers. After reproducing the full-sized TLC plates in your laboratory notebook, the TLC plates and the micropipettes should be put in the container for contaminated laboratory debris. Note that the plates from the DNPH analysis should be placed in the specially labeled container in the fume hood.

Discussion

- Discuss the results obtained from Plates 4 and 5.
- Explain how your group reached its conclusions about the adulterant. One way is to imagine that your group was hired by an herb/spice manufacturer to determine what adulterant was being used to contaminate their shipments of an expensive herb or spice.

Questions

1. Two students developed TLC plates using the same solvent system and each saw a single, UV-active spot with an $R_f = 0.42$. At first they assumed that their two samples must be the same compound. To confirm this, they co-spotted their samples on one TLC plate. Upon developing the plate and looking at it under a UV lamp, they observed the results shown below. Explain why they now see two spots, even though the R_f values on their previous plates were the same.

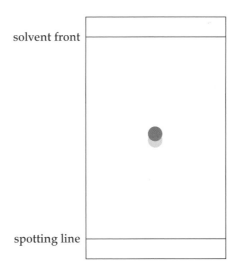

2. Two students analyzed the same spice by TLC. One of the students used hexane to extract the essential oil and the other student used diethyl ether. Would you expect their developed TLC plates to look the same if they used the same developing solvent? Explain your answer.

3. From the data recorded in your laboratory notebook, draw what a TLC plate would look like if it were spotted with the three compounds shown

below in different lanes and developed in 90% hexane:10% ethyl acetate. Next to this plate, draw another plate that was spotted with the same three compounds at identical starting points that was developed in 100% hexane to the same distance from the spotting line as the first plate.

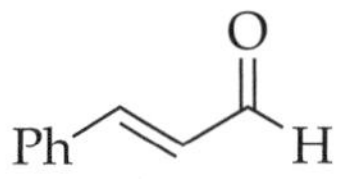

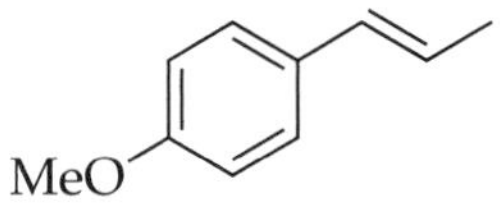

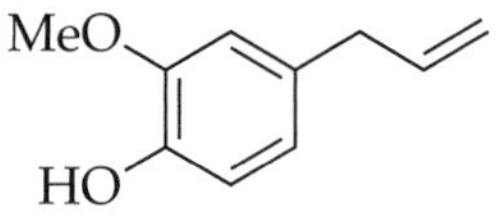

4. One of the known compounds you analyzed, 3,7-dimethyl-1,6-octadien-3-ol, is only available as a racemic mixture. Why did you not see two spots, one for each enantiomer, when you analyzed this sample?

5. Citral, a compound found in lemons and limes, is a mixture of two isomeric alkenes, *cis*- and *trans*-3,7-dimethyl-2,6-octadienal. Would you expect to see two spots, one for each isomeric alkene, when you analyzed this sample by thin layer chromatography? Explain.

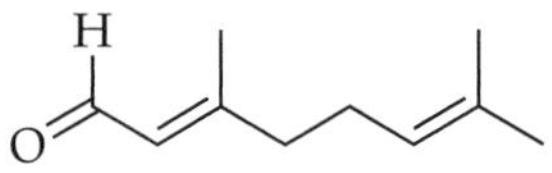

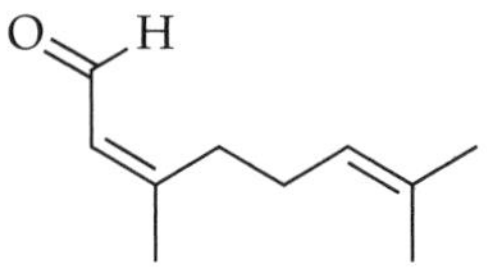

What Do You Take for Pain?

In this experiment, you will isolate naproxen from an over-the-counter (OTC) analgesic. In 1994, the FDA approved the OTC marketing of the nonsteroidal anti-inflammatory drug (NSAID) Aleve® (sodium naproxen) to compete with aspirin, acetaminophen, and ibuprofen. It works to relieve pain and inflammation by blocking the body's production of prostaglandins (Experiment 7). It is sold as the sodium salt of the *S*-(+)-enantiomer because the *R*-(-)-enantiomer is toxic to the liver. Each tablet contains 220 mg of the sodium salt in addition to some inactive ingredients that are insoluble in both aqueous and organic solvents. The pills are covered with Opadry®, a blue film coating that is soluble in water and some organic solvents. The solubility of naproxen is 1 g/25 mL in 95% ethanol; 1 g/40 mL in diethyl ether; and 1 g/15 mL in chloroform at room temperature. It is insoluble in cold water.

Did you know?

The stereoisomer of a chiral pharmaceutical that is pharmacologically active in the human body is called a **eutomer**. Stereoisomers that are either inactive or cause undesirable side effects are known as **distomers**.

Background Reading:

Before the laboratory period, review the discussion of acids and bases in Chapter 4, solubility in Section 6.1, recrystallization in Section 7.1, and thin layer chromatography in Section 7.8.3.

Prelab Checklist:

Include the structures and physical properties of the solvents and naproxen in your prelab write-up. Identify any stereocenters in the compound and include the specific rotation.

The ingredient in Aleve® is isolated as the free acid, not the sodium salt. Make sure that you understand the acid-base properties involved and the solubilities of the various forms of the compound. Include a flowchart of the procedure for the isolation of the free acid in your prelab write-up, clearly indicating which fraction contains the desired product. Write equations for any chemical reactions involved in the isolation of naproxen.

Experimental Procedure

Make sure you are wearing your safety glasses during the entire laboratory period.

Part I: Isolation and Purification of the Product

Weigh the tablet, put it in a 10-mL beaker, and add 0.5 mL of 95% ethanol. Swirl the beaker to loosen the blue covering on the tablet. Use a paper towel to wipe off any remaining blue covering. Let the tablet dry and weigh it. Put the tablet in a piece of weighing paper and fold the paper around the tablet. Enclose this package in a second piece of weighing paper and crush the tablet carefully by hitting it gently with a metal lab clamp. Transfer the pulverized material to a tared reaction tube/small test tube and record the weight.

Add 1 mL of water to the pulverized material in the reaction tube/small test tube. Heat the mixture gently for 5 minutes while stirring with a boiling stick to break up any clumps. Cap the tube and centrifuge the solution. Use a Pasteur pipette to transfer the supernatant to a clean reaction tube/small test tube. Add 0.5 mL hot water to the remaining residue and heat gently while stirring. Cap the tube, centrifuge, and transfer the liquid to the tube containing the supernatant from the first extraction. If the solution is not clear, it should be centrifuged again to remove all of the insoluble material.

TECHNIQUE TIP It is important to balance the centrifuge to prevent breakage of tubes. Use a second tube with water in it or share the "spin" with another student in the lab.

Cool the aqueous extract in an ice-water bath and add 1 N sulfuric acid dropwise with stirring until the solution is acidic (pH 1–2). Cap the tube, centrifuge the solution, and check that the solution is acidic. If it is not, add more sulfuric acid, stir, and recentrifuge. Decant the clear liquid into a reaction tube/small test tube (Section 7.2.4). Rinse the precipitate with a small amount of ice-cold water. After stirring, cap the tube, centrifuge, and add the rinse liquid to the acidic supernatant. *Warning:* Make sure that you understand whether the desired product is in the liquid or the precipitate and why.

CHEMICAL SAFETY NOTE Sulfuric acid is corrosive. If any solution containing sulfuric acid comes in contact with your skin, flush the affected area immediately with water for at least five minutes and have another student notify the instructor.

Recrystallize the material from ethanol/water (Section 7.1.3.a). Use the minimum volume of hot ethanol required to dissolve the crystals. Add hot water dropwise until the solution clouds. Then add hot ethanol slowly dropwise to obtain a clear solution (Section 7.1.3.a and Figure 7.5). Cool the solution to room temperature undisturbed and then put it in an ice-water bath. Remove the mother liquor using the square-tip Pasteur pipette method or filter the mixture on the Hirsch funnel. Rinse the crystals with a small amount of ice-cold water.

Transfer the crystals to a piece of filter paper and blot them dry. After the crystals have air dried, obtain the weight and melting point of the recrystallized material. Calculate the percent recovery of naproxen based on the weight of the sodium naproxen present in the tablet.

Part II: Thin Layer Chromatography

Dissolve a small amount of the product in acetone and spot it on a TLC plate. Develop the plate in ethyl acetate. Visualize the plate by UV light, mark the spot(s), and then develop the plate in I_2. Trace the TLC plate in your laboratory notebook and record the location of the visualized spot(s).

Save on Solvents: Share your TLC solvent chamber with a neighbor.

Part III: Polarimetry of the Product

Your instructor will demonstrate the operation of the polarimeter and indicate the concentration of the product needed to obtain a reasonable rotation in the polarimeter. It may be necessary to combine your product with that of other students to achieve the required concentration.

Dissolve the material in 95% ethanol. Zero the polarimeter using 95% ethanol in the cell for the blank reading. Empty the cell, transfer the ethanol solution of the product to the cell, and determine the rotation. Take five readings of the rotation and average them. Calculate the specific rotation using the average rotation, the weight of the material, the volume of the solution, and the length of the polarimetry cell.

Cleanup

Cool the acidic solution from the isolation of the crude product in ice water and neutralize it by adding sodium bicarbonate slowly while stirring. Discard the TLC solvents, the ethanol solution of the product, the neutralized acidic solution from the isolation, the mother liquor, and the washes from the recrystallization in the appropriate liquid waste containers. Weighing paper, TLC plates, contaminated Pasteur pipettes, and melting point and TLC capillaries should be put in the container for contaminated laboratory debris. Scrape the insoluble solid material out of the reaction tube/small test tube and put it in the container for solid chemical waste.

Discussion

- Discuss the acid-base chemistry used in this experiment.
- The solid product was recrystallized from a mixture of ethanol and water. Discuss the possible reason(s) for using this co-solvent system. Also discuss why diethyl ether or chloroform, the two other solvents mentioned in the introduction, were not used to recrystallize naproxen.
- Discuss what type of information the polarimetry studies provided. What are the inherent experimental errors associated with polarimetry measurements?

Questions

1. Why is the Opadry® coating removed using ethanol instead of water?
2. Do the results from the TLC prove that the product is pure? Explain.

Table E9.1 Solubility Data for Components of Analgesics

Compound	Solubility in Water	Solubility in Ethanol	Solubility in Other Solvents
aspirin	1 g/300 mL at 25°C 1 g/100 mL at 37°C	1 g/5 mL	1 g/10 mL of ether, 1 g/17 mL of dichloromethane
acetaminophen	0.1 g/100 mL at 22°C 1 g/20 mL at 100°C	soluble (~1 g/5mL at 25°C)	1 g/13 mL in acetone at 20°C, slightly soluble in ether, insoluble in hexane and dichloromethane
ibuprofen	<1 mg/mL	soluble (<1 g/2.5 mL at 25°C)	soluble in acetone
caffeine	1 g/46 mL at 25°C 1 g/1.5 mL at 100°C	1 g/66 mL at 25°C 1 g/22 mL at 60°C	1 g/500 mL in ether, 1 g/5.5 mL in dichloromethane

3. Extra-strength Excedrin is a combination of 250 mg acetaminophen, 250 mg aspirin, and 65 mg caffeine plus inactive ingredients that are insoluble in water and organic solvents. Using the data in Table E9.1, draw a flowchart to show how these three components could be separated using their solubility and acid-base properties. Hint: Caffeine is a nitrogen base similar in reactivity to diisobutylamine (Section 7.6.1 and Experiments 2 and 7).

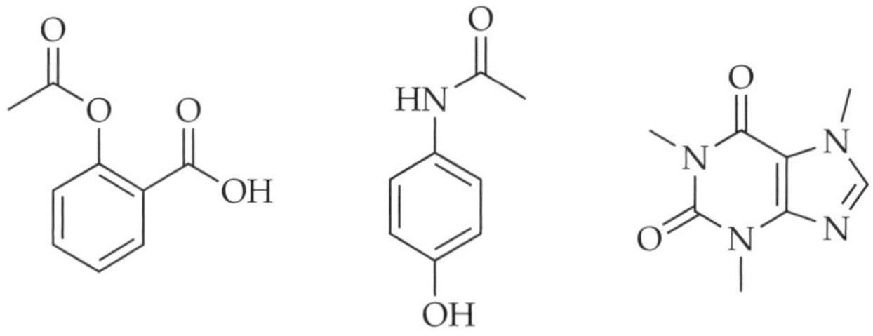

aspirin   acetaminophen   caffeine

4. You have isolated caffeine from the over-the-counter pharmaceutical called NoDoz and wish to purify it by recrystallization. Using the data in Table E9.1, explain which solvent you would use and why.

5. In this experiment, why was sulfuric acid used to neutralize the sodium salt of naproxen rather than acetic acid?

Nucleophilic Substitution Reactions of Alkyl Halides

In 1933, E.D. Hughes[1] and C.K. Ingold[2] proposed that two different reaction mechanisms accounted for the variation in the rate of substitution reactions for primary, secondary, and tertiary alkyl halides and for the differences in the reaction kinetics for the primary and tertiary compounds. They introduced the terms *nucleophile* and *electrophile* into the language of organic chemistry. From their kinetic studies, they defined the first-order two-step mechanism of substitution of tertiary alkyl halides as the **S_N1 reaction**. The second-order reaction of the primary compounds, which depended on the concentration of both the alkyl halide and the nucleophile, was explained by the one-step **S_N2 mechanism**. In 1935, using optically active compounds, they confirmed that inversion of configuration was observed for compounds reacting by the bimolecular S_N2 mechanism. A few years later, their experiments indicated that the unimolecular S_N1 pathway led to racemization with some accompanying inversion depending upon the reactants. Their work was widely criticized until about 1939 when these terms became the accepted terminology for both substitution and also the two elimination reaction mechanisms that they called **E1 and E2**.

Background Reading:

Before the laboratory period, read about nucleophilic substitution reactions in your lecture text.

Prelab Checklist:

In your prelab write-up, include the structures, molecular weights, and other relevant physical properties of the compounds and solvents being used in this experiment. Copy Table E10.1 into your laboratory notebook, leaving sufficient room to add your observations during the laboratory period. Make predictions for each substrate as to whether or not it will react with sodium iodide in acetone, silver nitrate in ethanol, or both and enter these predictions in the table.

Experimental Procedure

Wear your safety glasses at all times in the lab.

You will be testing the reactivity of 1-bromopentane, 1-chloropentane, 2-bromopentane, 3-bromopentane, 1-bromo-2-methylpropane, 2-chloro-2-methylpropane, 1-bromo-2,2-dimethylpropane, bromocyclopentane,

---

[1] JSTOR: Biographical Memoirs of Fellows of the Royal Society, **10**, 147
[2] JSTOR: Biographical Memoirs of Fellows of the Royal Society, **18**, 349

Table E10.1 Predictions and Observations for the Reactions of Alkyl Halides

Substrate		NaI/Acetone			AgNO₃/Ethanol			Name of collector
Name	Structure	Prediction	Observations		Prediction	Observations		
			RT	50°C		RT	50°C	

Continue for all eleven substrates.

bromocyclohexane, 1-bromoadamantane, and 1-chloroadamantane. The experiment is done in pairs with each student testing five or six substrates for a total of eleven halides with the two reagents. Make sure you record your partner's name and all results for the eleven substrates with the two reagents in Table E10.1 in your laboratory notebook.

It is important that you take care not to cross-contaminate any of the reactants. Note that 1-bromoadamantane and 1-chloroadamantane are solids and should be dissolved in a small amount of acetone or ethanol solvent before adding the test reagents.

CHEMICAL SAFETY NOTE   Alkyl halides are toxic if they are inhaled, ingested, or absorbed through the skin. You should avoid contact with them. In addition, avoid contact with the silver nitrate reagent, which will discolor your skin.

Did you know?

The antiviral drug amantadine (1-aminoadamantane), which is also used to treat Parkinson's disease, is synthesized from 1-bromoadamantane.

Part I: Reaction with Sodium Iodide in Acetone

Note: Acetone is a polar, aprotic solvent with a dielectric constant of 21. It does not promote the ionization of the alkyl halide substrates. Sodium iodide is soluble in acetone and the iodide ion is an excellent nucleophile. This reagent therefore favors S_N2 reactions of the substrates. The reaction can be detected because sodium bromide and sodium chloride are not as soluble as sodium iodide in acetone. A white precipitate forms when a reaction occurs. The formation of the insoluble product drives the reaction to completion.

Label eleven clean, *dry* reaction tubes/small test tubes with the identity of the alkyl halide substrates. Add 1 mL of an 18% solution of sodium iodide in acetone to each of the tubes (Experiment 1). Add four drops of one of the halides to one of these tubes and mix the reactants thoroughly with the stainless steel spatula by twirling to obtain a homogeneous solution. Record the time of addition.

Record the reaction time in Table E10.1 in your laboratory notebook next to the name of the compound if a precipitate forms at room temperature. If no visible change is detected within about 5 minutes, place the tube in a 50°C water bath and observe it for the next 5 to 6 minutes. The temperature of the water bath should be kept below 50°C because of the boiling point of acetone. Record your observations in Table E10.1

Repeat this procedure for the remaining substrates. In the case of the two solid substrates, dissolve 50 mg of the solid in 0.5 mL of acetone and add this solution to the reaction tube/small test tube with the sodium iodide/acetone test reagent.

Record the observations for all eleven substrates in the table in your laboratory notebook for your data as well as your partner's data and, in the last column, indicate the name of the person who collected the data.

Part II: Reaction with Ethanolic Silver Nitrate

Note: Ethanol is a polar, protic solvent with a dielectric constant of 24.3. The silver ion coordinates with the halogen of an alkyl halide in ethanol solution to weaken the carbon-halogen bond and promote ionization. The insoluble silver halide precipitates from solution when a reaction occurs. This reagent therefore promotes the S_N1 reaction of the substrates. The ethanol solvent is in much higher concentration than the nitrate ion and therefore ethanol behaves as the nucleophile to react with the resulting carbocation. Depending on the structure of the alkyl halide, elimination (E1) may compete with the substitution reaction.

Rinse any contaminated test tubes or reaction tubes/small test tubes with ethanol before proceeding. The tubes do not have to be dry.

Label the tubes as in the previous section and add 1 mL of the 1% ethanolic silver nitrate solution to each of the tubes. Add four drops of one of the halides to one of these tubes and mix the reactants thoroughly with the stainless steel spatula by twirling to obtain a homogeneous solution. Record the time of addition.

Record the identity of the halide and the time it takes for the formation of any precipitate at room temperature. If no precipitation occurs within 5 minutes, heat the tubes in a 50°C water bath for 5 to 6 minutes and record your observations.

Repeat for the remaining substrates. In the case of the two solid substrates, dissolve 50 mg of the solid in 0.5 mL of ethanol and add this solution to the 1% ethanolic silver nitrate reagent.

Record the observations for all eleven substrates in your laboratory notebook, using your data as well as your partner's data.

Cleanup

All of the reaction mixtures generated during this experiment, including all rinses, must be disposed of in the appropriate liquid waste containers. Make sure that no alkyl halides or solutions thereof are disposed of down the drain. All contaminated Pasteur pipettes must be disposed of in the container for contaminated laboratory debris.

Discussion

- For your laboratory report, analyze all of the results that you and your partner obtained for this experiment.

- Provide balanced equations for any reactions that took place.

- Write a brief explanation for the observed reactivity of each compound with each of the reagents.

- For the discussion, it is important to compare the reactivities of various substrates for both S_N1 and S_N2 reactions. For example, consider issues such as the nature of the halide leaving group (chloride versus bromide) and the structures of the alkyl groups. Include a discussion of the effect of the temperature on the reactions observed.

- If the reactivity differs from what you predicted, explain how the results helped you rethink your predictions.

- To aid in the analysis of your results, make models of bromocyclopentane, bromocyclohexane, and the intermediates you would expect to generate in the S_N1 reaction of 1-bromoadamantane or 1-chloroadamantane and of 2-chloro-2-methylpropane. Examine the models carefully. Explain why the results from the adamantane derivatives are surprising.

Questions

1. Why is it important to use dry tubes for the reactions involving sodium iodide in acetone?

2. The reaction of 1-bromopropane with sodium iodide to give 1-iodopropane is endothermic. Explain what strategy you would use to obtain a good yield of product from this reaction.

3. Explain why 2-bromopropane reacts with sodium iodide in acetone over 10^4 times faster than bromocyclopropane. Hint: Examine the transition state for each of the reactions.

4. Write rational arrow-pushing mechanisms for the following reaction to show how each of the products is formed.

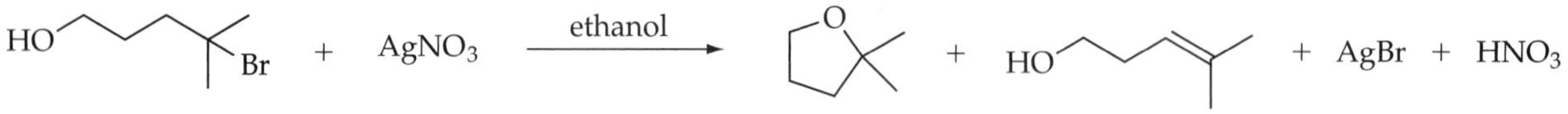

$$\text{HO}\diagdown\diagup\diagdown\!\!\overset{|}{\underset{\text{Br}}{\text{C}}}\!\!\diagup \quad + \quad \text{AgNO}_3 \quad \xrightarrow{\text{ethanol}} \quad \boxed{\text{O}}\!\!\diagdown \quad + \quad \text{HO}\diagdown\diagup\diagdown\diagup \quad + \quad \text{AgBr} \quad + \quad \text{HNO}_3$$

5. Propose two syntheses of *tert*-butyl ethyl ether. One of the synthetic schemes must use an S_N2 reaction and the other an S_N1 reaction.

Isolation of Trimyristin from Nutmeg

Nutmeg is an aromatic spice obtained from the seed inside the fruit of the evergreen tree, *Myristica fragrans*. Your goal in this experiment is to use liquid-solid extraction to separate and purify trimyristin, a naturally occurring triglyceride, from nutmeg.

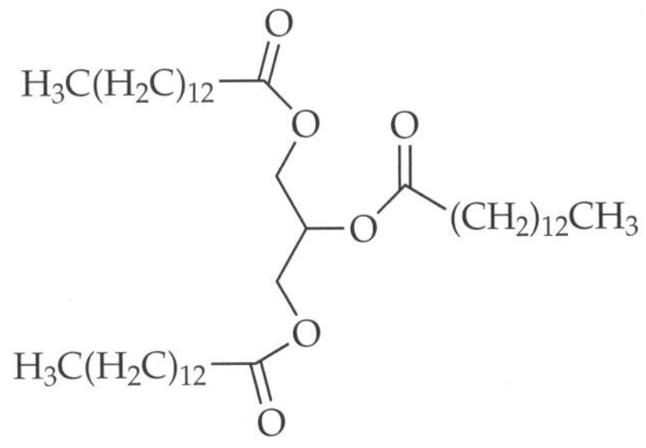

$$H_3C(H_2C)_{12}-C(=O)-O-CH_2-CH(-O-C(=O)-(CH_2)_{12}CH_3)-CH_2-O-C(=O)-(CH_2)_{12}CH_3$$

trimyristin

You will be introduced to the technique of distillation and will be working with larger amounts of materials than used in previous experiments. This experiment is done on a larger scale for two reasons. First, in order to obtain reasonable quantities of trimyristin, it is advantageous to use a greater amount of starting material. Second, it is important that you have the opportunity to work with larger quantities of chemicals to appreciate how it differs from microscale chemistry.

Because of the large amount of chemical waste that is generated, we typically don't work on a larger scale. However, in this experiment you will be minimizing the waste generated by recovering and recycling the solvent used for extraction.

Background Reading:

Before the laboratory period, you should read about liquid-solid extraction (Section 7.6.2), distillation (Section 7.3), and reflux (Section 7.4). Review the section on recrystallization in Section 7.1, including the section on removing colored impurities (Sections 7.1.4 and 7.1.5). Also refer to Experiment 6, Part II.

Prelab Checklist:

The prelab write-up should include a flowchart of the procedure. Include the structures and physical properties of trimyristin, *tert*-butyl methyl ether, and acetone.

Did you know?

Methyl *tert*-butyl ether (MTBE) has less of a tendency to form organic peroxides than diethyl ether and the cyclic ether, tetrahydrofuran (THF). High levels of these peroxides can cause explosions if, for example, solutions of diethyl ether or THF are distilled to dryness. Unlike diethyl ether, MTBE is not a suitable solvent for the Grignard reaction because it does not coordinate adequately with magnesium. Its use as a solvent is also limited by the fact that it decomposes in strong acid to form 2-methylpropene by way of an E1 reaction.

Laboratory Period 1: Extraction and Recrystallization of Trimyristin

Experimental Procedure

Don't forget to keep your safety glasses on during the entire laboratory period.

You will be working with a partner for the next two laboratory periods. All observations should be your own and all data must be recorded in each student's laboratory notebook. Work as a team, sharing the responsibilities in this detail-oriented experiment.

EXPERIMENTAL NOTE: It is important that you keep track of the amount of *tert*-butyl methyl ether that you use in this experiment. You will recover it at the end and will determine a percent recovery.

HELPFUL HINT: You will be collecting samples for TLC analysis during the experiment. You will be prompted with the notation *TLC Sample* when it is time to take the sample.

Part I: The Extraction

Weigh approximately 5 g of nutmeg powder and place it in a 100–mL round-bottom flask that is securely clamped to the lab rack. Add a 1–inch stir bar and 40 mL of *tert*-butyl methyl ether.

TECHNIQUE TIP   When using glassware with ground glass joints, it is important that you avoid getting any residue on the joint. A funnel should be used to transfer materials into the flask. If residue is present on the glass joint, try to wash it into the flask with the solvent being added to the flask.

Place the round-bottom flask in the well of the heating mantle. Apply a thin band of grease to the male ground-glass joint of a condenser column and attach the column to the 100–mL round-bottom flask, twisting it slightly to ensure that it rotates smoothly (Figure 7.15). Clamp the two pieces of glassware together to make sure that they do not separate during the extraction. The entire apparatus should be secured to the lab rack with a lab clamp.

TECHNIQUE TIP The condenser is fragile, so do not tighten a clamp around the condenser, but secure it loosely in the clamp.

Attach an appropriately-sized septum to the top of the condenser and then insert a needle into the septum.

EXPERIMENTAL NOTE A septum and needle are used to minimize the amount of vapor released from the system. It is critical to insert the needle into the septum in order to prevent having a closed system when heating.

Finish setting up the apparatus by attaching two water hoses to the condenser column. Attach the water inlet hose to the bottom of the condenser and the outlet hose to the nozzle at the top of the condenser as shown in Figure 7.15. Make sure the water inlet hose is securely attached to the water source and that the water outlet hose is in the drain. Turn the water on slowly. Adjust the water flow until a gentle stream of water is coming out of the outlet tube.

Start stirring and heating the solution. Adjust the heat control so that a gentle reflux of the solution is achieved. Once refluxing has begun, continue heating for another 45 minutes.

SAFETY NOTE If you detect any solvent escaping from the joint where the round-bottom flask and condenser are connected, turn off the heat control and notify your instructor.

TECHNIQUE TIP A gentle reflux typically means that the reflux ring (page 69) is lower than one-third of the height of the condenser column. If the reflux ring is higher than this, either increase the flow of water through your condenser or decrease the heat control setting.

After completing the 45-minute reflux, turn the heat control off and allow the solution to cool down.

After the solution has cooled to room temperature, remove the septum and rinse the inside of the condenser column with 2 mL of *tert*-butyl methyl ether. Use a Pasteur pipette to deliver the solvent from the top of the column.

Turn the water supply to the condenser off and then remove the condenser by gently twisting and then pulling. If the condenser does not come off easily, *do not attempt to force it off!* Contact your instructor for assistance.

Wipe off the grease on the female joint using a paper towel or tissue that has been wet with a small amount of acetone.

Prepare the filtration apparatus by placing a Büchner filter funnel in a 125–mL filter flask using the neoprene adapter between the funnel and the flask (Section 7.2.5 and Figure 7.10). Put a piece of filter paper in the funnel so that all of the holes in the funnel are covered. Wet the paper with a few drops of *tert*-butyl methyl ether. Attach a vacuum source to the nozzle of the filter flask. Slowly decant as much of the liquid as possible from the powder, passing this liquid through the filter (Section 7.2.4). The majority of the powder should remain in the 100–mL round-bottom flask. After all the liquid has passed through the filter paper, *remove the vacuum hose*.

Add another 10 mL of *tert*-butyl methyl ether to the powder in the 100–mL round-bottom flask and swirl or stir the contents of the flask for approximately 1 minute. The purpose of this step is to dissolve any crude trimyristin coating the powder. Decant the liquid, passing it through the same filter paper into the first liquid filtrate while briefly applying a vacuum.

Put a small stir bar into a clean 100–mL round-bottom flask and weigh it, making sure to record the weight of the flask plus the stir bar in your laboratory notebook. Transfer the contents of the filter flask to this 100–mL round-bottom flask using a funnel to aid with the transfer. If necessary, rinse the filter flask with a small amount of *tert*-butyl methyl ether and add these rinses to the 100–mL flask. Record how much *tert*-butyl methyl ether you used. Temporarily cap this flask.

TLC Sample #1: Put a few drops of the liquid from the round-bottom flask in a small test tube. Cap the test tube and label it #1.

Part II: The Distillation

Set up the apparatus for a simple distillation of the solution from Part I (Figure 7.11). Apply a thin band of grease to the male ground-glass joint(s) except for the male joint at the receiving end of the distillation apparatus. Secure all of the connections so that they do not separate during the distillation. Make sure that the bottom of the thermometer probe is placed at the fork of the distillation head. Wrap cotton around the neck of the distillation head up to the point where the cooling arm is attached. Weigh a capped, clean 100–mL round-bottom flask. Uncap the flask, saving the cap, and attach the flask to the ungreased joint at the receiving end of the distillation apparatus. Do not twist the ungreased joint.

Begin stirring and heating the solution. The heat control setting should be positioned at approximately the level that resulted in a gentle reflux in Part I. Keep an eye on the cooling arm of the distillation head for the first signs of distillate. Be sure to record the temperature at this time. The temperature should continue to climb to approximately the boiling point of *tert*-butyl methyl ether. Continue to monitor the cooling arm and the collection flask. If the distillate comes over too quickly, turn the heat control down. Make sure that no ether leaks from the joint between the distillation head and the flask containing the original extracts. If leaking occurs, turn the power off and notify your instructor.

When the temperature of the distillate starts to decrease, monitor the rate of liquid dripping into the collection flask. When the amount of distillate coming over into the receiving flask has decreased dramatically, turn the heat control off. It is critical that you do not overheat the residue left behind in the flask.

SAFETY NOTE   Overheating of the residue in the distillation pot may result in a fire. Therefore, it is very important that you watch your distillation at all times. In particular, look for any leaks and monitor the level of the solution being distilled carefully. You will never see all of the liquid in this flask disappear. (Think about why this is the case.) You must therefore keep an eye on the collection flask to determine when it is time to stop heating.

Allow the system to cool to room temperature, remove the collection flask, and cap it. Weigh the capped flask and record the weight in your laboratory notebook.

TLC Sample #2: Put a few drops of the distillate from the round-bottom flask in a small test tube. Cap the test tube and label it #2.

Remove the distillation head from the other 100–mL round-bottom flask and, if necessary, slowly spin the flask in the palm of your hand to help remove the last traces of *tert*-butyl methyl ether. After the residue has solidified, use a paper towel and a small amount of acetone to wipe the grease off of the inside of the joint. Weigh the flask and record the weight of the crude product obtained from the extraction.

Part III: The Recrystallization

Fit a Hirsch funnel with glass-fiber filter paper in order to remove activated charcoal (Section 7.2.3) and place the funnel on a 25–mL side-arm filter flask (Section 7.2.5 and Figure 7.10).

Add 12 mL of acetone, approximately 700 mg of activated carbon, and a boiling stick to the crude extract. Bring the solution to a boil. After boiling for a couple of minutes, pour the solution through the filter. It may be necessary to apply a vacuum briefly to help with the filtration process.

SAFETY NOTE   If any acetone is spilled onto the heating mantle, be sure to wipe it off before you begin heating.

After filtering the solution, rinse the round-bottom flask with another 2–3 mL of acetone and pass this rinse through the filter using a Pasteur pipette.

Transfer the filtrate to a 25–mL Erlenmeyer flask. Use another 1 mL of acetone to rinse the filter flask, adding the rinse to the Erlenmeyer flask along with a boiling stick.

TLC Sample #3: Put a few drops of the liquid from the Erlenmeyer flask in a small test tube. Cap the test tube and label it #3.

Heat the acetone solution to boiling in a hot water bath on the sand bath and reduce the volume to approximately 10 mL. Remove the boiling stick and cap the flask. Enclose the flask in cotton and place it in a beaker. Cover the beaker and label it with your name, the name of the experiment, and the date. It should be stored for a minimum of 24 hours. It is important that it not be moved around much during this time.

Spot a TLC plate with the three TLC samples. If any of your samples have evaporated to dryness, add three drops of either *tert*-butyl methyl ether or acetone and proceed with the analysis. Develop the plate in 80:20 hexane:ethyl acetate and visualize it with UV light and iodine.

Save On Solvents: Share your TLC solvent chamber with a neighbor.

Cleanup for Laboratory Period 1

Pour the recovered *tert*-butyl methyl ether into the appropriate container. The solvent will be recycled, so be conscientious and do not add anything else to this container.

To clean the 100–mL round-bottom flask containing the powder residue, add some water to the flask and swirl the contents to suspend the powder in the water. Retrieve the stir bar using a stir-bar retriever or other type of magnet and then quickly pour the suspension into the appropriately labeled waste container. If necessary, repeat the procedure. Finish cleaning the flask with soap and water. Place any other solutions in the appropriately labeled waste container. Be careful not to add any of these to the *tert*-butyl methyl ether recycling container. All used weigh paper, Pasteur pipettes, TLC plates, filter paper, and boiling sticks should be discarded in the container for contaminated laboratory debris.

Laboratory Period 2: Isolation of the Trimyristin

Experimental Procedure

Make sure you are wearing your safety glasses during the entire laboratory period.

After standing a minimum of 24 hours, there should be a considerable quantity of crystals in the Erlenmeyer flask. Place the flask in an ice-water bath for at least 20 minutes. In the meantime, put 3 mL of acetone in a test tube and cool it in an ice-water bath. It is important that the acetone be ice-cold before using it to wash the crystals.

Isolate the crystals by brief vacuum filtration (Section 7.2.5 and Figure 7.10) and wash them with approximately 2 mL of ice-cold acetone. Try to get the acetone to come into contact with the majority of crystals. Do not discard the mother liquor.

Dry the product to a constant weight and measure the melting point range. Calculate the percent recovery of trimyristin based on the amount of nutmeg powder used.

Cleanup for Laboratory Period 2

Pour the mother liquor into the appropriate liquid waste container. Additional material can be recovered from this solution if desired. Place your product in a suitable container. Any used weigh paper, Pasteur pipettes, or boiling sticks should be discarded in the container for contaminated laboratory debris.

Discussion

- Now that you have performed a series of larger-scale laboratory techniques, compare this approach with the microscale techniques you used in earlier experiments.

- Calculate the percent recovery of *tert*-butyl methyl ether based on the total amount used. Discuss why you did not recover 100% of the *tert*-butyl methyl ether. In particular, consider where the losses occurred. If you were to run this experiment again, how could you improve the recovery of *tert*-butyl methyl ether?

- Discuss the TLC plates you developed, including the answers to the following questions. Did you identify trimyristin by TLC? Were there other spots present? Did you observe any changes after decolorizing the solution?

- Approximately 25% by weight of nutmeg powder is trimyristin. Compare your yield with the amount expected based on this figure. If you were to run this experiment again, how could you improve your yield?

Questions

1. In Part III of Laboratory Period 1, after treating the crude extract with activated carbon, the solution was somewhat lighter in color, but still quite dark. Knowing that trimyristin is in fact a white crystalline compound, why wasn't more activated carbon used in an attempt to remove most of the colored impurities?

2. Gas chromatographic analysis of the *tert*-butyl methyl ether collected from the distillation in Part II of Laboratory Period 1 showed no impurities in the distillate. The oil of nutmeg is known to consist of many other organic compounds besides trimyristin. What does the purity of the distillate indicate about these other compounds?

3. What would be the outcome of the recrystallization of trimyristin if the following instructions were provided at the beginning of Laboratory Period 2? "After standing a minimum of 24 hours, there should be a considerable quantity of crystals in the Erlenmeyer flask. Place the flask in an ice-water bath for at least 20 minutes. Isolate the crystals by brief vacuum filtration and wash them with 2 mL of acetone. Try to get the acetone to come into contact with the majority of crystals."

4. What probably occurred if, after the recrystallization of trimyristin, the crystals had black particles dispersed throughout? What could be done at this point to correct the situation?

5. A student was in a hurry to leave the lab and recorded the melting point of the recrystallized trimyristin right after washing the crystals with cold acetone. The melting point range was 46–50°C. Is this a good result? If not, explain what happened.

The Magtrieve™ Oxidation of 4-Chlorobenzyl Alcohol, A Solvent-Free Reaction

The oxidation of primary alcohols to aldehydes and secondary alcohols to ketones is an important transformation in the synthetic organic chemist's toolbox. Historically, such oxidations have generally been carried out using chromium (VI) or manganese (VII, IV) reagents. One of the most important of the chromium reagents is pyridinium chlorochromate, often called PCC, that was introduced to the synthetic chemistry community by Professor E.J. Corey[1] of Harvard University in 1975. In the absence of water, this reagent oxidizes a primary alcohol to an aldehyde without over-oxidation to a carboxylic acid, which is a common problem with other chromium(VI) oxidants. This was an important step forward in the development of selective reagents for organic synthesis. Although such oxidants continue to be used in research labs around the world, they have some disadvantages. First, the chromium(VI) reagents are harmful to humans if inhaled in the form of dust and several are suspected carcinogens. Second, the chromium(III) products resulting from these oxidation reactions are generally not recyclable.

Recent advances in the area of oxidation chemistry are focusing on developing reagents and/or catalysts that are either recyclable or that use relatively benign oxidants such as bleach (NaOCl), hydrogen peroxide (H_2O_2), or, ideally, oxygen (O_2). In this experiment, you will be working with a recyclable chromium oxidant that does not have the problems mentioned previously.

One stage of the production of magnetic recording tape employs solid chromium (IV) dioxide, CrO_2. In 1997, researchers at DuPont[2], one of the manufacturers of CrO_2, reported the use of CrO_2 as a selective oxidant with properties superior to MnO_2. Important features of this reagent include:

- The oxidation of primary alcohols stops at the aldehyde so that over-oxidation is not a problem.

- The reaction mixture is heterogeneous because the oxidant is not soluble in organic solvents. The organic materials are therefore easily separated from the oxidant and its reduced form.

---

[1] E.J. Corey was awarded the Nobel Prize in 1990 "for his development of the theory and methodology of organic synthesis." See http://nobelprize.org/chemistry/laureates/1990/index.html.

[2] Lee, R.A.; Donald, D.S., *Tetrahedron Lett.* **1997** *38*, 3875.

- The reduced chromium(III) product from this reaction, CrO(OH), is readily converted back to the CrO_2 oxidant by passing air over the solid at 350°C.

DuPont named this reagent Magtrieve™ because it can be retrieved from a reaction by using a magnet. In the original paper describing this oxidant, all of the reactions were carried out in a solvent such as dichloromethane or toluene. In order to take this environmentally-friendly reaction one step further, no solvent is used in the reaction or in the workup (Equation E12.1).

Because the product from this reaction, 4-chlorobenzaldehyde, has a high vapor pressure relative to the starting alcohol, 4-chlorobenzyl alcohol, the two can be separated without the use of solvent. The reaction is carried out at a temperature above the melting point of the 4-chlorobenzyl alcohol. As this compound liquefies, it is oxidized by the Magtrieve™ to form the 4-chlorobenzaldehyde. The aldehyde also has a melting point below the reaction temperature and, because of its high vapor pressure, it evaporates. The gaseous aldehyde solidifies on a cold surface. The product can be scraped off this surface. This a rare example of a completely solvent-free oxidation reaction.

A relatively large amount of Magtrieve™ is used compared with the amount of starting material because only the surface layer of CrO_2 in each particle of Magtrieve™ is available to react (Figure E12.1). Although this may seem wasteful, recall that the Magtrieve™ is recyclable and is therefore not wasted by using an excess of the reagent.

Figure E12.1   **Cross Section of a Magtrieve™ Particle**

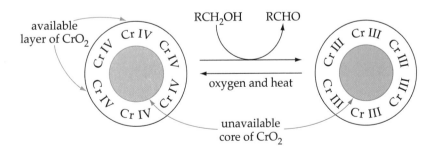

Did you know?

Audio recording tape was invented in Germany in 1926, but was kept secret from the rest of the world until after World War II. Magnetic tape used for audio and video recording consists of three layers. The plastic or metal base material is coated with a magnetic oxide, such as ferric oxide or chromium dioxide, that is held to the base with binder glue. A square inch of the tape contains about a billion of these particles that function like microscopic bar magnets. The ferromagnetic oxide of chromium is ideal for this purpose because it is a submicron-size powder with an acicular (needle) shape. The size required is determined by the frequency being recorded with long particles used to record low frequencies and short particles used for high frequencies. Chromium dioxide is preferable to ferric oxide because it permits higher recording levels at high frequencies and higher density storage, resulting in greater fidelity.

Background Reading:

Before the laboratory period, read about sublimation/evaporation in Sections 6.5 and 7.5. Also read about the oxidation of alcohols in your lecture text.

Prelab Checklist:

In your prelab write-up, include the balanced equation for the reaction (Section 4.2). Include the structures and physical properties of 4-chlorobenzyl alcohol and 4-chlorobenzaldehyde. Calculate the number of mg of 4-chlorobenzyl alcohol to be used and the theoretical yield expected (Section 4.2).

Experimental Procedure

Don't forget to keep your safety glasses on during the entire laboratory period.

Place 1 mmol of 4-chlorobenzyl alcohol and 700 mg of Magtrieve™ in a 30–mL beaker.

SAFETY NOTE   To prevent inhaling Magtrieve™ dust, use your spatula to scoop it out of the container rather than pouring it.

Mix the contents thoroughly with a glass stirring rod, breaking up any large clumps of the alcohol. Spread the mixture out evenly over the bottom of the beaker and place it in a sand bath. The beaker should be approximately 1/8 to 1/4 inch below the top of the sand. Place a watch glass (75 mm diameter) over the top of the beaker and put some ice in the watch glass (Figure 7.18).

Start heating the system using a low setting. Measure the sand bath temperature at the bottom of the beaker with a digital thermometer to make sure it does not go above 110°C. Use a Pasteur pipette to remove water from the watch glass as the ice melts and add more ice as necessary. Be careful that water does not overflow into the beaker.

Watch for condensation of the product around the top of the beaker. When this occurs, crystals will start to appear on the underside of the watch glass. After approximately 30 minutes of heating, carefully remove the entire set-up from the sand bath and let it cool to room temperature. While leaving the watch glass on the beaker, pipette away the water and remove any remaining ice. Blot the top of the watch glass with a paper towel to remove any traces of water. Gently lift the watch glass off the beaker and scrape the solid product onto a second tared watch glass. If there is any uncontaminated crystalline material near the top edges of the beaker, scrape this onto the tared watch glass also. Record the weight of the product in your laboratory notebook.

Level out the remaining reaction mixture in the beaker with a glass stirring rod, place the first watch glass on the beaker, and add ice. Put the set-up back on the sand bath to isolate a second crop of solid. You can repeat this a third time if you wish. Make sure to record the combined weight of the product and calculate a percent yield. Measure the melting point of the product. Dissolve a small amount of the starting material and the product in acetone and spot solutions of each of these compounds on a TLC plate. Develop the plate in 60:40 hexane:ethyl acetate. Use UV light and iodine to visualize the spots.

Save On Solvents: Share your TLC solvent chamber with a neighbor.

Cleanup

Place your combined product in an appropriate container. The used Magtrieve™ should be placed in the jar labeled "Magtrieve™ for Recycling." The TLC developing solvents should be placed in the appropriate liquid waste container. Contaminated Pasteur pipettes, weigh paper, and melting point capillaries should be disposed of in the container for contaminated laboratory debris.

Discussion

- Analyze the ^{1}H NMR and the ^{13}C[^{1}H] NMR spectra of the starting material and the product (Chapter 8). Your instructor will let you know how or where to obtain the spectra.

- Describe the phase changes occurring in the beaker that allow the collection of the purified product. Make sure that you consider the temperature of the sand bath and the temperature of the watch glass (Section 6.2).

- Based on the balanced equation you included in your prelab write-up, how much of an excess of Magtrieve™ did you use? Express your answer in both mmol and mg (Section 4.2). Discuss why using an excess of the reactant was necessary in the experiment.

Questions

1. Explain the relative R_f values for the starting material and the product.
2. Write the structure of the product that would be obtained if potassium dichromate in aqueous acid were used for the oxidation.

3. In this experiment, you collected two crops of crystals and were given the option of collecting a third crop. The continued heating of the reaction mixture to collect the third and additional crops could be necessary for one of two reasons. First, not all of the starting material had been oxidized after 30 minutes and therefore a longer reaction time was needed. Second, the reaction was complete after 30 minutes; however, the product had not completely evaporated from the reaction mixture. Propose a simple experimental method that would determine which scenario is correct.

4. Explain why the vapor pressure of 4-chlorobenzaldehyde is greater than that of the alcohol (Chapter 5).

5. Find an experimental procedure in the chemical literature that makes use of a traditional chromium or manganese oxidation reagent (that is, PCC, CrO_3, $K_2Cr_2O_7$, MnO_2, or $KMnO_4$) (Chapter 3). Compare and contrast the environmental consequences of the experimental procedure from the literature with the one in this experiment using Magtrieve™. Include the title of the article, the author(s), the journal name, the publication year, the volume, and page number(s).

The Sodium Borohydride Reduction of Benzil and Benzoin[1]

Before the availability of sodium borohydride and lithium aluminum hydride, aldehydes and ketones were reduced using active metals such as sodium or zinc in acetic acid or alcohols. These reactions required high temperatures and long reaction times. The synthesis of the hydride reagents in the 1940s provided an inexpensive and easy method for reducing polarized double bonds. These and other hydride reducing agents have been the focus of Professor Herbert Charles Brown's research career. His interest in boron chemistry began in 1936 quite by accident. His future wife gave him a book on boron chemistry as a graduation gift when he received his undergraduate degree. It was the Great Depression and she chose the least expensive chemistry book ($2.06) in the University of Chicago bookstore.[2] He concentrated his research efforts on sodium borohydride when the Army Signal Corps needed a method of generating hydrogen in the field. The synthesis of sodium borohydride was reported in 1942 and, in an attempt to purify the compound using acetone as a solvent, its ability to reduce aldehydes and ketones at room temperature was discovered. Now thousands of tons of sodium borohydride are produced per year. Not only did it provide a quick and easy method of reducing aldehydes and ketones, but it also solved a major ecological problem in the paper pulp industry. A zinc compound used in the bleaching of paper pulp was contaminating rivers and killing fish. By using sodium borohydride for the reduction process, the problem was eliminated.[3] Professor Brown was awarded a Nobel Prize in 1979[4] for his contributions to the chemistry of organoborane compounds and hydride reducing agents. The compounds that he synthesized provide a range of reducing agents with different chemoselectivities that allow selective reductions of specific functional groups.

Sodium borohydride is a much milder reducing agent than lithium aluminum hydride. Although both sodium borohydride and lithium aluminum hydride reduce aldehydes, ketones, and acid chlorides at 25°C,

[1] Rowland, Alex T., *J. Chem Ed.* **1983** *60*, 1085.
[2] Brown, H.C. *Science* **1980** *210* 4469 485.
[3] Brown, H.C. *Chem.Br.* **1980** *16* 11 606.
[4] http://nobelprize.org/chemistry/laureates/1979/index.html.

sodium borohydride does not reduce carboxylic acids, epoxides, nitro groups, nitriles, amides, or alkyl halides. Both reagents provide hydride ions that behave like hydrogen nucleophiles and attack polarized multiple bonds at the electrophilic atom. They do not react with nonpolar carbon-carbon multiple bonds. While lithium aluminum hydride reacts violently with protic solvents, sodium borohydride reactions can be run in protic solvents. In fact, it can be purchased as a 12% aqueous solution in 14 M sodium hydroxide. Its solubility in water is 55 g/100 mL at 25°C; in methanol 16.4 g/100 mL at 20°C; in ethanol 3.2 g/100 mL at 20°C; and in 2-propanol 0.37 g/100 mL at 25°C and 0.88 g/100 mL at 60°C.

Originally, it was thought that the tetraalkoxyborate salt of the product alcohol was produced in the reaction and that this had to be hydrolyzed to give the free product alcohol. Mechanistic studies in protic solvents indicate that the solvent is involved in the transition state. One solvent molecule provides a proton to the carbonyl oxygen while the oxygen atom of another solvent molecule bonds to the boron atom[5].

Did you know?

Foxing is a term used to describe the brown spots that are found in older books and manuscripts. The term, first used in 1848, is thought to be derived either from the color of the spots that resemble those of a fox or from their shape that often resemble their paws. The discoloration is caused by the formation of chromophores by oxidation. Sodium borohydride has been used to reduce the color-producing aldehydes and ketones back to the colorless alcohols.

Background Reading:

Before the laboratory period, read about the reduction of aldehydes and ketones in your lecture text.

Prelab Checklist:

In addition to the physical properties of all starting materials and products, include a balanced equation for the reaction in your prelab write-up and identify the limiting reactant (Section 4.2). Calculate the weight of the starting materials from the number of mmol specified.

Experimental Procedure

Wear your safety glasses at all times in the lab.

SAFETY NOTE   There should be no flames in the lab when the sodium borohydride is being used. It is also important to keep the top on the container to minimize the exposure of the sodium borohydride to moisture.

---

[5] Wigfield, Donald C. *Tetrahedron* **1979** *35* 449.

Part I: The Reduction of Benzil and Benzoin in 2-Propanol

This experiment is to be done in pairs with one person using benzil as the starting material and the other using (±)-benzoin. Record the name of your partner in your laboratory notebook.

Weigh approximately 0.35 mmol of the compound to be reduced in a reaction tube/small test tube. Add approximately 0.5 mL of 2-propanol. Record the physical characteristics of the mixture in your laboratory notebook. Spot a small amount of the solution on a TLC plate that can accommodate three lanes. Add approximately 15 mg of sodium borohydride and stir the mixture occasionally at room temperature for 15 minutes. Follow the reaction by thin layer chromatography by spotting the reaction mixture on a TLC plate after 2, 5, 10 minutes, etc. Put three spots on each TLC plate. Develop the plates in 60:40 hexane:ethyl acetate and visualize them using UV light and iodine. Be sure you make careful observations and record them in your laboratory notebook: What color is the solution? Is it homogeneous? Is any heat produced? Are any bubbles generated? Record the time it takes for any physical changes to occur and any differences between your reaction and your partner's reaction.

Save On Solvents: Share your TLC solvent chamber with a neighbor.

Add 1.0 mL of water and a boiling stick and heat the solution gently for approximately 5 minutes. Be careful that the solution doesn't bump. Record your observations and explain what is occurring. Add hot water to give a total volume of approximately 3.0 mL. Allow the solution to cool slowly to room temperature before putting it in an ice-water bath.

Remove the solvent using the square-tip Pasteur pipette filtration method (Section 7.2.1). Rinse the crystals with approximately 0.5 mL of ice-cold water and remove the rinse with the Pasteur pipette. Scrape the wet crystals onto a piece of filter paper and blot them dry. Allow them to dry completely before determining the weight and melting point. Measure the mixed melting points of the product with benzoin and with your partner's product.

Before leaving the laboratory, determine the theoretical yield based on the amount of starting material actually used and calculate the percent yield. Also record your partner's results in your laboratory notebook.

Part II: The Effect of Solvent on the Reaction of Benzil

Each partner should weigh approximately 50 mg of benzil in a reaction tube/small test tube. One partner should add 0.5 mL of methanol, the other partner should add 0.5 mL of 95% ethanol to their reaction tube/small test tube. Cool the resulting suspensions in an ice-water bath.

Add approximately 10 mg of sodium borohydride and allow the mixture to warm to room temperature. Stir occasionally at room temperature for approximately 10 minutes. Follow the reaction by TLC using 60:40 hexane:ethyl acetate to develop the plates and visualize them using UV light and iodine. Record the differences among the reactions in the three solvents: methanol, ethanol, and the 2-propanol from Part I. Record your observations and those of your partner.

Add 0.5 mL of water and heat gently. Add hot water to give a total volume of approximately 3.0 mL. Allow the solution to cool slowly to room temperature before putting it in an ice-water bath.

Remove the solvent using the square-tip Pasteur pipette filtration method. Rinse the crystals with approximately 0.5 mL of ice-cold water and remove the rinse with the Pasteur pipette. Scrape the wet crystals onto a piece of filter paper and allow them to dry before determining the weight and melting point. Before leaving the laboratory, determine the theoretical yield based on the amount of starting material actually used and calculate the percent yield. Also record your partner's results in your laboratory notebook.

Cleanup

Put your product in an appropriate container. Neutralize the filtrate with 1 N sulfuric acid to destroy any remaining reducing agent. Put the aqueous filtrate in the appropriate liquid waste container. Dispose of all contaminated Pasteur pipettes, melting point and TLC capillaries, and TLC plates in the container for chemically contaminated laboratory debris.

CHEMICAL SAFETY NOTE   Sulfuric acid is corrosive. If any solution containing sulfuric acid comes in contact with your skin, flush the affected area immediately with water for at least five minutes and have another student notify the instructor immediately.

Discussion

- Discuss the fact that there is more than one possible stereoisomeric product from the reduction of (±)-benzoin.

- Discuss the reduction of benzil and identify the products that are possible from this reaction.

- Based on your and your partner's results, which product(s) was formed? If one product appeared to be favored, speculate as to why this was the case.

- Why did you need to make two mixed melting point determinations? Did these help you identify the major product? If so, explain how.

- Discuss and explain the similarities and differences between the reactions run in 2-propanol, methanol, and 95% ethanol.

Questions

1a. The product from your reaction has a sharp melting point, indicating that it is either *meso*-hydrobenzoin or a racemic mixture of the (1R, 2R) and (1S, 2S) enantiomers. Why is it not possible that your product is either the (1R, 2R) or the (1S, 2S) pure enantiomer?

1b. What is the enantiomeric excess (ee) of a racemic mixture of enantiomers?

1c. If the racemic mixture of the hydrobenzoins were formed in the reaction of benzil with sodium borohydride, describe a set of experimental conditions for which you might get a non-racemic mixture of these two products. You can change anything except the two starting materials.

2. Calculate the theoretical weight of sodium borohydride needed to reduce 50 mg of benzil to hydrobenzoin. Show your work.

3. Draw all possible products, including stereoisomers, for the reaction shown below.

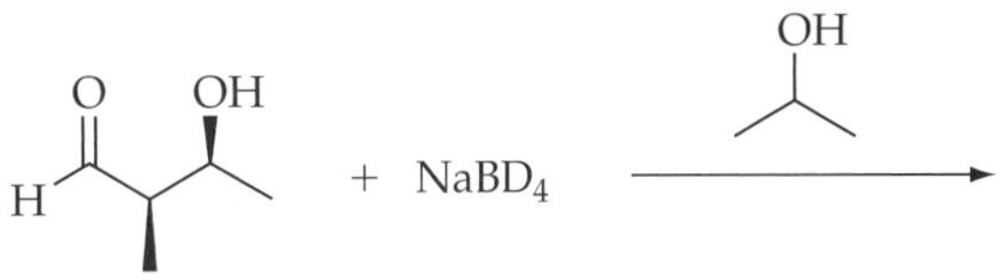

4. Explain why the ^{1}H and ^{13}C[^{1}H] NMR spectra of the meso and the racemic forms of 1,2-diphenyl-1,2-ethanediol are not informative in distinguishing these stereoisomers (Chapter 8).

5. A common strategy used to distinguish the diastereomeric meso and racemic forms of 1,2-diaryl-1,2-ethanediols is to prepare the acetonides and then examine the ^{1}H and ^{13}C[^{1}H] spectra of these compounds. An acetonide is a ketal derived from the reaction of a vicinal diol with acetone in the presence of an acid catalyst.

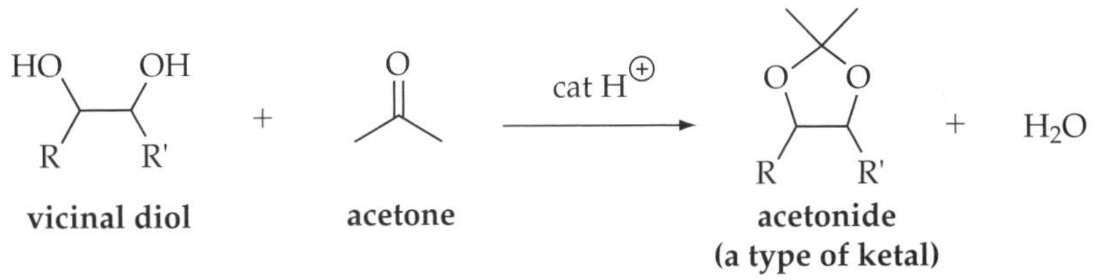

The spectral data for the acetonides of the two possible vicinal diols you prepared in this experiment are given below. Draw the structures of both acetonides. Match each set of ^{1}H and ^{13}C[^{1}H] data with an acetonide (Chapter 8). Explain the rationale behind each choice.

Acetonide 1: ^{1}H NMR (400 MHz, CDCl$_3$) δ 7.08–7.00 (m, 10H), 5.53 (s, 2H), 1.84 (s, 3H), 1.62 (s, 3H); ^{13}C [^{1}H] NMR (100 MHz, CDCl$_3$) δ 137.7, 127.5, 127.1, 126.9, 108.8, 85.5, 26.7, 24.5.

Acetonide 2: ^{1}H NMR (400 MHz, CDCl$_3$) δ 7.33–7.23 (m, 10H), 4.77 (s, 2H), 1.70 (s, 6H); ^{13}C[^{1}H] NMR (100 MHz, CDCl$_3$) δ 136.8, 128.4, 128.2, 126.7, 109.3, 85.4, 27.2.

The Grignard Reaction: The Preparation of 1,1-Diphenylethanol

The Grignard reaction is one of the most common methods used to form a carbon-carbon bond. The reaction is named after François Auguste Victor Grignard (1871–1935) who was a doctoral student in the laboratory of François Phillipe Antoine Barbier. In 1899, Barbier discovered that the reaction of magnesium with alkyl halides was superior to using zinc, which resulted in pyrophoric organometallic compounds. Barbier had reacted methyl iodide with 6-methyl-5-hepten-2-one in the presence of magnesium and isolated an alcohol after aqueous workup (Equation E14.1).

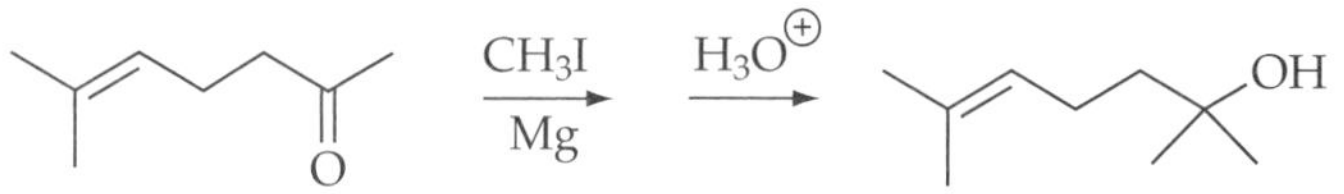

$$\text{(E14.1)}$$

Grignard altered the synthesis by first reacting the alkyl halide with the magnesium in ether and then treating this solution of the alkyl magnesium halide with other reactants. By 1901, Grignard had authored seven papers on these versatile reagents and was awarded the doctorate. During his academic career, he continued to focus on the alkyl magnesium compounds and their applications. In 1912, he was a co-recipient of the Nobel Prize[1] for his extensive investigations of their reactions. At the time of his death there were over 6000 references to Grignard reagents in the literature.

In this experiment, you will generate the Grignard reagent phenyl magnesium bromide from bromobenzene and magnesium. Reaction of this reagent with acetophenone followed by an aqueous acidic workup yields 1,1-diphenylethanol (Equation E14.2).

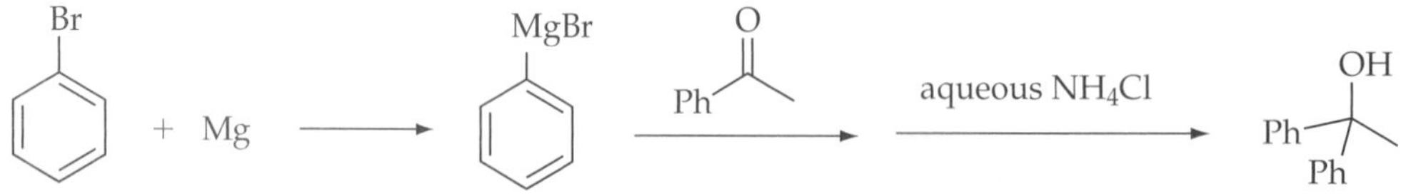

$$\text{(E14.2)}$$

---

[1] For additional information see: http://orac.sunderland.ac.uk/~hs0bcl/h_vg.htm and http://www.nobel.se/chemistry/laureates/1912/.

Did you know?

Acetophenone used to be called hypnone. In 1886, it was recommended as a drug to induce sleep.

Background Reading:

Before the laboratory period, read about the Grignard reaction in your lecture text.

Prelab Checklist:

The prelab write-up should include the balanced equation for the reaction and the physical properties of the starting materials and products. Calculate the weight of each starting material from the number of mmol specified (Section 4.2).

Experimental Procedure

Wear your safety glasses at all times in the lab.

Part I: Preparation of the Apparatus

TECHNIQUE TIP   Be very careful when removing the cover from the syringe needle. First put the needle on the syringe with the cover intact. Then hold the cover with it positioned down and remove it by exerting a slight downward motion while twisting the cover off. In this way you have less of a chance of sticking yourself.

SAFETY NOTE   Do not dispose of the syringes or needles unless they are clogged. If a needle is clogged, it must be discarded in the container for syringe needles. A clogged syringe should be put in the container for chemically contaminated laboratory debris. Never put a syringe needle in this container.

Place a magnetic stir bar in a 5–mL long-neck round-bottom flask and attach the air condenser (Figure 7.15). Flame dry the flask and air condenser using a microburner, being careful not to burn the connector. To minimize the amount of water vapor reentering the apparatus and provide pressure relief, attach a septum loosely to the top of the air condenser and insert a syringe needle in the septum. *Warning: Hot!* After the equipment is cool, secure the septum. Do not use any ether until the instructor has indicated that there are no flames in the lab. While the apparatus is cooling, go on to Part II.

TECHNIQUE TIP   When flame drying glassware, hold the apparatus using a clamp. Start heating the set-up at the bottom of the round-bottom flask, working your way up to the opening at the end of the air condenser. Rotate the set-up by 180° about the rod and repeat the process. Flame drying should not take longer than 30 seconds.

Part II: Preparation of the Grignard

Prepare an ice-water bath in a 50-mL beaker for use in the case of a run-away reaction.

EXPERIMENTAL NOTE   A run-away reaction is one that becomes extremely exothermic. This situation can lead to the contents of the reaction flask being ejected or, in the worst case, catching on fire.

The approximate amounts of the starting materials in mg should be in your prelab write-up. Check the calculations with your instructor before proceeding.

Weigh approximately 4 mmol of magnesium powder on a piece of weighing paper. Remove the air condenser from the round-bottom flask long enough to add the magnesium. Reattach the round-bottom flask to the air condenser. Based on the amount of magnesium you used, weigh an equimolar amount of bromobenzene in a small dry test tube labeled #1. Do not use a syringe to measure the amount of bromobenzene by volume.

Put 1 mL of anhydrous ether in a separate small dry test tube, label it #2, and cap it. Dilute the bromobenzene in test tube #1 with an additional 1 mL of anhydrous ether. Draw the solution into the syringe and expel it a couple of times to ensure that the contents are mixed. Use the syringe to transfer approximately 0.25 mL of the bromobenzene solution (test tube #1) to the reaction flask through the septum at the top of the air condenser. *Caution: Do not add any more than this amount at this time*. Keep test tube #1 capped.

Remove the air condenser from the reaction flask and grind the magnesium gently for less than one minute with a clean, dry glass stirring rod. Immediately replace the air condenser, making sure that the septum contains a needle for pressure relief. Keep the air condenser and septum in place to prevent moisture from entering the reaction flask.

SAFETY NOTE   Diethyl ether, the solvent for the reaction, is very volatile and flammable. It is heavier than air and creeps along bench tops and along the floor. No flames will be allowed in the lab once diethyl ether is being used. Ether absorbs moisture and oxygen from the air that interfere with the Grignard reaction and that can also lead to the formation of explosive peroxides.

SAFETY NOTE   The magnesium metal has an oxide coating that can be removed by *gently* grinding with a glass stirring rod to expose the clean metal surface. Great care must be exercised so that the stirring rod does not break the bottom of the round-bottom flask.

Stir the reaction mixture, looking for signs of initiation, often referred to as *catching*. The solution will become cloudy and eventually turn a reddish-brown color. Bubbles will be visible on the surface of the magnesium. The reaction is extremely exothermic so the solution will feel warm to the hand and the ether will begin to reflux.

Once the reaction has started, use the syringe to deliver the 1 mL of anhydrous ether from test tube #2 through the septum, keeping the pressure relief needle in place. Use the syringe to remove the remaining bromobenzene solution from test tube #1 and add it dropwise to the stirred reaction through the septum so that the reflux ring does not go higher than the tip of the syringe needle. Rinse test tube #1 with about 0.3 mL of anhydrous ether and use the syringe to deliver the rinse to the reaction flask. If refluxing becomes too vigorous (a run-away reaction), use an ice-water bath to control the reaction. If the reaction slows down, warm the flask in a beaker of warm tap water (45–50°C). Do not use a sand bath. Continue stirring and reflux the solution for an additional 10 minutes until most of the magnesium has disappeared. There may be bits of magnesium left, but these will not interfere with the subsequent steps.

The Grignard reagent cannot be stored. It must be reacted with the acetophenone during the same laboratory period.

Part III: Formation of the Alkoxide

Based on the amounts of magnesium and bromobenzene used, weigh an equimolar amount of acetophenone in a clean, dry reaction tube/small test tube. Do not use a syringe to estimate the amount of acetophenone by volume. Dilute this with 1 mL of anhydrous ether and mix the solution thoroughly. Remove the contents with the syringe and, while stirring the reaction, add the acetophenone solution dropwise to the Grignard reagent through the septum. Make sure that the pressure relief needle is in place. The rate of addition should be adjusted to maintain a gentle reflux. Use an ice-water bath to control a run-away reaction, that is, if reflux becomes too vigorous. After the addition is complete, reflux the reaction mixture for an additional 5–10 minutes using a beaker of warm tap water (45–50°C). The reaction mixture will be thick.

Part IV: Hydrolysis of the Alkoxide

Cool the reaction mixture in an ice-water bath. Take the septum off the air condenser and add about 1 mL of a 4.5 M aqueous ammonium chloride solution dropwise to the stirred reaction mixture. Hold a piece of moistened pH paper over the flask during the addition and record any changes in its appearance in your laboratory notebook. The mixture should begin to stir more easily and two layers will appear.

EXPERIMENTAL NOTE   Aqueous ammonium chloride is a weak acid used to neutralize the reaction mixture (Table 4.1). Employing stronger acids, such as sulfuric acid, promotes reactions of the desired product and lowers the yield.

TECHNIQUE TIP   If any clumps of solid remain attached to the walls of the round-bottom flask after addition of the ammonium chloride, try breaking these up by holding the flask directly on top of the stir plate. Rotate the flask so that the stir bar is directly on top of the clump and allow the stir bar to break it up.

Pour the reaction mixture into a centrifuge tube. Rinse the reaction flask with 1 mL of ether and add this to the centrifuge tube. Any sticky residue remaining in the reaction flask should not be transferred to the centrifuge tube. Cap the tube and shake it. Release any pressure buildup in the tube by slowly removing the cap. Replace the cap and shake the centrifuge tube again. Transfer the ether layer to a large test tube using a Pasteur pipette. Add another 1 mL of ether to the centrifuge tube, cap, and shake. Remove the ether layer and combine this with the previous ether extracts. Dry the ether extracts with anhydrous sodium sulfate (Section 7.7).

EXPERIMENTAL NOTE   About 1.5% by weight of water dissolves in diethyl ether and must be removed before the product is isolated (Section 6.1). The granular form of anhydrous sodium sulfate is preferred for this purpose because it can easily be removed by decanting the solution away or by filtering using the square-tip Pasteur pipette method. Calcium chloride is not an appropriate drying agent (Section 7.7).

TECHNIQUE TIP   It is important to keep the caps on the bottles of drying agents so they remain anhydrous. Please be a good citizen and always replace the tops on bottles of all chemicals used in the lab.

While the solution is drying, spot a TLC plate with a small amount of the ether solution of the product and solutions of acetophenone and bromobenzene in three separate lanes. Develop the plate in 90:10 hexane:ethyl acetate. Visualize the developed plate by UV light and iodine. Record the results in your notebook.

Save On Solvents: Share your TLC chamber with a neighbor.

Part V: Isolation of the Product

Cool about 5 mL of hexane in a large test tube in an ice-water bath.

Decant the ether solution from the sodium sulfate into a 25–mL Erlenmeyer flask. Wash the sodium sulfate with 0.5–1 mL of ether and add the washes to the flask. Remove the ether in the fume hood using a *warm* sand or water bath and a stream of air. The mixture should solidify to a yellow-white solid or an oily material.

TECHNIQUE TIP If you have difficulty removing the ether in the Erlenmeyer flask, the solution can be transferred to a small beaker to expedite evaporation.

Add approximately 2 mL of cold hexane to the solid or oily material and stir it with a spatula, breaking up any clumps. Filter the mixture on a Hirsch funnel using gravity filtration (Section 7.2.3 and Figure 7.8). Wash the solid with another 1 mL of cold hexane. Allow the white solid to dry and then weigh it. Recrystallize the solid from hot hexane and allow the solution to cool slowly to room temperature before cooling the solution in an ice-water bath. Pipette the mother liquor away from the crystals, rinse with cold hexane, dry, and weigh.

Calculate the percent yield and determine the melting point.

Dissolve some of the crystals in a small amount of ethyl acetate and spot this solution on a TLC plate along with a solution of acetophenone. Compare these results with the TLC plate of the crude ether extracts and the starting materials.

Cleanup

The product should be put in an appropriate container. It may be used in Experiment 20. All liquids should be disposed of in the appropriate waste containers. The aqueous layer from the Grignard reaction should be neutralized before it is placed in the appropriate container for liquid waste. The round-bottom flask can be cleaned with a small amount of 4.5 M ammonium chloride and the wash discarded in the appropriate liquid waste container. Pasteur pipettes, weigh paper, TLC and melting point capillaries, TLC plates, and the septa should be placed in the container for chemically contaminated laboratory debris. Clogged syringe needles must be disposed of in the container for syringe needle disposal. Under no circumstances should syringe needles be placed in the container for laboratory debris.

Discussion

- Analyze the ^{1}H and the ^{13}C[^{1}H] NMR spectra and the EI mass spectrum of the product (Chapter 8). Your instructor will let you know how or where to obtain the spectra.

- Write a rational arrow-pushing mechanism for the reaction of the Grignard reagent with acetophenone.

- Formation of Grignard reagents is best accomplished using clean, dry glassware and, under optimum conditions, an inert atmosphere such as nitrogen or argon gas. Explain why, being specific and using balanced equations.

- Dropwise addition of the acetophenone to the Grignard reagent is recommended. What did you observe that warrants this precaution? Explain.

- On the first TLC plate you developed, were there any other compounds present in the ether extracts in addition to the product? If so, comment on what these compounds could be and why they are present. If you don't know what these compounds are, make an informed guess based on their position on the TLC plate relative to the product and the starting materials (Chapter 5).

- Why was it necessary to wash the crude product with cold hexane before recrystallizing it from hexane?

Questions

1. What product would you obtain from the experiment if you used a strong acid such as sulfuric acid to protonate the alkoxide? Write a rational arrow-pushing mechanism for the formation of this product.

2. What product might you obtain from the experiment if you used hydrochloric acid in methanol to protonate the alkoxide? Write a rational arrow-pushing mechanism for the formation of this product.

3. A student has a dirty syringe and rinses it with acetone just before using it to transfer the ether to the Grignard that has just been initiated. What compound will contaminate the product?

4. Propose three different syntheses of the alcohol shown below starting with a ketone and a Grignard reagent. In all cases, first show how you would make the Grignard reagent.

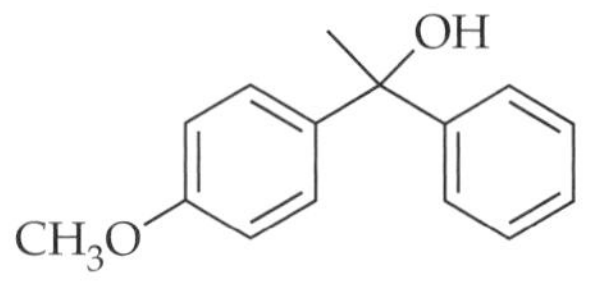

5. A student proposed the synthesis shown in Equation PE 14.5.

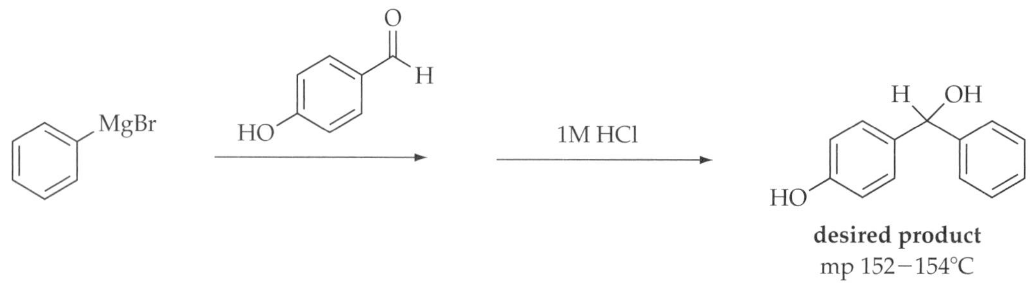

(PE 14.5)

desired product
mp 152−154°C

After extraction and recrystallization, a white, flaky, crystalline product was obtained that had a melting point of 117–119°C. The desired product, shown in Equation PE14.5, has a melting point of 152–154°C. Write the structure of the compound that the student isolated and briefly explain what happened. Write a rational arrow-pushing mechanism for the formation of this product.

Hint: If you were actually doing this reaction, you would have prepared a prelab write-up that included all of the physical constants for your starting materials.

The Asymmetric Dihydroxylation of trans-Stilbene [1]

One of the most rapidly growing and economically important areas of organic chemistry is enantioselective catalysis. It is desirable to perform many of the reactions you have learned about in organic chemistry lecture in a manner that leads primarily or exclusively to one of two possible enantiomeric products. For example, if a standard homogenous hydrogenation catalyst such as Wilkinson's catalyst, $RhCl(PPh_3)_3$, is used to add hydrogen across the double bond of the *N*-acylated unsaturated α-amino acid, a racemic mixture of *N*-acylated phenylalanine is obtained (Equation E15.1).

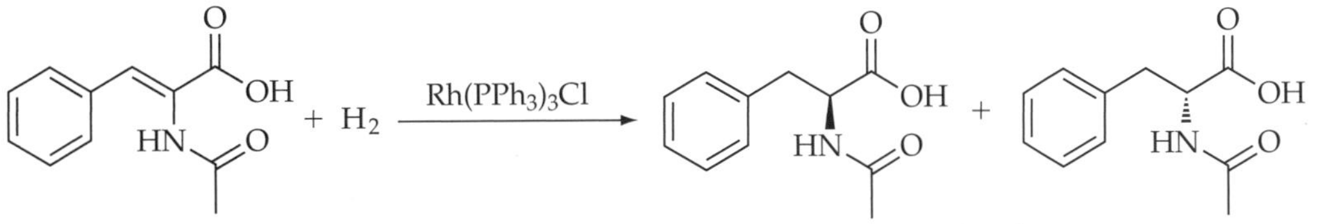

$$\text{(E15.1)}$$

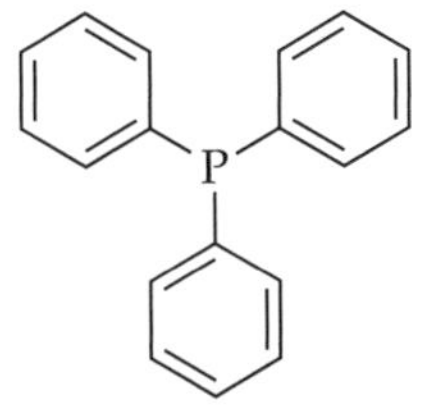

racemic mixture

The two hydrogen atoms add to the same face of the alkene and there is an equal chance of addition to either face. The transition states leading to the enantiomers are enantiomeric and, by definition, they must be equal in energy. To favor the addition of the hydrogen atoms to only one face, the two transition states must be different in energy. In other words, the transition states need to be diastereomeric. This has been accomplished by using chiral phosphine ligands in place of the achiral triphenylphosphine. Because the alkene attaches to the rhodium catalyst during the hydrogenation reaction, the complexes resulting from addition to the two alkene faces are diastereomeric. If the energies between these two diastereomeric complexes are sufficiently different, one of the transition states is preferred, leading predominantly to one of the enantiomers.

triphenylphosphine

You will perform another type of alkene addition called a dihydroxylation using osmium(VIII) tetroxide (OsO_4) as a catalyst for the reaction. Two of the oxygen atoms of the osmium species add to the same side of the

---

[1] Adapted from Wang, Zhi-Min and Sharpless, K.Barry *J.Org.Chem.* **1994** *59* 8302.

76 190.2

Os

2.2 $6s^25d^6$

Osmium

alkene to give a racemic osmium(VI) diolate product (Equation E15.2). This complex is hydrolyzed by water to give the racemic diol and an osmium(VI) intermediate that is reoxidized to OsO_4 using a variety of oxidants, including 4-methylmorpholine *N*-oxide (Equation E15.2).

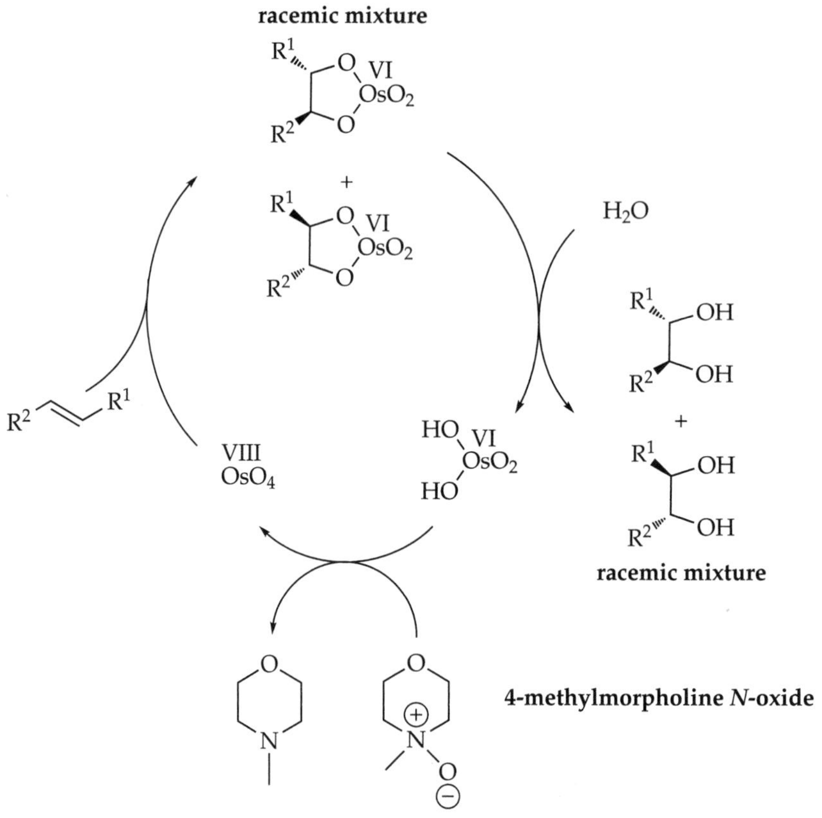

(E15.2)

An enantioselective version of this catalytic reaction was developed by Professor K. Barry Sharpless[2] (Scripps Research Institute) using chiral amine ligands that bind to the OsO_4. These ligands are derived from a naturally occurring class of amines known as the cinchona alkaloids. The most famous member of this class of compounds is quinine, which saved much of the world from malaria in the 19th century. Analogous to the hydrogenation reaction using the rhodium catalyst, by attaching a chiral amine to the osmium, the transition states for the approach of the osmium to the alkene become diastereomeric. Because the transition states have different energies, it is possible to obtain different amounts of the two enantiomeric diol products.

---

[2] K. Barry Sharpless won the Nobel Prize in Chemistry for 2001 for his work in asymmetric synthesis using chiral reagents to catalyze oxidation reactions. See http://www.nobel.se/chemistry/laureates/2001/.

Figure E15.1   The Structures of (DHQD)₂PHAL and (DHQ)₂PHAL

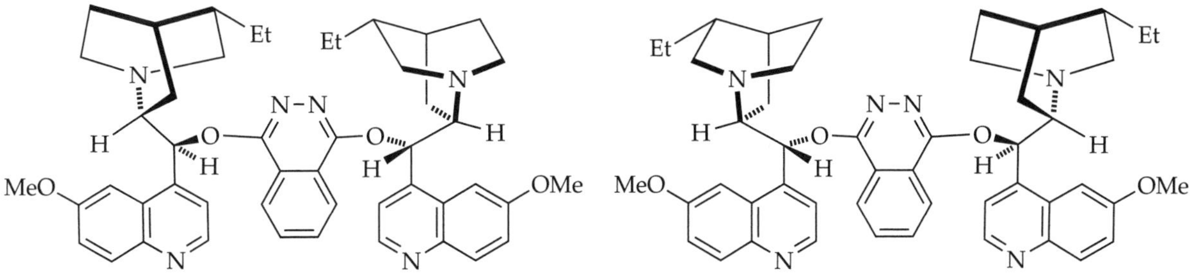

(DHQD)₂PHAL   (DHQ)₂PHAL

Did you know?

The gland secretions that result in fingerprints are unsaturated organic compounds. Because osmium tetroxide oxidizes double bonds, the fingerprints turn dark gray-black when exposed to it. It is also used for staining fatty tissue on microscope slides where it cross-links the lipids that contain double bonds both staining and fixing them.

Background Reading:

Before the laboratory period, read about the dihydroxylation of alkenes in your lecture text.

Prelab Checklist:

The structures of (DHQD)₂PHAL and (DHQ)₂PHAL are shown in Figure E15.1. When the (DHQD)₂PHAL ligand is used in the reaction, the product is (R,R)-hydrobenzoin. The (DHQ)₂PHAL ligand gives the enantiomeric hydrobenzoin. The reason for the effectiveness of these ligands is not clearly understood.

You will be using either (DHQD)₂PHAL or (DHQ)₂PHAL, whichever is available in the lab. In the prelab write up, include the equation for the reaction of the alkene with osmium tetroxide and the mmol amounts for all reactants. Write the structure of one of the enantiomers of the product in a Fischer projection and specify the stereochemistry of the enantiomer you have drawn. The physical properties of the starting materials, solvents, and product and the specific rotation of the expected product should be included in your prelab write-up.

Experimental Procedure

Wear your safety glasses at all times in the lab.

Part I: The Dihydroxylation Reaction

Each student should do the experiment individually.

Weigh 350 mg of *trans*-stilbene in a 5–mL long-neck round-bottom flask. Using the 5–mL syringe provided, add 2.0 mL of the acetone stock solution of either the (DHQD)$_2$PHAL or (DHQ)$_2$PHAL ligand (concentration 2 mg/mL). Add 0.6 mL of the aqueous 4-methylmorpholine *N*-oxide (NMO) stock solution (492 mg/mL) in the fume hood using the 1 mL syringe provided. Be careful to expel any air bubbles from the syringe so that an accurate amount of the NMO is used in the reaction. Do not remove the reagents or the syringes from the fume hood. Clamp the reaction flask about 2–3 cm above the stir plate. Introduce a stir bar and stir the heterogeneous solution. Your instructor will add approximately 0.04 mL (40μL) of 2.5 wt% OsO$_4$ in *tert*-butyl alcohol to the reaction mixture.

> **CHEMICAL SAFETY NOTE**   Osmium tetroxide is a catalyst in the reaction; therefore the exact amount used is not critical except for the fact that it affects the rate of the reaction. The catalyst is very expensive and it is important that it not be wasted. Care should be exercised in handling osmium compounds. Osmium(VIII) tetroxide is a volatile solid and is toxic, particularly to the eyes and respiratory tract; however the volatility of this material is reduced significantly by dissolving it in *tert*-butyl alcohol.

Put a cap loosely in the neck of the round-bottom flask and continue to stir, making sure that the reaction flask is about 2–3 cm above the stir plate. Remember to record any physical changes. Put a few drops of acetone in a small test tube. Dip an open-end capillary tube that is attached to a boiling stick in the reaction mixture to obtain a drop of the solution and put the capillary tube in the small test tube containing acetone. Spot this solution on a TLC plate along with an acetone solution of *trans*-stilbene. Develop the plate in 60:40 hexane:ethyl acetate. Use UV light and iodine to visualize the spots.

Save On Solvents: Share your TLC chamber with a neighbor.

Use new open-end capillary tubes that are attached to boiling sticks to take samples of the reaction solution every 30 minutes. Place the capillary tube with a small amount of the reaction solution in a clean small test tube and dilute the reaction mixture with a few drops of acetone. Spot the dilute acetone solution on a TLC plate.

After the TLC indicates that all the stilbene has reacted or after the reaction has run for 90 minutes, add 0.2 mL of the aqueous stock solution of 4,5-dihydroxy-1,3-benzenedisulfonic acid, disodium salt monohydrate (Tiron®) (concentration 17.5 mg/mL). Stir the reaction for 5 minutes.

EXPERIMENTAL NOTE   The Tiron® reacts with all of the osmium in solution providing a water soluble, non-volatile complex that is easily separated from the product.

Add 2.5 mL of water dropwise with continued stirring. Isolate the product by gravity filtration on the Hirsch funnel, using a 10–mL Erlenmeyer flask to collect the mother liquor. Pour the contents of the Erlenmeyer (the mother liquor) into the appropriate waste container in the fume hood immediately after filtration. The mother liquor contains the osmium complexed by the Tiron® and a tertiary amine, which gives it a fishy odor.

Fit the Hirsch funnel to a filter flask and use vacuum filtration to rinse the product two or three times using 0.6 mL of water each time.

Recrystallize the crude product from the minimum volume of hot 95% ethanol in a 10–mL Erlenmeyer flask. Collect the product on the Hirsch funnel and rinse it twice with 0.5 mL of ice-cold ethanol.

EXPERIMENTAL NOTE   If the solution is colored, consult with your instructor. It may be necessary to add a small amount of activated charcoal (Section 7.1.4) and a boiling stick and to heat the solution to boiling on the sand bath. The charcoal can be removed by gravity filtration using the Hirsch funnel that has been fitted with a glass-fiber filter.

Transfer the product to a piece of filter paper and blot it dry. Determine the weight, calculate the yield, and assess the purity of the product by TLC. Measure the melting point of the dry recrystallized product.

Part II: Polarimetry of the Product

Your instructor will demonstrate the operation of the polarimeter (Experiment 9) and indicate the concentration of material necessary to obtain a reasonable rotation on the polarimeter. You may have to combine your product with that of other students to achieve the required concentration.

Record the weight of the material used and dissolve it in 95% ethanol. Zero the polarimeter using 95% ethanol. Transfer the combined ethanol solution of the product to the polarimeter cell carefully. Take five readings of the rotation of the sample and average them. Calculate the specific rotation from the average rotation, the concentration used, and the length of the cell.

Cleanup

Discard all solutions in the appropriate waste containers. Dispose of all contaminated Pasteur pipettes, TLC and melting-point capillaries, and TLC plates in the container for contaminated laboratory debris.

Discussion

* The dihydroxylation reaction you performed was carried out in a solvent mixture. Determine the approximate composition of the mixture and discuss the possible reasons for using a solvent mixture.

* Tiron® is used to remove the osmium in this reaction by turning it into a water-soluble complex. Propose a structure for the osmium-Tiron® complex and discuss why it is important to make this complex water soluble.

* Discuss the results of your polarimetry measurements. Based on the specific rotation of your product, which enantiomeric hydrobenzoin did you synthesize? Calculate the enantiomeric excess (ee) of the

product. What are the inherent errors associated with polarimetry measurements?

Questions

1. Based on your experimental results, draw the major enantiomer expected from the following two reactions.

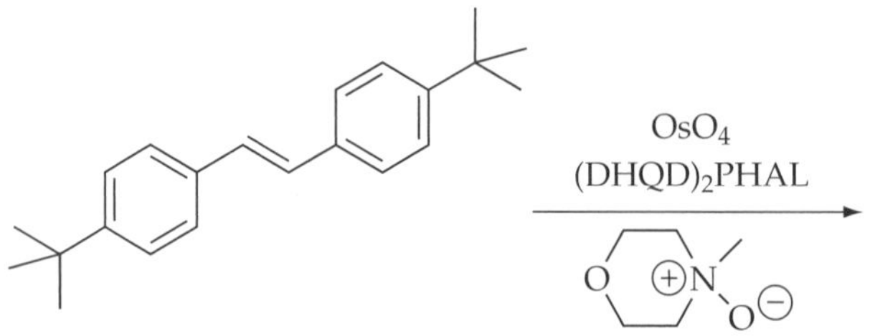

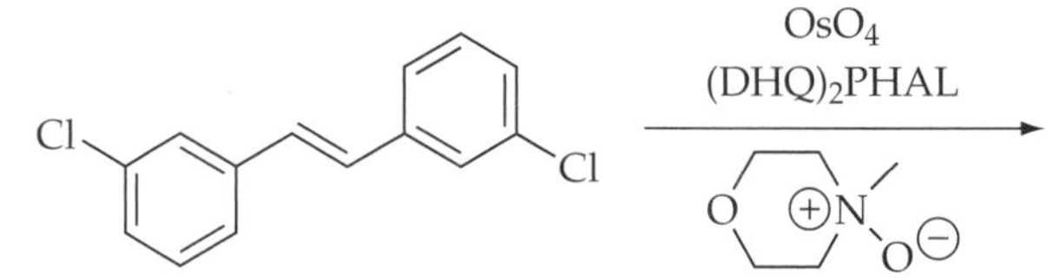

2. When a student ran an asymmetric dihydroxylation reaction using *cis*-stilbene and the ligand (DHQ)$_2$PHAL, the hydrobenzoin showed an optical rotation of 0°. Explain this result.

3. Are the ligands (DHQ)$_2$PHAL and (DHQD)$_2$PHAL enantiomers? Explain.

4. Search the recent chemical literature for an example in which an osmium-catalyzed dihydroxylation was used in the synthesis of an organic compound (Chapter 3). Briefly discuss the importance of dihydroxylation in accomplishing the synthetic goal of the article. Show the structure of the final product and indicate where the oxygen atoms from dihydroxylation were introduced into the product. Include the title of the article, the author(s), the journal name, the publication year, the volume, and the page number(s).

Some Chemistry of α-Pinene Oxide

α-Pinene (Figure E16.1) is one of the most ubiquitous monoterpenes. It is found in the essential oils of a variety of plants and is the major constituent of oil of turpentine, which is obtained from various species of *Pinus* (pine trees).

Figure E16.1: α-Pinene

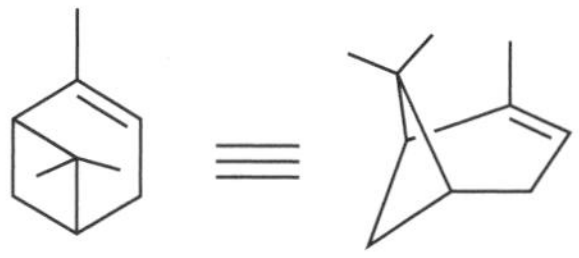

α-pinene

Did you know?

The α-pinene isolated from oil of turpentine from North American pines is the 1*R*,5*R*-isomer, which is dextrorotatory, whereas that from most European pines is the levorotatory 1*S*,5*S* isomer.

The ring strain of the bicyclic skeleton that contains a four-membered ring plays a significant role in the reaction chemistry of α-pinene. For example, interaction of α-pinene with dry hydrogen chloride gas gives the bicyclic chloride shown in Equation E16.1. Conversion of α-pinene to limonene, another common monoterpene, can be accomplished using clay, a strong acid catalyst (Equation E16.2). These unusual reactions take place by way of the formation of carbocations and are often referred to as Wagner-Meerwein rearrangements. Carboxylic acids are weak acids and do not normally promote such rearrangements.

(E16.2) 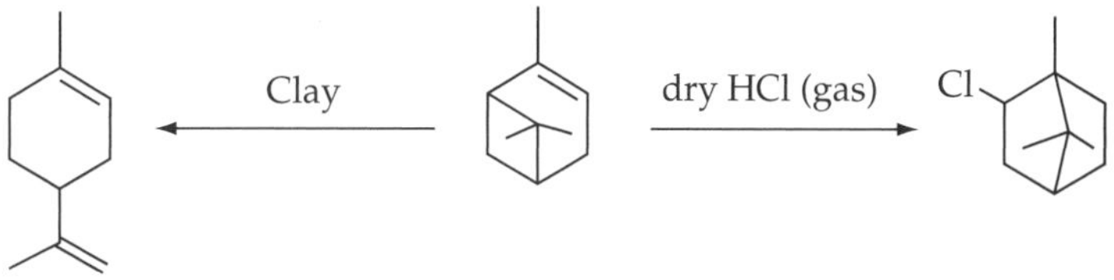 (E16.1)

limonene

Under non-acidic conditions, the double bond of α-pinene reacts with reagents in a more predictable way. For example, epoxidation with *meta*-chloroperbenzoic acid (MCPBA) yields the epoxide known as α-pinene oxide, which is the starting material in this experiment (Equation E16.3). Even though the epoxide is formed with no surprises, this does not mean that the α-pinene oxide does not undergo some rather bizarre chemistry.

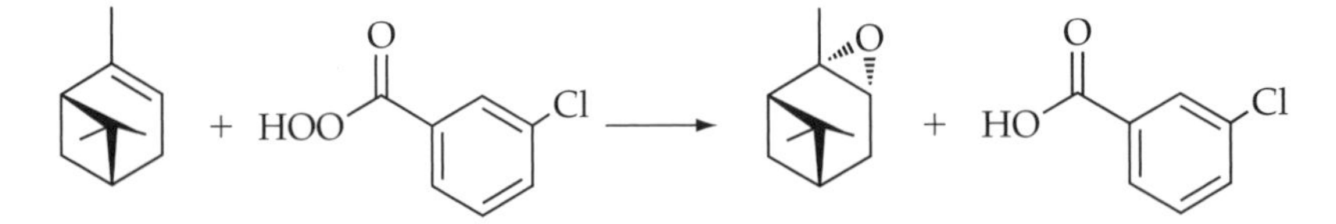

$$(\pm)\text{-}\alpha\text{-pinene} \quad\quad \textbf{MCPBA} \quad\quad\quad (\pm)\text{-}\alpha\text{-pinene oxide} \quad\quad\quad (E16.3)$$

The experimental procedure you will follow in this experiment will not always point out the exact physical characteristics of the reaction mixture and products so it is important to record complete and careful observations in your laboratory notebook.

Background Reading:

Before the laboratory period, read about carbocation rearrangements in your lecture text.

Prelab Checklist:

The prelab write-up should include the structures and physical constants for the starting materials and any solvents used in this experiment.

Experimental Procedure

Don't forget to keep your safety glasses on during the entire laboratory period.

Place a stir bar in a 5–mL short-neck, round-bottom flask. Deliver 0.8 mL of 0.1 M H_2SO_4 directly into the flask. Add 1 mL of hexane and start stirring this mixture vigorously. Using a syringe, add 0.25 mL of α-pinene oxide dropwise over a two-minute period.

EXPERIMENTAL NOTE   Rapid addition of the α-pinene oxide will result in low or no yield of product.

CHEMICAL SAFETY NOTE   Sulfuric acid is corrosive. If any solution containing sulfuric acid comes in contact with your skin, immediately flush the affected area with water for at least five minutes and have another student notify the instructor.

After 20 minutes, filter the reaction mixture on a Hirsch funnel and wash the resulting solid twice with 1–mL portions of hexane. Spot the hexane filtrate on a TLC plate, leaving enough room on the plate for three more spots. Save the hexane filtrate until the end of the laboratory period in case something happens to your TLC plate.

Wash the solid in the Hirsch funnel twice with 1–mL portions of water. Briefly dry the solid and determine the weight of the crude product.

Recrystallize the crude solid from a minimum amount of ethyl acetate, starting with 0.5–1 mL. Isolate the product and wash it with 0.5 mL of cold ethyl acetate. Determine the weight and melting point of the product. Dissolve a small amount of the product in ethyl acetate and spot this solution on the TLC plate along with the mother liquor from the recrystallization. Also spot the plate with a solution of α-pinene oxide in hexane. Develop the plate in 90:10 hexane:ethyl acetate. Visualize the plate by UV light and iodine.

Save On Solvents: Share your TLC solvent chamber with a neighbor.

TECHNIQUE TIP　It is important to consider data from both visualization techniques before drawing conclusions about the presence or absence of compounds.

EXPERIMENTAL NOTE　The α-pinene oxide starting material is pure and is stable in hexane and in ethyl acetate.

Before continuing, present your instructor with the following data:

* The weight of recrystallized product.
* The TLC plate you developed.

Your instructor will ask you the following about the TLC results.

* Is your crystalline product more or less polar than the starting material?
* Explain your observations for the TLC of α-pinene oxide.

After this discussion, run another TLC of α-pinene oxide and visualize the plate using a solution of dinitrophenylhydrazine (DNPH) (Section 7.8.3.d and the *Procedure for Students Using DNPH* in Part II of Experiment 8). After performing this task, present your instructor with the results and explain what you think is occurring.

EXPERIMENTAL NOTE　Before dipping the plate into the DNPH solution, examine it with the UV lamp. Outline any spots that are visible and sketch the plate in your laboratory notebook before continuing. Do not visualize the plate with iodine.

CHEMICAL SAFETY NOTE　Do not touch the plate after it has been dipped in the DNPH solution. The reagent solution is made with 85% phosphoric acid, an extremely corrosive material. Do not take the plates to your laboratory bench or place them in your laboratory notebook. Sketch the plates as best you can while they are in the fume hood.

Cleanup

Place your recrystallized product in an appropriate container. Place any remaining solvents, including TLC chamber solvents, in the appropriate liquid waste containers. Dispose of the contaminated Pasteur pipettes, used weigh paper, TLC plates (except those visualized with the dinitrophenylhydrazine solution), and melting-point and TLC capillaries in the container for contaminated laboratory debris. Dispose of the plates used with the dinitrophenylhydrazine solution in the specially labeled container in the fume hood. Clogged syringe needles must be placed in the container for syringe needles.

Discussion

- Analyze the ^{1}H and ^{13}C[^{1}H] NMR spectra of α-pinene oxide, the solid product isolated from the reaction of α-pinene oxide with aqueous sulfuric acid, and the decomposition product that you observed when analyzing α-pinene oxide by thin layer chromatography (Chapter 8). Your instructor will let you know how or where to obtain the spectra.

  Analyze the EI mass spectrum to determine if the molecular ion and any major fragment ions help you to determine or confirm the structure of your product.

- Discuss your strategy for determining the structure of the solid you isolated from the reaction of α-pinene oxide with aqueous sulfuric acid. Was it necessary to analyze the mass spectrum of the product in order to establish the identity of the compound? Explain your answer.

- Write a balanced equation for the formation of the product and calculate the percent yield.

- Write a rational arrow-pushing mechanism for the formation of this product.

- Discuss your strategy for determining the structure of the decomposition product you observed when analyzing α-pinene oxide by thin layer chromatography.

- Write a balanced equation for the formation of the decomposition product.

- Write a rational arrow-pushing mechanism for the formation of the decomposition product.

- Discuss the two TLC plates you developed. What did each lane in the first TLC plate tell you about the overall reaction?

Questions

1. Draw the two enantiomers of α-pinene (Figure E16.1) and assign the stereocenter(s) for each isomer as *R* or *S*.

2. Write arrow-pushing mechanisms for the reactions shown in Equations E16.1 and E16.2.

3. Write an arrow-pushing mechanism for the reaction shown in Equation E16.3.

4. Write rational arrow-pushing mechanisms leading to three additional products that might have been formed in the reaction of α-pinene oxide with aqueous sulfuric acid that are different from the product you isolated and the decomposition product you identified.

5. Identify the product from the reaction shown below[1]. The ^{1}H NMR data for the product is: (CDCl$_3$) δ 0.65 (s, 3H), 1.28 (s, 3H), 1.63–2.55 (m, 6H), 4.42 (d, 1H), 4.82 (m, 1H), 5.00 (m, 1H) (Chapter 8). Note that hydroxyl hydrogens are sometimes not visible in ^{1}H NMR spectra (Section 8.1.4.b).

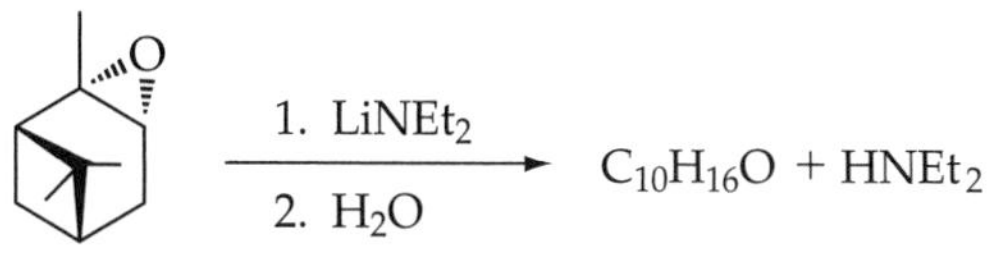

$$\xrightarrow[\text{2. H}_2\text{O}]{\text{1. LiNEt}_2} \quad \text{C}_{10}\text{H}_{16}\text{O} + \text{HNEt}_2$$

---

[1] Crandall, J.K. and Crawley, L.C. *Org. Syn.* **1988** *Coll. Vol. 6* 948.

A Dehydrogenation/ Hydrogenation Reaction

The hydrogenation of alkenes with hydrogen gas and a metal catalyst is a synthetic tool familiar to all students who study organic chemistry. In addition, the energy released upon hydrogenating a carbon-carbon double bond, the heat of hydrogenation, is sometimes used to determine the relative stability of isomeric alkenes. In many cases, these reactions require the use of specialized equipment capable of containing hydrogen gas at elevated temperatures and pressures. Large reactors are commonly used in industry and smaller versions are often available in academic research laboratories. Some hydrogenation reactions can be carried out at room temperature and atmospheric pressure in a simple set-up consisting of a balloon filled with hydrogen gas attached to the reaction vessel. However, special safety precautions must be observed in either case due to both the flammability of hydrogen gas and the fact that the pressurized systems can fail, leading to explosive release of their contents.

An alternate method of hydrogenating carbon-carbon double bonds relies on the fact that all reactions are reversible if one can determine the necessary conditions to drive the equilibrium in the desired direction. With a suitable choice of reactants, a transfer-hydrogenation reaction can be accomplished where one molecule is dehydrogenated and donates hydrogen to another molecule that is consequently hydrogenated. For example, if cyclohexene is heated in the presence of palladium on carbon (Pd/C), a mixture of cyclohexane and benzene is obtained (Equation E17.1). In the presence of the catalyst, two equivalents of hydrogen are lost from cyclohexene to form benzene. The driving force for this reaction is the stability of the aromatic product. The two equivalents of hydrogen then react with two equivalents of the cyclohexene to form the more stable saturated compound, cyclohexane.

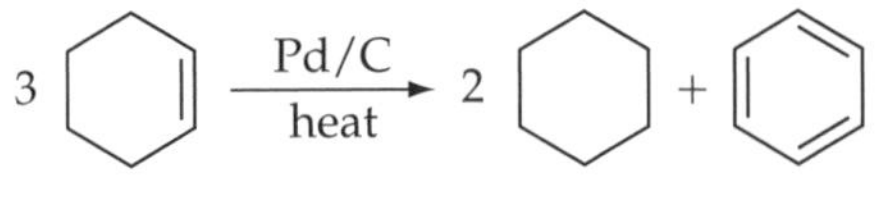

$$3 \; \bigcirc \!\!\!\!= \xrightarrow[\text{heat}]{\text{Pd/C}} 2 \; \bigcirc + \bigcirc \qquad \text{(E17.1)}$$

Transfer-hydrogenation reactions are very useful in two areas of synthesis: 1) in hydrogenation, as a method of generating stoichiometric amounts of hydrogen for transfer to another molecule to obtain a saturated molecule, and 2) in dehydrogenation, to remove hydrogen from a molecule to obtain aromatic compounds with defined substitution patterns.

In the first case, molecules that are easily coaxed into giving up one or more molecules of H_2 from their skeleton to form stable unsaturated compounds are used as sources of hydrogen for the hydrogenation reactions. Formic acid is one example. In the presence of a catalyst such as Pd/C, this carboxylic acid decomposes to give H_2 and CO_2. The reaction is rendered irreversible because the CO_2 bubbles out of the reaction solution. Salts of formic acid, such as ammonium formate, have also been used because both the ammonia and carbon dioxide products are gases (Equation E17.2 and Experiment 36).

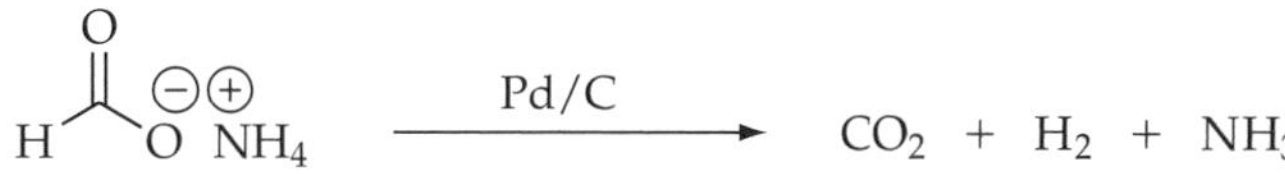

$$
\underset{\substack{\text{ammonium} \\ \text{formate}}}{\text{H}\overset{\overset{\text{O}}{\|}}{\text{C}}\text{O}^{\ominus}\text{NH}_4^{\oplus}} \xrightarrow{\text{Pd/C}} CO_2 \;+\; H_2 \;+\; NH_3 \qquad \text{(E17.2)}
$$

If the compound being hydrogenated has functional groups that are sensitive to acids or ammonia, there are several cyclic hydrocarbons that can serve as hydrogen donors in the presence of a metal catalyst because they form stable aromatic compounds upon dehydrogenation. In the example given in Equation E17.1, the cyclohexene gives up two equivalents of hydrogen to yield benzene. The hydrogen is transferred to cyclohexene to furnish two equivalents of cyclohexane; however a different unsaturated compound can serve as the hydrogen acceptor as shown in Equation E17.3.

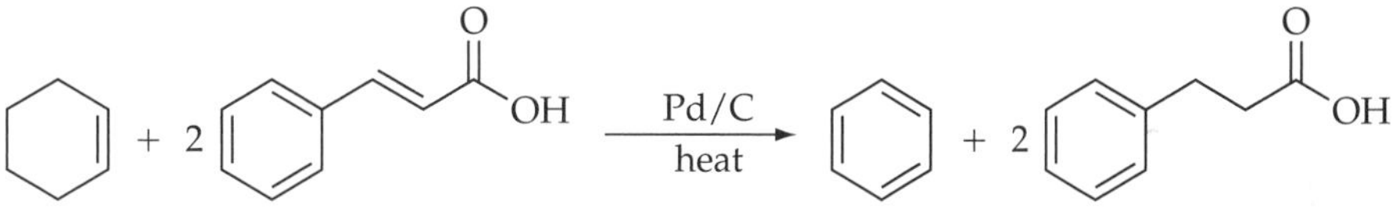

$$
\bigcirc\!\!\!\| \;+\; 2\; \text{(cinnamic acid)} \xrightarrow[\text{heat}]{\text{Pd/C}} \bigcirc \;+\; 2\; \text{(hydrocinnamic acid)} \qquad \text{(E17.3)}
$$

Another useful source of hydrogen is 9,10-dihydroanthracene, which provides one equivalent of hydrogen and yields anthracene (Equation E17.4).

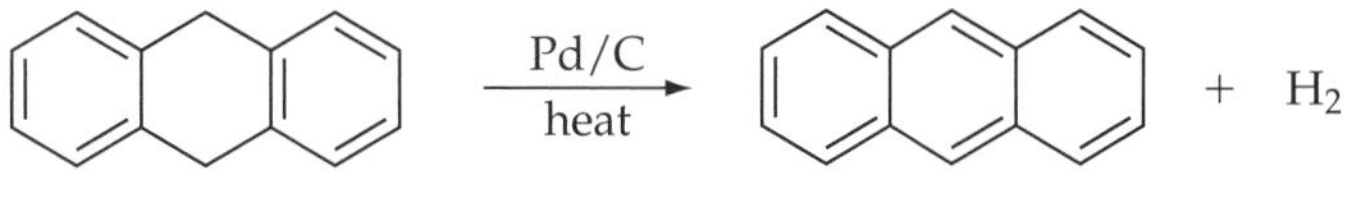

$$
\text{(9,10-dihydroanthracene)} \xrightarrow[\text{heat}]{\text{Pd/C}} \text{(anthracene)} \;+\; H_2 \qquad \text{(E17.4)}
$$

The main criteria to consider in choosing transfer-hydrogenation substrates other than formic acid and its salts is that they be cyclic and capable of forming aromatic molecules upon the loss of hydrogens from at least two sp^3 hybridized carbons in the ring.

Are these reactions limited to hydrocarbon starting materials? For example, can a cyclohexanone be used to obtain a phenol? It is easier to see the analogy by starting with the enol tautomer as shown in Equation E17.5.

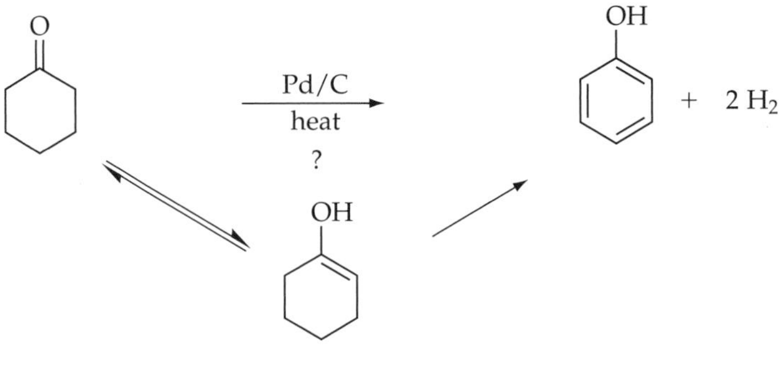

$$\text{(E17.5)}$$

In the experiment, you will explore this question by examining the reaction of the naturally occurring monoterpene ketone, carvone with Pd/C and heat (Equation E17.6).

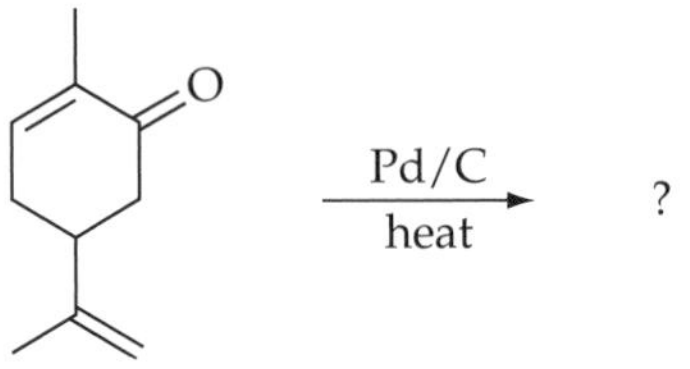

$$\xrightarrow[\text{heat}]{\text{Pd/C}} \quad ? \qquad \text{(E17.6)}$$

Did you know?

Our olfactory receptors contain chiral groups and are able to differentiate between the enantiomers of some compounds. In the case of carvone, the *S*-(+) compound smells like caraway and is the principle component of caraway seeds and dill-seed oil. The *R*-(−) isomer is found in spearmint oil and has a minty odor. A study done in 1971 found that about 8 to 10% of the population exhibited carvone *anosmia* or carvone odor blindness, that is, they were unable to smell carvone. This effect was greater for the *R*-(−) compound than for its enantiomer.

Russell, G.F. and Hills, J.I. *Science* **1971** *172* 1043.
Friedman, L. and Miller, J.G. *Science* **1971** *172* 1044.

Background Reading:

Before the laboratory period, read about aromaticity, keto-enol equilibria, and hydrogenation in your lecture text.

Prelab Checklist:

Include the structures and physical constants for all reactants and solvents used in this experiment in the prelab write-up.

Experimental Procedure

Don't forget to keep your safety glasses on during the entire laboratory period.

Add 26 mg of 10% Pd/C to a reaction tube/small test tube followed by 0.9 mmol of (*R*)-carvone delivered by volume from a syringe. Cap the reaction tube/small test tube with a septum and insert a needle into the septum. Heat the reaction to 180°C on a sand bath.

SAFETY NOTE   Palladium on carbon can be pyrophoric when exposed to air, especially when it is dry. The weigh paper used for the Pd/C should be put in the common container of water for Pd/C contaminated materials that is located in the fume hood. Also, make sure that you put the top back on the container of Pd/C.

EXPERIMENTAL TIP   Monitor the temperature of the sand bath frequently by placing the digital thermometer probe at the same depth as the bottom of the reaction tube/small test tube. When the temperature is close to 180°C, turn the power down a little.

SAFETY NOTE   Reaction temperatures above 200°C can lead to ignition of any hydrogen generated in this reaction.

After the reaction has been at 180°C for 15 minutes, remove the septum briefly and use a TLC capillary attached to a boiling stick to remove a small drop of the reaction mixture. Because no solvent is used in this reaction, the reaction sample must be diluted. Dip the TLC capillary containing the reaction mixture up and down a few times in 0.2 mL of ethyl acetate in a small test tube. Spot this mixture in lanes 2 and 3 of a TLC plate with three lanes. Spot a solution of carvone in lanes 1 and 2 (Section 7.8.3.f). Elute the plate with 95:5 hexane:ethyl acetate, and visualize the plate with UV light and iodine. Using this procedure, monitor the reaction every fifteen minutes until the reaction is complete or has proceeded for 1 hour, whichever comes first.

TECHNIQUE TIP   In order to see the difference between the R_f of the starting material and that of the product, the spots must be small. This means the solutions must be applied sparingly. For example, lightly spot each solution twice, allowing the solvent to evaporate between dabs. If no difference in the R_f values for the starting material and product can be seen, dilute the TLC reaction mixture even more or spot it less. Cospot lane 2 with both the starting material and the reaction mixture (Section 7.8.3.f).

TECHNIQUE TIP   If the product and starting material are not clearly resolved in the TLC of the reaction mixture, place the developed plate back into the TLC chamber and develop it a second time. Although this technique is not often used, it is analogous to spotting the starting material and product at two slightly different starting points on the plate; consequently, spots with similar R_f values can sometimes be separated.

Save On Solvents: Share your TLC solvent chamber with a neighbor.

Remove the reaction tube/small test tube from the sand bath and allow it to cool to room temperature. After allowing the mixture to cool to room temperature, dilute the reaction mixture with 1 mL of hexane. Filter the mixture into a centrifuge tube using the glass-fiber Pasteur pipette filtration method (Section 7.2.2). Apply gentle pressure to the pipette bulb to help force the solution through the filter paper. Rinse the reaction tube/small test tube with an additional 0.5 mL of hexane, and pass this mixture through the pipette filter as well.

TECHNIQUE TIP If all of the reaction mixture does not go through the filter after completely depressing the pipette bulb, equalize the pressure in the pipette by first removing the bulb from the pipette before releasing your grip on the bulb (Section 7.2.2).

SAFETY NOTE Palladium on carbon can be pyrophoric when exposed to air, especially when it is dry. Immediately after use, the pipettes containing the Pd/C should be submerged in water in a common container located in the fume hood.

Add 1 mL of 2N (10%) aqueous sodium hydroxide to the hexane solution, cap the centrifuge tube, and shake well. Allow the organic and aqueous layers to separate. Centrifugation should not be necessary. Remove the aqueous layer with a pipette and put it in a second centrifuge tube. Add another 1 mL of 2N sodium hydroxide to the hexane layer and repeat the above procedure. Combine the aqueous layers and check the pH of the aqueous solution.

CHEMICAL SAFETY NOTE Sodium hydroxide is corrosive. If any solution containing sodium hydroxide comes in contact with your skin, immediately flush the affected area with water for at least 5 minutes. Sodium hydroxide is particularly harmful to the eyes. If any solution contacts your eyes, proceed immediately to the eyewash station and flush your eyes for at least 5 minutes. In either case, have another student notify the instructor.

Acidify the combined aqueous layers by slowly adding 1 mL of 6N aqueous hydrochloric acid. Shake the tube gently from side to side while adding the hydrochloric acid to ensure rapid mixing. Check the pH of the solution after the addition is complete. Add 1 mL of diethyl ether to this solution, cap the tube and shake it to mix the layers. Allow the two layers to separate. If necessary, centrifuge the mixture. Remove the ether layer and place it in a large test tube. Repeat this process with two more 1–mL portions of diethyl ether. All three ether extracts should be combined in the large test tube.

CHEMICAL SAFETY NOTE Hydrochloric acid is corrosive. If any solution containing hydrochloric acid comes in contact with your skin, immediately flush the affected area with water for at least five minutes and have another student notify your instructor.

SAFETY NOTE Be sure to balance your centrifuge tube with another tube or a tube from another student that has approximately the same amount of liquid.

Dry the combined ether extracts over sodium sulfate. Transfer the dry ether solution to a tared vial. Rinse the sodium sulfate with another 1 mL of ether and add these rinses to the vial. Remove the diethyl ether with a gentle stream of air in the fume hood.

Record the amount of product obtained. Carefully smell the contents of the vial by waving your hand over the top of the vial while holding it at a distance equivalent to the width of your hand from your nose and in the flow of air being generated. Record your observation.

Cleanup

Place the product in an appropriate container as specified by your instructor. Place all organic solvents, including TLC solvents, and aqueous waste in the appropriate waste containers. The Pasteur pipettes and weigh paper that contain Pd/C should already be in the bucket of water in the fume hood. Dispose of all contaminated Pasteur pipettes, used TLC plates, and melting point and TLC capillaries in the container for contaminated laboratory debris. Clogged syringe needles must be placed in the container for syringe needles.

Discussion

- Analyze the ^{1}H NMR and ^{13}C[^{1}H] NMR spectra of the starting material and the product of the reaction (Chapter 8). Your instructor will let you know how or where to obtain the spectra.

- Briefly discuss how you determined the structure of your product.

- Calculate the percent yield of the product.

- Discuss the overall energetics of the reaction. Is the reaction exothermic or endothermic? Assume you are starting with the enol form of carvone. You will need to consult your lecture text to find bond energies for the various bonds being broken and formed.

- Discuss the steps used to isolate and purify the product from this reaction.

Questions

1. In this experiment you used *R*-carvone. Would the outcome of the reaction be different if you used *S*-carvone? Explain your answer.

2. In addition to TLC, how do you know a reaction occurred in the experiment?

3. What feature(s) in the ¹H NMR spectra of molecules A and B could be used to distinguish between the two (Chapter 8)? Be specific in your explanation.

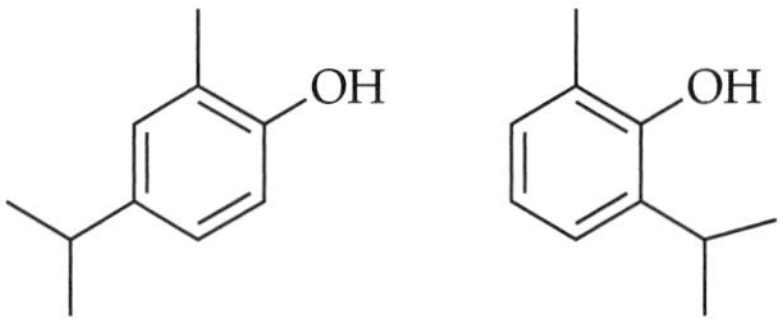

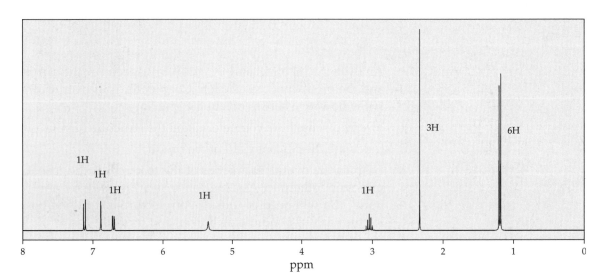

A   **B**

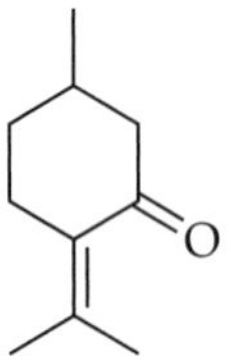

4. Pulegone, a cyclic monoterpene, can be isolated from the essential oil of pennyroyal, a member of the mint family. Reaction of pulegone with Pd/C at 180°C, followed by the same work-up used in the reaction of carvone with Pd/C, generated two products in equal quantities. One product was isolated from the aqueous layer. The ¹H NMR of this material is shown below. The other product was isolated from the hexane layer by evaporating the hexane. The EIMS of this compound showed a molecular ion at $m/z = 154$. The IR spectrum had a sharp band at 1736 cm⁻¹. In the ¹³C[¹H] NMR spectrum there was a peak at 198 ppm. Identify the compound isolated from each phase and explain how you reached your conclusions (Chapter 8). Write a balanced equation for the reaction of pulegone and Pd/C.

Pulegone

5. The reaction of the ketone with Pd/C at elevated temperatures shown below led to no reaction. Explain this result.

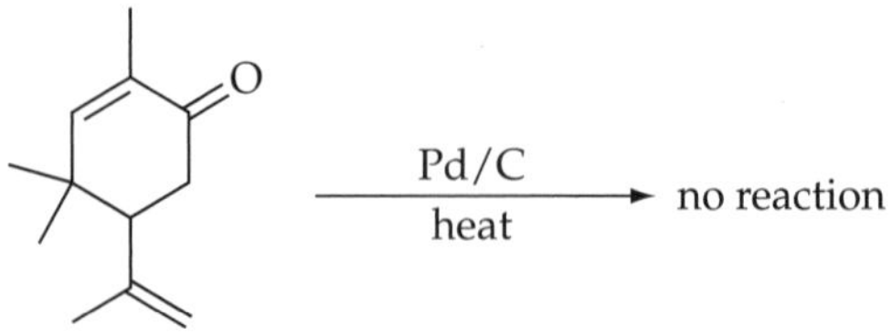

Pd/C
heat
→ no reaction

The Friedel-Crafts Reaction

The formation of a carbon-carbon bond between an aromatic hydrocarbon and a suitable electrophile, known as the Friedel-Crafts reaction, is an important synthetic organic reaction. Between 1874 and 1891, Charles Friedel[1] and James Crafts[2] collaborated to develop the reaction between benzene and alkyl and acyl halides using a Lewis acid catalyst (Figure E18.1). The aluminum or iron trichloride Lewis acid polarizes the carbon-halogen bond to facilitate its cleavage to form a carbocation. The reaction, which was published in 1877[3, 4], is referred to as an **Electrophilic Aromatic Substitution**.

The usual procedure for the Friedel-Crafts reaction uses commercially available $AlCl_3$. This material is extremely hygroscopic and difficult to handle without the appropriate equipment needed to weigh it under an inert atmosphere. Often this is accomplished using an atmosphere of argon or nitrogen. These are referred to as inert atmospheres because these gases are unreactive with most organic compounds. In this experiment, the $AlCl_3$ is generated directly from aluminum metal in the form of aluminum foil. Samples of t-butyl chloride contain small amounts of HCl that react with aluminum metal to produce $AlCl_3$ (Equation E18.1).

$$Al \ + \ 3\,HCl \ \longrightarrow \ AlCl_3 \ + \ 1.5\,H_2 \tag{E18.1}$$

Figure E18.1   Friedel-Crafts Alkylation and Acylation Reactions

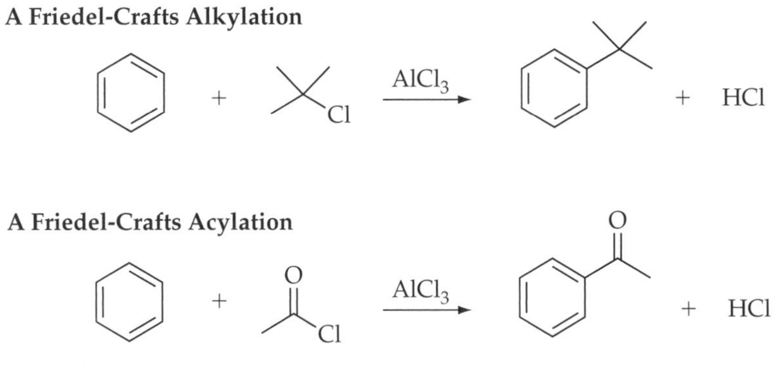

A Friedel-Crafts Alkylation

A Friedel-Crafts Acylation

---

[1] Charles Friedel, a French mineralogist, is credited with achieving the first synthesis of 2-propanol, also known as isopropyl or rubbing alcohol. From 1879 to 1887, he concentrated on making artificial diamonds and successfully created them using heat and pressure.

[2] Crafts, who never received a Ph.D., was appointed the first chemistry professor at Cornell University in 1868. In 1870, he moved to MIT and, in 1874, took a leave of absence to work with Friedel in Paris. He returned to MIT in 1891 and became its fifth president in 1898.

[3] Friedel, C. and Crafts J. M. *Compt. Rend.* **1877** *84* 1392.

[4] Friedel, C. and Crafts J. M. *Compt. Rend.* **1877** *84* 1450.

Background Reading:

Before the laboratory period, read about the Friedel-Crafts reaction in your lecture text.

Prelab Checklist:

In your prelab write-up, provide all of the physical constants for the starting materials and solvents.

Experimental Procedure

Don't forget to keep your safety glasses on during the entire laboratory period.

Set up the hydrogen chloride gas trap as shown in Section 7.4.1 and 7.4.1.a. Make sure that the cotton is loose so that the pressure does not build up. Do not attach the septum to your reaction tube/small test tube yet.

> **SAFETY NOTE**   Do not use an old septum when setting up the gas trap.

Caution: Make sure there are no flammable solvents on the laboratory bench.
Cap a reaction tube/small test tube with a septum and insert a needle into the septum. Use your microburner to flame dry the tube. After heating, remove the needle and let the system cool down. Do not start with a visibly wet reaction tube/small test tube.

After your tube has cooled down, add 0.16–0.17 g of biphenyl. Dissolve this in 0.5 mL of dichloromethane and shake or swirl it to mix the contents thoroughly. Spot some of this solution on a TLC plate in lanes 1 and 4. To the test tube, add 0.35 mL of 2-chloro-2-methylpropane (*tert*-butyl chloride) and a small piece of aluminum foil that is approximately 2 mm × 2 mm. That means a piece this big: □!

EXPERIMENTAL NOTE   Do not use a large excess of foil. Larger amounts of foil result in an oily mixture of products.

Put the stainless steel spatula in the reaction mixture and scratch the surface of the foil in several places until a red color appears on the metal surface.

Immediately cap the tube with the septum attached to the HCl trap. Periodically swirl the reaction mixture. Allow the solution to stand until it is red-orange in color and bubbling has subsided (approximately 15 min). Spot some of this mixture in lane 2 of the same TLC plate as used above.

> **SAFETY NOTE**   When removing the septum for TLC sampling, do not place your face directly over the reaction tube/small test tube. Also, be sure to keep the system capped while you are spotting the TLC plate. Otherwise some of the HCl may escape.

Wash the dichloromethane solution twice with 1.5–mL portions of water, making sure that the layers have been thoroughly mixed (Section 7.6.3 and Figure 7.21).

Discard the aqueous layer in the appropriate liquid waste container. Do *not* pour this aqueous waste down the drain.

Pipette the dichloromethane layer away from any remaining foil and water drops and place it in a clean reaction tube/small test tube. Add 1.0 mL of 95% ethanol and a boiling stick and heat to boiling. If some of the solid material does not dissolve, continue adding small portions of 95% ethanol until the solution is clear. Let the solution cool slowly to room temperature and then cool it in an ice-water bath before collecting the product by vacuum filtration. The filtrate from this recrystallization contains dichloromethane and should be discarded in the appropriate liquid waste container.

Determine the melting point and a mixed melting point of the product with biphenyl. Spot an acetone solution of the product in lanes 3 and 4 of the TLC plate used previously (Section 7.8.3.f) and develop the plate in hexane. Visualize the plates using UV light and iodine.

Save On Solvents: Share your TLC solvent chamber with a neighbor.

Did you know?

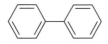

biphenyl

Biphenyl is used as a fungistat (antifungal agent) for citrus fruits. A pad of paper with a prescribed amount of biphenyl impregnated in it is placed either around each individual fruit or on the top and bottom of a larger box of fruit. The vapor of this compound helps to prevent the growth of fungi.

Cleanup

Place the product in an appropriate container. Dispose of all liquids in the appropriate waste containers. Do not put any solutions containing chlorinated materials down the drain. The wet cotton used in the gas trap will contain HCl. It should be discarded in the special waste container in the fume hood. Dispose of contaminated Pasteur pipettes, used TLC plates, and melting point and TLC capillaries in the container for contaminated laboratory debris. Any syringe needles that have become clogged must be disposed of in the special container for syringe needles.

Discussion

- Analyze the ^{1}H and the ^{13}C[^{1}H] NMR spectra of the product of the reaction (Chapter 8). Your instructor will let you know how or where to obtain the spectra.

- Discuss how you used the NMR spectra to determine the structure of your product relative to other possible isomers that might have been formed in this reaction. To answer this question, draw the structures of other possible isomers and consider what their NMR spectra would look like.

- Assume the product you obtained is the major product formed in the reaction. Rationalize this outcome relative to other possible isomers using words and arrow-pushing mechanisms. Make sure that the mechanism shows the function of the $AlCl_3$.

- Comment on the TLC behavior of the starting material and product. Are the results what you would have expected (Chapter 5)? Suggest another chromatographic technique that might accomplish separation of these compounds. Explain your choice (Chapter 5 and Section 7.8).

- Calculate the percent yield of the product. Based on the yield and TLC results, can you state with certainty that the product you isolated was the only product formed in this reaction (Section 7.8.2 and 7.8.3.f)? Explain.

Questions

1. In the introduction to this experiment, the text states that small amounts of HCl are present in samples of *tert*-butyl chloride. Write an arrow-pushing mechanism that explains the source of the HCl.

2. When monitoring the reaction between benzene and 2-chloro-2-methyl-propane using $FeBr_3$ as the catalyst, researchers observed the formation of a new alkyl halide. Draw the structure of this compound and show how it was formed using an arrow-pushing mechanism.

3. Why is it important that there be no significant quantity of water present in the reaction?

4. Considering the mechanism of the reaction, suggest other possible sources of the *tert*-butyl group. Include any reagents needed to activate these compounds.

5. The reaction shown below of biphenyl with two equivalents of the alkyl chloride in the presence of $AlCl_3$ gave two products with very similar R_f values by TLC. One of these products was visualized under the UV lamp, while the other was not. Both products were observed when the plate was stained with iodine. Write the structures of both products.

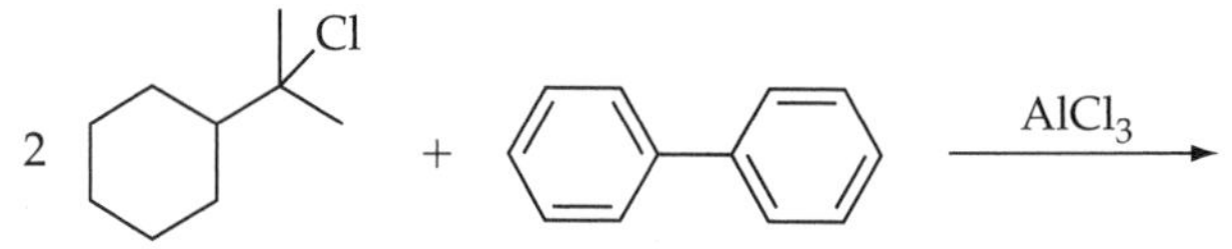

Microwave Heating of Organic Compounds

Using microwave heating in synthetic chemistry is a relatively new methodology. Although the first reports of heating organic reactions using microwaves appeared in 1986, it wasn't until several years later that this method became widely used by synthetic chemists.

Microwave heating is often referred to as dielectric heating. A domestic microwave oven operates at a set frequency of 2.45 GHz. This frequency corresponds to a wavelength of 12.2 cm, which places microwaves between infrared and radio waves in the electromagnetic spectrum. When molecules with molecular dipole moments (Section 5.2.1) are exposed to microwave radiation, they rotate to align themselves with the sign of the wave. Because the waves change direction 2.45×10^9 times per second, rotation of the molecules in solution cannot keep pace with the changing external field. Therefore, the rotating molecules generate heat through frictional encounters with other molecules. Basically, the molecules get out of step with the "music", and keep bumping into one another, similar to what you might observe in a country-line dance class for beginners. This generation of heat is known as dielectric loss (ε''). The ratio of ε'' to the dielectric constant, ε', is a measure of the ability of a molecule to transform microwave energy into thermal energy. Molecules with molecular dipole moments absorb microwaves and are heated while molecules with little or no molecular dipole moment are not.

One of the advantages of microwave heating is that microwaves do not heat many solids, such as porcelain, Teflon, or Pyrex glass, even though all three materials have molecular dipole moments. The molecules are locked in place and are not able to follow the phase of the alternating microwaves. With no molecular rotation, heat is not generated, enabling such materials to be used as containers for heating materials in a microwave. Even though these materials are not heated by microwave radiation, they can become hot because of heat transfer from their contents, so care must be used when handling the containers.

Background Reading:

Before the laboratory period, read the information on the polarity of molecules in Sections 5.1 and 5.2.

Prelab Checklist:

In your prelab write-up, provide the structures and relevant physical constants for all of the compounds being used in this experiment.

Based on the discussion about how microwaves heat, predict whether or not each of the compounds being used in this experiment will exhibit an increase in temperature when they are heated in a microwave oven.

Experimental Procedure

Please make sure you are wearing your safety glasses during the entire laboratory period.

Part I: Microwave Heating of Pure Liquids

In this part of the experiment, you will work in a group of four students so that you can compare your data. Each student is responsible for obtaining her/his own results for each of the compounds. The compounds you will be working with are water, decane, 1-octanol, 1-tetradecanol, 1,2-dichlorobenzene, and 1,4-dichlorobenzene.

CHEMICAL SAFETY NOTE   It is prudent to assume that any of the organic chemicals used in this experiment can act as skin irritants. Therefore, if any part of your skin comes in contact with any of the compounds, thoroughly wash the affected area with soap and water and have another student report the incident to your instructor. If any material comes in contact with your eyes, immediately proceed to the eyewash and flush the affected eye(s) with water for at least five minutes. Be sure that your instructor is aware of the situation.

Use a different *clean* and *dry* reaction tube/small test tube for each of the compounds. Add 0.5 mL of a liquid or 250 mg of a solid sample to the reaction tube/small test tube. Label each tube, indicating what the sample is and to whom it belongs.

Your instructor will give you instructions on the use of the microwave and the number of student samples that can be heated at the same time. Place a beaker containing the sample in the microwave oven.

EXPERIMENTAL NOTE   Do not run more than one of your samples at the same time because the temperature measurements must be made as soon as the microwave heating is over.

When the specified time is up, remove your sample and immediately measure the temperature of the liquid. Record the maximum temperature observed as the final temperature T_f. Record your results in a tabular format as shown in Table E19.1, using room temperature as T_i. Repeat this procedure for the remaining compounds.

Table E19.1 Experimental Data for the Pure Liquids

Compound	T_i	T_f	$\Delta T (T_f - T_i)$
Water			
decane			
1-octanol			
1-tetradecanol			
1,2-dichlorobenzene			
1,4-dichlorobenzene			

Compare your results with those from the other members of your group. If there are any significant differences in the results, repeat the experiment.

EXPERIMENTAL NOTE

- Be sure not to cross-contaminate samples.
- Make sure that you record the temperature of the room.
- After you are finished measuring the T_f of the sample, be sure to rinse the probe tip of the digital thermometer with a little acetone and dry it with a paper towel.

Did you know?

The common name for 1-tetradecanol is myristyl alcohol. The name is derived from *Myristica fragrans*, the source of nutmeg. The myristic acid (tetradecanoic acid) from trimyristin (Experiment 11) can be reduced with lithium aluminum hydride to give the alcohol. Both the acid and the alcohol are used to make ingredients for cosmetics.

Part II: Microwave Heating of Mixtures

In Part II, you will examine the effects of an additive on the microwave heating of decane. The four additives are 1-octanol, 1-tetradecanol, 1,2-dichlorobenzene, and 1,4-dichlorobenzene. Each member of the group is responsible for making up one mixture and measuring the ΔT.

Prepare a mixture of each additive in 0.5 mL of decane. Calculate the weight of 0.5 mL of decane and multiply this number by 0.3 to determine the weight of the additive to use. Add this weight of additive to a *clean* and *dry* reaction tube/small test tube. If the additive is a solid, be sure to crush up any large crystals. Then add the 0.5 mL of decane and mix thoroughly with your metal spatula using a twirling motion until you have a homogenous solution. Label the tube with your name and the name of the additive.

Your instructor will give you instructions on the use of the microwave and the number of student samples that can be heated at the same time. Place the reaction tube/small test tube containing the sample in a small beaker in the microwave oven.

 When the specified time is up, remove your sample and immediately measure the temperature of the liquid. Record the maximum temperature observed as the final temperature T_f. You should record the data from all four mixtures in a tabular format in your laboratory notebook as shown in Table E19.2, using room temperature as T_i.

Table E19.2 Experimental Data for the Mixtures

Mixture	T_i	T_f	$\Delta T \; (T_f - T_i)$
decane and 1-octanol			
decane and 1-tetradecanol			
decane and 1,2-dichlorobenzene			
decane and 1,4-dichlorobenzene			

Cleanup

None of the samples should be disposed of in the sink drains. Place all liquid samples in the appropriate liquid waste containers. Rinse any tube that contained the samples with two 1-mL portions of acetone and discard the rinses in the appropriate liquid waste containers. At this point, you may wash out the tube with soap and water in the sink. Place the solid samples in the bottles labeled 1-tetradecanol and 1,4-dichlorobenzene for recycling. Be careful not to cross-contaminate these samples. Dispose of all contaminated Pasteur pipettes, paper towels, and weigh paper in the container for contaminated laboratory debris.

Discussion

- Explain the results from Part I based on what you know about microwave heating. Are the results consistent with what you predicted in the prelab? If not, explain.

- Discuss the data in Part II by comparing it with your observations in Part I.

- What overall lessons have you learned about heating organic compounds using microwave radiation?

Questions

1. Why is it important to use clean and dry reaction tubes/small test tubes in this experiment?

2. Some polymerization reactions are carried out using microwave energy. Although the reaction temperatures can be relatively high at the beginning of the reaction, they decrease as a function of time. Explain this observation assuming that the monomer used in the polymerization reaction has a significant molecular dipole moment. An example of such a polymerization reaction is shown in Equation PE19.2.

$$Nu^{\ominus} + \text{excess} \quad H_3CO\overset{O}{\underset{}{\parallel}}\diagdown \longrightarrow H_3CO\overset{O}{\underset{}{\parallel}}\left(\diagdown\right)_n \quad\text{(PE19.2)}$$

3. Sometimes silica gel is dried in a microwave oven. Explain why the silica gel is hot to the touch when removed from the oven even though it does not absorb appreciable amounts of microwave radiation.

4. In each of the following lists, which compound would you expect to heat most rapidly in a microwave oven? Explain.

   List 1: benzene, chlorobenzene, toluene
   List 2: carbon tetrachloride, 1,1-dichloroethene, 1,4-difluorobenzene
   List 3: *trans*-1,2-dichloroethene, hexadecane, *cis*-1,2-dichloroethene

5. Below are three molecules commonly found in many types of beef. Explain which of these three would contribute least to the heating of beef in a microwave oven.

H_2O

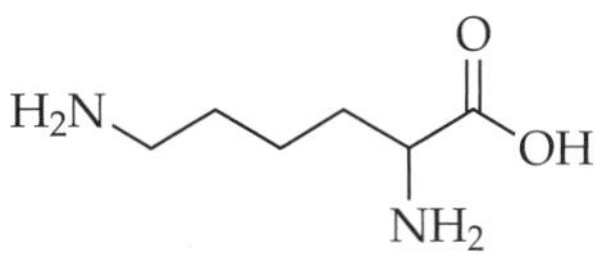

H₂N‚‚‚‚‚‚‚OH

lysine

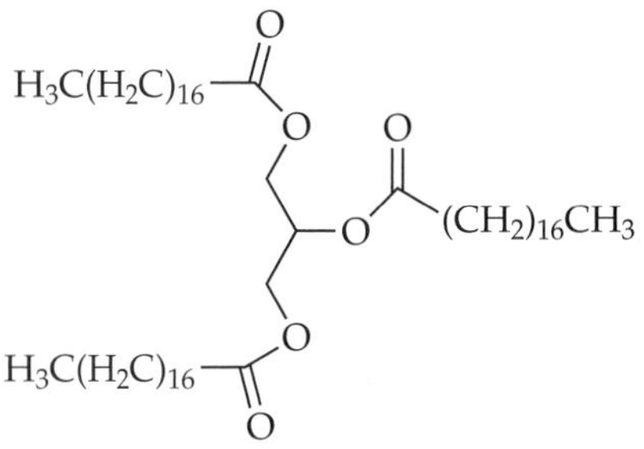

tristearin

20 EXPERIMENT

The Reaction of 1,1-Diphenylethanol on Clay in the Presence of Microwave Radiation

The thought of purposely pouring dirt into your reaction flask may at first seem strange. However, there are various naturally occurring clays that are quite useful to the synthetic chemist.

Clays are sedimentary minerals, composed of layered oxides. They are often referred to as aluminosilicates after the two most abundant oxides found in their structures. The two-dimensional sheets of oxides that make up the three-dimensional layers are composed of SiO_4 tetrahedra and AlO_6 octahedra (page 125). Protons and a variety of other ions, including Li^+, Na^+, K^+, and Ca^{2+}, balance the excess charge. The arrangement of these sheets leads to differences among the clays.

The material used in this experiment is known as Montmorillonite and is made up of layers consisting of one sheet of octahedral units sandwiched between two sheets of tetrahedral units. The interlayer distances are quite variable and can be changed dramatically by the addition of water or other solvents, a process known as *swelling*, or by the addition of various transition metal or organic ions. The later process is known as *pillaring* and results in very selective heterogeneous catalysts.

Most clays are strong Brønsted acids. Their acidity results from the terminal and bridging hydroxyl groups. The acidity of these materials is enhanced by washing them with a mineral acid such as HCl, which replaces ions such as Na^+ and K^+ with H^+. The surface acidity of acid-washed clays is comparable to concentrated sulfuric acid. Therefore, these materials are more acidic than resins, such as Amberlyst-15 (Experiment 30), whose acidity is attributed to benzene sulfonic acid residues ($pK_a = 1$) bound to a polymer backbone.

This experiment investigates the product formed when 1,1-diphenylethanol is heated in the presence of a strong acid (Equation E20.1). The experiment is environmentally friendly because it uses clay, a naturally occurring material, and no solvent is used during the reaction. Microwave

312

radiation is used to heat the reaction instead of a sand bath or other heat source.

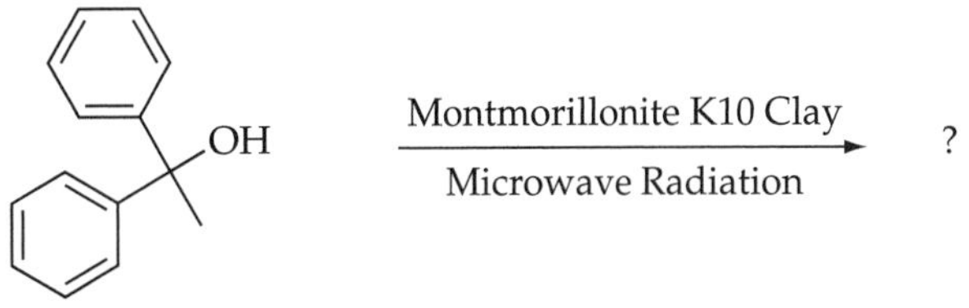

$$\underset{\textbf{1,1-diphenylethanol}}{\text{Ph}_2\text{C(OH)CH}_3} \xrightarrow[\text{Microwave Radiation}]{\text{Montmorillonite K10 Clay}} \quad ? \quad \text{(E20.1)}$$

Did you know?

Animals and the indigenous cultures in North and South America, Africa, and Australia have used clay for centuries as an external and internal healing agent. Animals lick selective clays as part of their diet or roll in it to relieve pain from injuries. In humans, it was applied externally to open wounds to reduce inflammation and pain. Humans have eaten it, called geophagy, to obtain calcium, iron, magnesium, potassium, manganese, silica, and trace amounts of many other minerals that are essential for good health. Because it carries a negative charge, clay adsobs positively charged impurities and toxins and then absorbs them into its structure so that they can be eliminated from the body as waste. NASA has even added a special calcium montmorillonite clay, called Terramin, to the diet of astronauts to counter calcium loss during weightlessness.

Background Reading:

Before the laboratory period, review information about dehydration and electrophilic addition in your lecture text. Also, review the information about microwave heating in Experiment 19.

Prelab Checklist:

The prelab write-up should include the structures and physical constants for the starting materials and solvents used in this experiment.

Experimental Procedure

Don't forget to keep your safety glasses on during the entire laboratory period.

Combine approximately 1 mmol of 1,1-diphenylethanol and 200 mg of Montmorillonite K10 clay in a one-dram vial.

SAFETY NOTE   To prevent inhaling clay dust, use your spatula to scoop out clay from the container rather than pouring it out.

Mix the solids thoroughly by twirling the tip of a spatula throughout the mixture. Stir to ensure that the 1,1-diphenylethanol is evenly distributed throughout the clay. Label the vial with your name.

Warning: Do Not Cap the Vial.

Your instructor will demonstrate the use of the microwave oven. After heating the vial for the time specified by your instructor, carefully remove it from the oven.

Caution: Be careful, the vial may be hot.

Allow the contents of the vial to cool to room temperature. Using a spatula, transfer as much of the solid mixture as possible to a clean centrifuge tube. Add 1 mL of acetone to the vial and pour the solution into the centrifuge tube. Repeat this one more time. The majority of the solid mixture should now be in the centrifuge tube. Briefly stir the mixture in the centrifuge tube with a spatula and then centrifuge it for 2 minutes.

> **SAFETY NOTE** Be sure to balance your centrifuge tube with another tube that has approximately the same amount of liquid.

Weigh a 10–mL beaker and record the weight in your laboratory notebook. Use a Pasteur pipette to remove the acetone from the clay. Filter the solution into the tared beaker using the glass-fiber Pasteur pipette filtration method (Section 7.2.2). Extract the clay with acetone two more times using 1 mL of acetone each time, centrifuge, and then pass the extract through the filtration pipette into the tared beaker.

Prepare a TLC plate to accommodate four lanes. Spot a solution of the starting alcohol in lane 1 and the filtrate from the reaction in lane 2.

Place a boiling stick in the acetone filtrate and heat the solution on the sand bath. Evaporate approximately 60–80% of the acetone, leaving a total volume of 1–2 mL.

> **CHEMICAL SAFETY NOTE** Acetone is a flammable solvent. Do not overheat your sand bath. Do not use a burner during the laboratory period.

Remove the beaker from the sand bath and allow the solution to cool to room temperature. Then place it in an ice bath until crystallization has ceased. Remove the mother liquor with a Pasteur pipette and place it in a test tube. Rinse the crystals with 1 mL of ice-cold acetone. To assist with drying the crystals, pass a stream of air over the product.

Weigh the product and measure the melting point. Dissolve a few of the crystals in acetone and spot the solution in lane 3 of the previously prepared TLC plate. Also, spot the mother liquor in lane 4. Develop the plate in 100% hexane and visualize it with UV light and iodine.

Save On Solvents: Share your TLC solvent chamber with a neighbor.

Cleanup

Scrape the product out of the beaker and put it in an appropriate container. Place any remaining solvents, including the TLC solvents, in the appropriate liquid waste containers. Dispose of your microwave reaction vial, contaminated Pasteur pipettes, used weigh paper, TLC plates, and melting point and TLC capillaries in the container for contaminated laboratory debris. After the acetone has evaporated from the clay, dispose of the clay in the container for solid chemical waste.

Discussion

- Analyze the ^{1}H and ^{13}C[^{1}H] NMR spectra of the product. In the EI mass spectrum of the product, assign the ions at m/z 360, 345, and 283. Show reasonable structures for these ions and rational arrow-pushing mechanisms that lead to them (Chapter 8). Your instructor will let you know how or where to obtain the spectra.

- Discuss your strategy for determining the structure of the product of the reaction. Was it necessary to use the mass spectrum to establish the identity of the product? Explain your answer.

- Write a balanced equation for the formation of the product and calculate the percent yield.

- Write a rational arrow-pushing mechanism for the formation of the product.

- Discuss the TLC plate you developed. What did each lane indicate about the overall reaction?

- Assume that this reaction could be carried out using hot sulfuric acid. Compare the procedure using sulfuric acid with the one in this experiment using clay. Be sure to compare each step of the reactions from beginning to end noting which one is more environmentally friendly.

Questions

1. Is the product you obtained in this experiment chiral? If so, draw both enantiomers and label all stereocenters as *R* or *S*.

2. In the first step of the reaction, an alkene is formed. Referring to the mechanism you wrote in your discussion, draw the structure of the alkene and provide its melting and boiling points. Is it relevant to the success of this solvent-free reaction that the alkene is a liquid at room temperature? Explain.

3. Devise a synthesis of the product you obtained from this experiment starting with 3,3,3-triphenylpropanal.

4. If the reaction in this experiment is run at lower temperatures, a significant amount of an alkene with the molecular formula $C_{28}H_{24}$ is formed. Draw the structure of this product and explain why it is formed at the lower temperatures. Hint: Examine the mechanism you wrote for the discussion.

5. When 2-methylpropene is reacted with Montmorillonite K10 clay, a large number of products are formed including the one shown in Equation PE20.5. Write a rational arrow-pushing mechanism leading to this product.

excess ⟍⟋ $\xrightarrow{\text{Montromillonite K10 Clay}}$ ⟍⟋⟍⟋ + many other products   (PE20.5)

The Wittig Reaction[1]

The Wittig reaction and its variants are some of the most useful carbon-carbon bond forming reactions in organic chemistry (Equation 21.1).

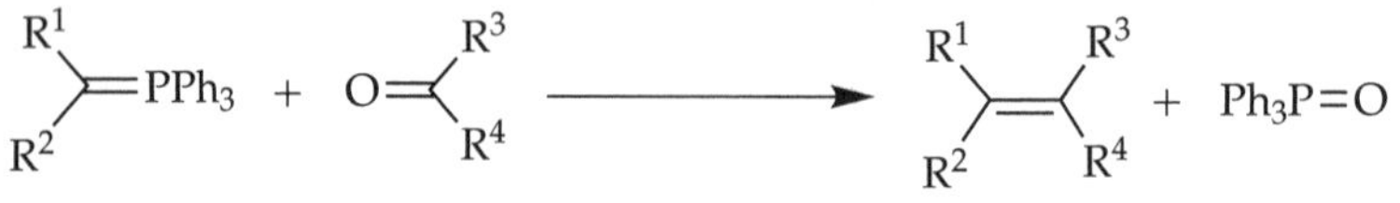

$$\begin{array}{c} R^1 \\ {\Large\diagup}\!\!=\!PPh_3 \\ R^2 \end{array} + \begin{array}{c} R^3 \\ O\!=\!\!{\Large\diagdown} \\ R^4 \end{array} \longrightarrow \begin{array}{c} R^1 \quad R^3 \\ {\Large\diagdown}\!\!=\!\!{\Large\diagup} \\ R^2 \quad R^4 \end{array} + Ph_3P\!=\!O \qquad (E21.1)$$

The starting materials for these reactions are readily available and the carbon-carbon double bond formed affords numerous opportunities for further synthetic transformations. In this experiment, you will react a stable phosphorane, methyl (triphenylphosphoranylidene)acetate (R^1 = C(O)OMe and R^2 = H in Equation E21.1) with 2-nitrobenzaldehyde. Because at least two products are formed in this reaction, an alkene and triphenylphosphine oxide, you will use column chromatography to separate them (Section 7.8.4).

Although the Wittig reaction often proceeds at room temperature, it is not uncommon to apply heat, for example, from a sand bath. In this experiment, you will use microwave radiation (Experiment 19). In addition, you will not use any solvent for the reaction.

Did you know?

Phosphorane is the term used for pentavalent phosphorous compounds. For example, $P(C_6H_5)_5$ is known as pentaphenylphosphorane. A Wittig reagent goes by many names, including phosphorus ylide and phosphorane. In the latter case, the name refers to the fact that the phosphorus atom in a Wittig reagent is formally phosphorus (V).

Background Reading:

Before the laboratory period, read about chromatography in Sections 7.8.1 and 7.8.2 and about column chromatography in Section 7.8.4. Review the information about microwave heating in Experiment 19. Also, read about the Wittig reaction in your lecture text.

Prelab Checklist:

The prelab write-up should include the structures and physical constants for the starting materials and solvents used in this experiment and a balanced equation for the reaction.

---

[1] Georg Wittig, who shared the Nobel Prize in 1979 with H.C. Brown, discovered the reaction in 1954. See http://nobelprize.org/nobel_prizes/chemistry/laureates/1979/.

Experimental Procedure

Don't forget to keep your safety glasses on during the entire laboratory period.

Part I: Synthesis of the Product

Combine 0.50 mmol of 2-nitrobenzaldehyde, 0.52 mmol of methyl (triphenylphosphoranylidene)acetate, and 100 mg of silica gel in a 1-dram vial.

> **CHEMICAL SAFETY NOTE**   Silica gel is a very flocculent powder. To prevent inhaling silica gel, use a spatula to scoop the powder out of the container rather than pouring it.

Mix the solids thoroughly by twirling a spatula throughout the mixture. Stir it well enough to ensure that all three solids are evenly distributed in the vial. Some of the aldehyde may melt during mixing, but this will not affect the experiment. Label the vial with your name.

Warning: Do not cap the vial.

Your instructor will demonstrate the use of the microwave oven. After heating for the specified time, remove the vial from the oven.

Caution: Be careful, the vial may be hot.

Mix the solid mass with a spatula before it cools to room temperature.

EXPERIMENTAL NOTE   If the solid is not mixed before it cools, it hardens and you will not be able to transfer the mixture to the chromatography column.

Part II: Packing the Chromatography Column

Place 10 mL of a 50:50 ethyl acetate:hexane solution in a graduated cylinder. Secure a chromatography column in the upright position with a clamp and leave the stopcock at the bottom of the column open. Put a funnel in the top of the column and place a 30–mL beaker underneath the nozzle of the column. Weigh 0.65 g of silica gel in a 25–mL Erlenmeyer flask and add 2 mL of the solvent mixture. Swirl the mixture and quickly pour it into the column while tapping the sides of the column to help distribute the silica gel evenly and to aid in the removal of air bubbles. Before all of the solvent drips out, wash the Erlenmeyer flask twice using 1 mL of solvent each time and pour the washes into the column. Use an additional 1 mL of solvent to wash down any silica gel sticking to the inside of the column. Before closing the stopcock, allow the solvent to drain until the meniscus is 2–3 mm above the top of the silica gel.

Do not let the solvent level drop below the top of the column of silica gel.

The silica gel column should be approximately 2.5 cm in height.

Part III: Isolation of the Product

Add the reaction mixture directly to the column using a spatula. Use 0.5 mL of solvent to wash down any of the reaction mixture that sticks to the inside of the column. Tap the sides of the column to release any trapped air bubbles. Open the stopcock and drain the column until the meniscus is just above the top layer of material. Add another 0.5 mL of solvent and repeat the process to ensure that all of the products are adsorbed onto the silica gel. This process ensures that none of the products are dissolved in solution above the column. Replace the beaker underneath the column with a reaction tube/small test tube and carefully pour the solvent in the beaker back into the column. In order to prevent disturbing the top of the silica gel column, pour the solvent in slowly, letting it run down the inside of the column. Also, pour the remaining solvent from the graduated cylinder into the column. By recycling the solvent used to pack the column, the amount of solvent waste generated in this experiment is minimized.

Open the column stopcock and begin collecting drops in a reaction tube/small test tube. When 1.5 mL of solution has been collected, close the stopcock, label the tube, and replace it with another tube. Repeat this procedure until four 1.5 mL fractions have been collected. Use a small Erlenmeyer flask for the fifth fraction and drain the column until the meniscus is just above the top of the silica gel.

Spot a TLC plate with solutions of triphenylphosphine oxide, fraction 1, and fraction 2 as shown in Figure E21.1. Develop this plate in 50:50 hexane:ethyl acetate and visualize it using a UV lamp. While the plate is developing, spot a second plate with the remaining fractions as shown in Figure E21.1. From the TLC plates, determine which fractions contain the alkene product and combine these fractions in a *tared* 25-mL filter flask. Do not combine any fractions that show more than one spot on the TLC plate.

Save On Solvents: Share your TLC Solvent Chamber with a neighbor.

Figure E21.1   TLC Plates of the Fractions from Column Chromatography

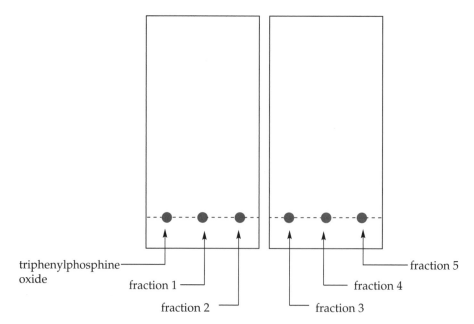

EXPERIMENTAL NOTE The desired alkene product from this reaction is the spot with the highest R$_f$ value. Triphenylphosphine oxide is not the desired product and should not be isolated from any of the fractions.

Evaporate the solvent in the fume hood using a stream of air and heating gently. If necessary, remove the last traces of solvent by capping the flask and pulling a vacuum on the system for a couple of minutes.

Determine the weight of the product and the melting point and calculate the percent yield.

Cleanup

Scrape the product out of the filter flask and place it in an appropriate container. The product may be used as a starting material in Experiment 36. Place any remaining chromatography solvents and TLC chamber solvent in the appropriate liquid waste containers. Dispose of the microwave reaction vial, contaminated Pasteur pipettes, used weigh paper, TLC plates, and melting point and TLC capillaries in the container for contaminated laboratory debris.

Discussion

- Make reasonable assignments for the ^{1}H and ^{13}C[^{1}H] NMR spectra of the starting aldehyde, the phosphorus ylide, and the product (Chapter 8). Your instructor will let you know how or where to obtain the spectra.

  Where relevant, look for any coupling to ^{31}P and identify it (Section 8.1.4.c).

- Discuss your strategy for determining the major alkene isomer formed in this reaction.

- Write a rational arrow-pushing mechanism for the formation of this product.

- Simply heating the two reactants together also results in product. Discuss the advantages of performing the reaction on silica gel compared with just heating.

- Discuss the two TLC plates you developed. Were there any fractions that showed more than one spot on the TLC plate? If so, what might you have done differently when separating these components by column chromatography?

Questions

1. Most Wittig reagents must be generated *in situ* from a phosphonium salt and base. The phosphorane used in this experiment is one of the few Wittig reagents, known as stabilized ylides, that can be stored in a bottle. Explain what makes this particular phosphorane stable relative to, for example, Ph$_3$P=CH$_2$.

2. After the reaction mixture was added to the top of the silica gel column, the instructions stated: "Open the stopcock and drain the column until the meniscus is just above the top layer of material. Add another 0.5 mL

of solvent and repeat the process to ensure that all of the desired products are adsorbed onto the silica gel. This process ensures that none of the products are dissolved in solution above the column." Why is this procedure important?

3. A researcher used column chromatography to separate two compounds. The desired compound, which was quite valuable, had an R_f approximately equal to 0.5. Two-mL fractions were collected, giving the TLC results shown below. Based on these results, explain what the researcher needs to do to maximize recovery of the desired compound, including the use of additional separations.

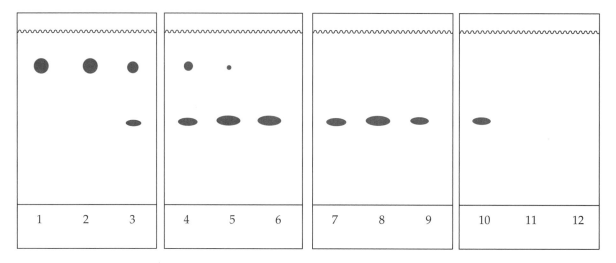

4. The Horner-Wadsworth-Emmons reaction is a commonly used variant of the Wittig reaction. What feature of this reaction, shown below, is particularly useful compared with the Wittig reaction? Hint: Consider the structure of the product(s).

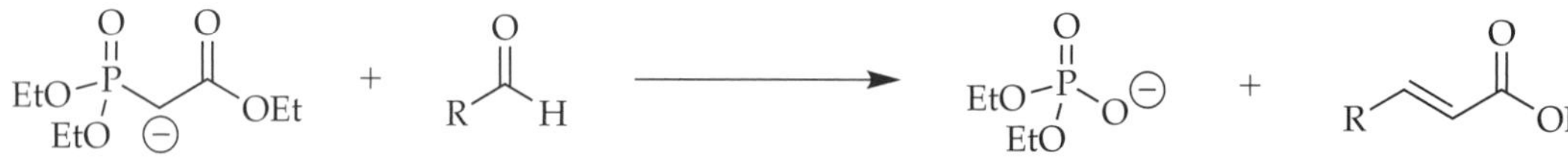

5. Based on your background reading about the Wittig reaction, propose a synthesis of the phosphorane used in this experiment, starting with triphenylphosphine and whatever other commercially available reactants and reagents you choose.

The Suzuki Reaction

The Suzuki reaction is named after Akira Suzuki who, along with Norio Miyaura, first reported the coupling of aryl halides and alkenyl halides with alkenylboranes using a palladium catalyst in 1979[1,2]. The reaction was extended to the coupling of aryl halides and arylboronic acids in 1981[3].

Because the palladium is used in catalytic quantities, the active form of this metal must be continually regenerated. A proposed catalytic cycle is shown in Figure E22.1. This cycle relies on two fundamental reactions, an oxidative addition and a reductive elimination. In the oxidative-addition step, palladium(0) is oxidized to palladium(II) by insertion into a carbon-bromine bond, analogous to the generation of a Grignard reagent from magnesium metal and an organohalogen. After the oxidative addition, two ligand-exchange reactions occur. One of them involves the exchange of a phenyl group on boron for a hydroxide ligand on palladium. The resulting diaryl palladium complex undergoes a reductive-elimination reaction. This process uses two of the four electrons in the palladium-carbon bonds

Figure E22.1 Proposed Catalytic Cycle for the Suzuki Reaction

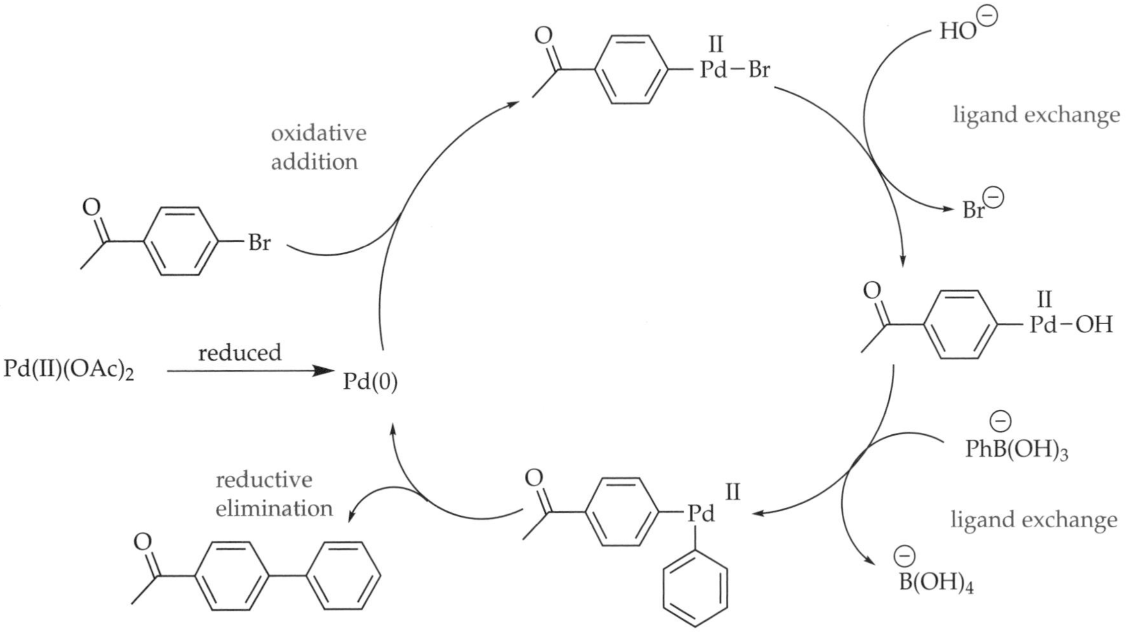

[1] Miyaura, N., Suzuki, A. *J.C.S. Chem. Comm.* **1979,** 866.

[2] Miyaura, N., Yamada, K., Suzuki, A. *Tet. Lett.* **1979,** 3437.

[3] Miyaura, N., Yanagi, T., Suzuki, A. *Syn. Comm.* **1981,** *11,* 513.

to form a carbon-carbon bond between the two aryl groups. Two electrons remain on palladium, formally reducing it back to palladium(0). At this point, the palladium is available for another oxidative-addition reaction.

Unlike the highly polarized nature of the metal-carbon bond in a Grignard reagent, analogous bonds involving transition metals such as nickel and palladium are not significantly polarized. Because of this lack of polarity, reactions like the Suzuki coupling can be run in the presence of protic solvents, such as water, and reactive functional groups, such as ketones and aldehydes. Consequently, unlike Grignard reagents, a variety of functional groups are tolerated under the reaction conditions and do not need to be protected. For example, the acetyl group of 4-bromoacetophenone remains unaffected in this carbon-carbon bond-forming reaction.

The reaction is environmentally friendly because it uses a small amount of catalyst (2 mg) and water is the solvent. Tetrabutylammonium bromide is added to the reaction as a phase-transfer agent. The purpose of this material is to aid in the transfer of the hydrophobic 4-bromoacetophenone to the aqueous phase. The four butyl groups provide a hydrophobic environment and help solvate the 4-bromoacetophenone, whereas the positive charge on the nitrogen atom is attracted to the water by ion-dipole interactions, so that the entire reaction mixture is soluble in water (Section 6.1). Potassium carbonate reacts with water to provide the hydroxide ions that are involved in the ligand exchange with bromine in the catalytic cycle. The potassium hydroxide also reacts with the boronic acid produced to form a water-soluble potassium salt, facilitating separation of this material from the product after the reaction is complete.

The equation for the reaction in this experiment is shown in Equation E22.1.

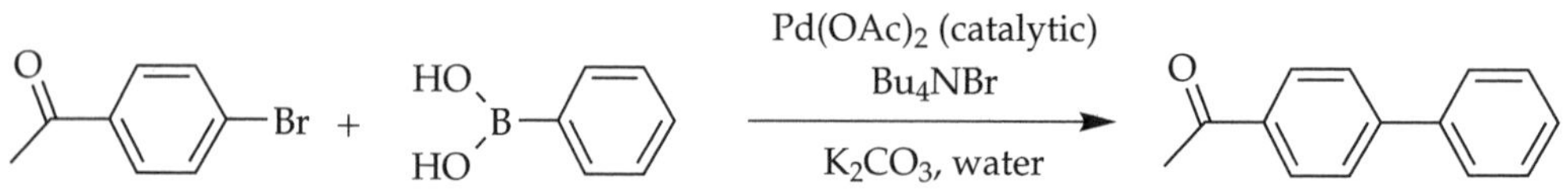

$$\text{(E22.1)}$$

Background Reading:

Before the laboratory period, find an example of the Suzuki reaction in the literature (Chapter 3). In your prelab write-up, cite the complete reference and write the equation for the reaction reported in the article.

Prelab Checklist:

Include structures and physical properties for all starting materials and solvents used in this reaction in your prelab write-up.

Did you know?

Phenylboronic acid is used as a wood preservative to prevent damage by wood-destroying organisms such as termites and fungi. It kills the protozoa found in the gut of the termite that are necessary to convert wood cellulose into the absorbable sugars that serve as the termites' food. It is relatively non-toxic to humans when compared with other wood preservatives that contain arsenic, chromium, and copper.

Experimental Procedure

Don't forget to keep your safety glasses on during the entire laboratory period.

Weigh the following materials separately and combine them in a long-neck 5–mL round-bottom flask: 1 mmol of 4–bromoacetophenone, 1.1 equivalents of phenylboronic acid, 2 mg of palladium(II) acetate, 1 equivalent of tetrabutylammonium bromide, and 2.5 equivalents of potassium carbonate.

Technique Tip   Calculate the weight of 1 mmol of 4-bromoacetophenone. Convert the exact amount weighed to mmol and calculate the weights of the other reactants needed to satisfy the above stoichiometry (Section 4.2).

Add a stir bar followed by 1 mL of distilled water, being careful to rinse down any materials that are clinging to the walls of the flask. Attach a condenser (Figure 7.15) and cap the condenser with a septum. Insert a needle into the septum to allow for any pressure buildup to be vented from the reaction while minimizing the exposure of the palladium to the oxygen in the air.

Heat the reaction at reflux for 45 minutes on a sand bath and stir the system using the magnetic stirrer. The temperature of the sand bath should be approximately 120°C. The reaction will turn black. Allow the reaction mixture to cool to room temperature.

Filter the mixture on a Hirsch funnel. Break up any clumps of the solid with a spatula and wash the resulting gray solid with 10 mL of hot water.

Transfer the solid to a centrifuge tube. Add enough acetone to bring the volume to approximately 3 mL and cap the tube. Shake the tube vigorously until the solid mass dissolves. The black particulate matter will remain suspended in the mixture. Centrifuge the mixture for 2 minutes.

SAFETY NOTE   Be sure to balance your centrifuge tube with another centrifuge tube that has approximately the same amount of liquid.

Being careful not to disturb the black sediment, pipette the clear, dark-yellow supernatant into a large test tube. Add a boiling stick and reduce the volume of the solution in the test tube to approximately 2 mL by heating on a sand bath.

Add water dropwise. During this addition, some material will precipitate. Shake the tube after each addition of water to see if the solid will dissolve. Add water until a very small amount of crystals remains after shaking. After crystallization has begun, cool the mixture to 0°C.

Isolate the crystals on a Hirsch funnel and wash them with cold water. Dry the solid thoroughly while it is on the funnel by pulling air through the filter flask using a vacuum source. Weigh the product and obtain a melting point. Calculate the percent yield.

Cleanup

Place the product in an appropriate container. Put the filtrates in the appropriate liquid waste containers. Contaminated pipettes, melting point capillaries, and used weigh paper should be disposed of in the container for contaminated laboratory debris.

Discussion

- Analyze the ^{1}H and ^{13}C[^{1}H] NMR spectra of the product. For the EI mass spectrum, assign the ions at m/z of 196, 181, and 152 (Chapter 8). Your instructor will let you know how or where to obtain the spectra.

- Explain each of the steps used to separate the final product from the other compounds present, including tetrabutylammonium bromide, potassium carbonate, any boron containing salts, and the palladium catalyst.

Questions

1. Identify each of the following reactions as an oxidative addition or reductive elimination.

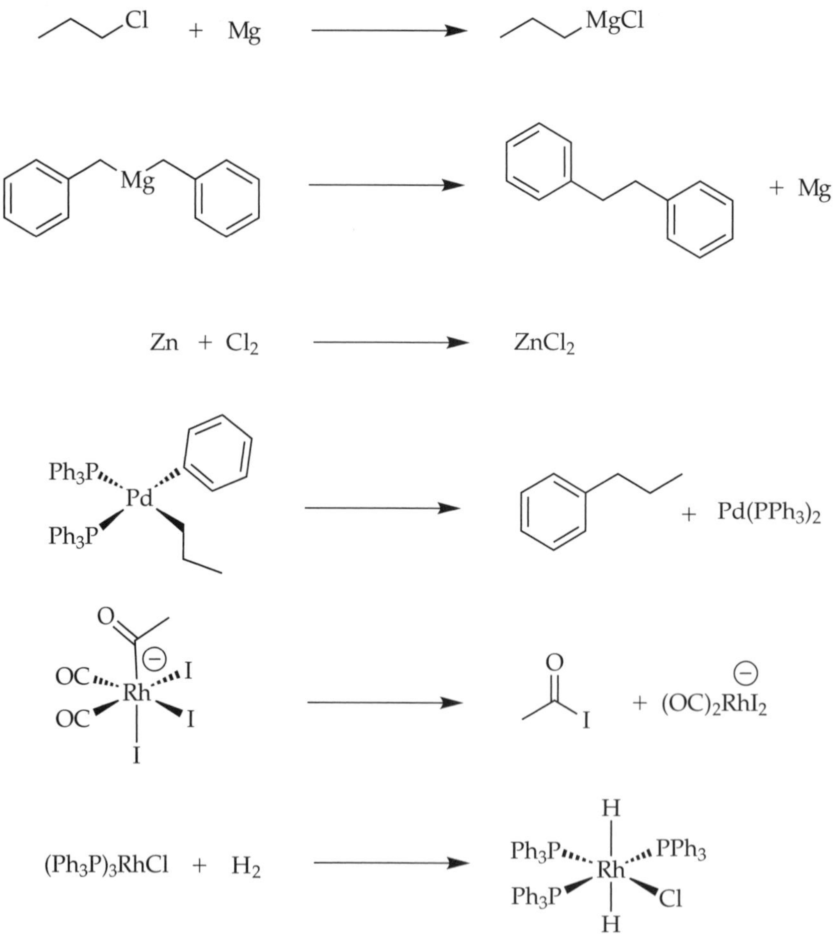

2. Provide an example of a phase-transfer catalyst other than the one used in this experiment.

3. List three features associated with this experiment that make it environmentally-friendly.

4. The reactions shown below are found in the first two papers that Suzuki published on cross coupling. What are the organic products of these reactions?

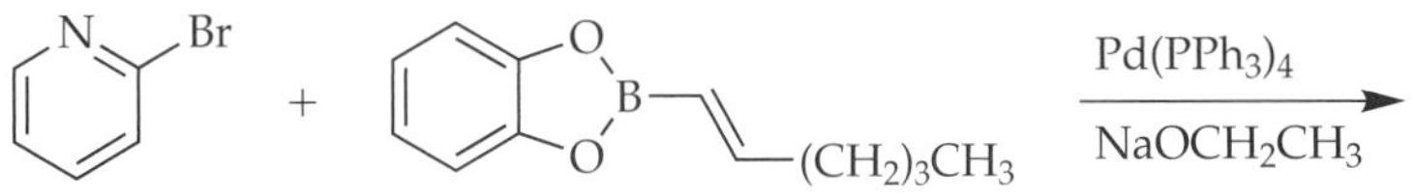

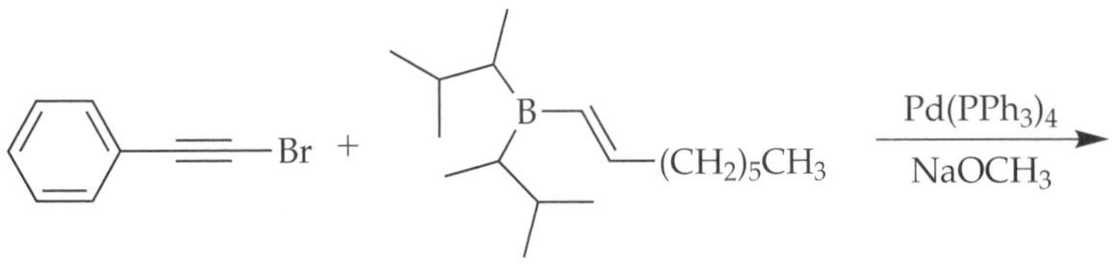

5. Propose a synthesis for molecule A, shown below, using benzene, phenyl-boronic acid, *tert*-butyl chloride, and any inorganic reagents needed (Experiment 18). A retrosynthetic scheme is shown. Put the pieces together showing the necessary reactants and reagents.

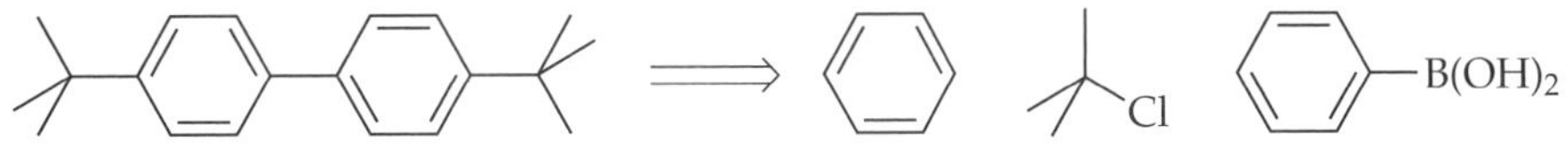

A

The Crossed-Aldol Condensation

The aldol reaction is a valuable method for constructing carbon-carbon bonds from aldehydes, ketones, or a combination of both. The reaction requires a hydrogen atom on the α-carbon of one of the reactants and an acid or base catalyst. The latter is more commonly used to remove the α-hydrogen, forming an enolate anion (Equation E23.1). This anion adds to the carbonyl carbon of the second reactant to give a β-alkoxycarbonyl (Equation E23.2). Protonation of this intermediate gives the aldol product, a β-hydroxycarbonyl. When an aldehyde enolate is used (R = H), the product is referred to as an aldol, a term that blends the names of the two functional groups present, an **ald**ehyde and alcoh**ol**. When a ketone enolate is reacted with another aldehyde or ketone, the product is sometimes referred to as a ketol. However, it is more common to call the product an aldol also, emphasizing the general nature of the reaction.

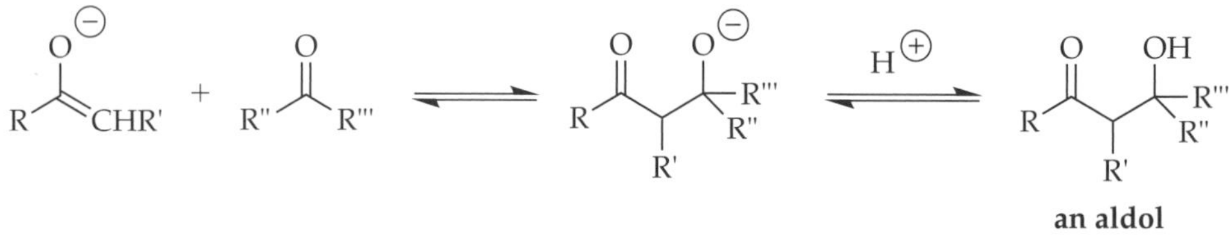

$$\text{R}-\underset{\text{CH}_2\text{R}'}{\overset{\text{O}}{\|}}\quad +\quad \text{Base} \quad \rightleftharpoons \quad \text{R}-\underset{\text{CHR}'}{\overset{\text{O}^{\ominus}}{|}}\quad +\quad \text{H-Base} \qquad (E23.1)$$

an enolate

$$\text{R}-\underset{\text{CHR}'}{\overset{\text{O}^{\ominus}}{|}} + \text{R}''-\underset{\text{R}'''}{\overset{\text{O}}{\|}} \rightleftharpoons \text{R}-\overset{\text{O}}{\|}-\underset{\text{R}'}{\overset{}{\underset{}{}}}-\underset{\text{R}''}{\overset{\text{O}^{\ominus}}{|}}\text{R}''' \quad \xrightarrow{\text{H}^{\oplus}} \quad \text{R}-\overset{\text{O}}{\|}-\underset{\text{R}'}{\overset{}{\underset{}{}}}-\underset{\text{R}''}{\overset{\text{OH}}{|}}\text{R}''' \qquad (E23.2)$$

an aldol

The aldol reaction is a versatile reaction used by both academic and industrial chemists. For example, pentaerythritol, an important intermediate in resins, is prepared by three aldol reactions between acetaldehyde and formaldehyde to give the aldol product that is then reduced by a Cannizaro reaction. All of these reactions take place in one reaction vessel making the product suitable for large-scale synthesis.

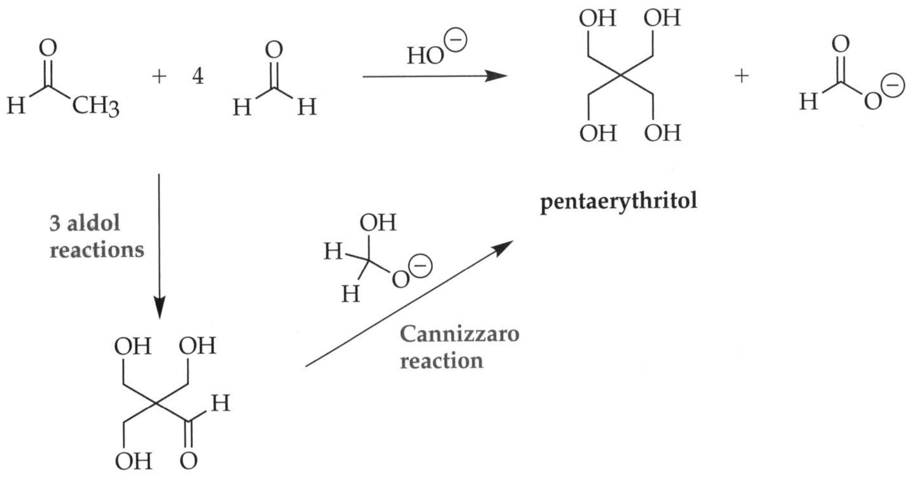

pentaerythritol

In nature, enzymes in some cells catalyze aldol reactions to produce a variety of important molecules. For example, the addition of glyceraldehyde-3-phosphate to dihydroxyacetone phosphate forms fructose 1,6-bisphosphate, a precursor to glucose.

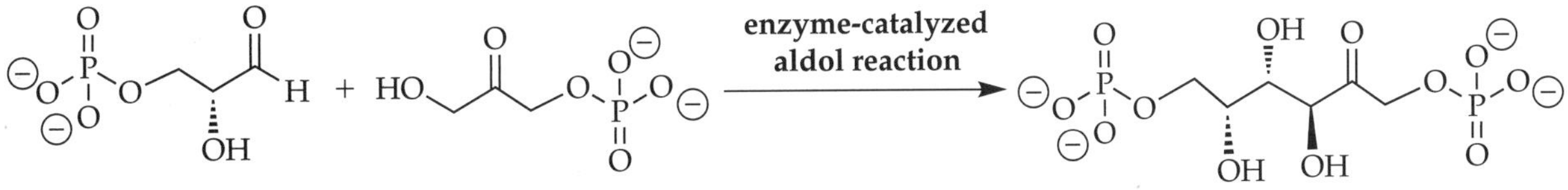

If the product of an aldol reaction has at least one remaining α-hydrogen between the carbonyl and the β-hydroxy group, dehydration leading to an α,β-unsaturated carbonyl is possible. Synthesis of an α,β-unsaturated carbonyl from the aldol reaction between two carbonyls followed by the dehydration of the β-hydroxycarbonyl product is known as an aldol condensation reaction (Equation E23.3). Both the aldol and the aldol condensation reactions are reversible and care must be taken to remove the base or acid catalyst before purification of the product.

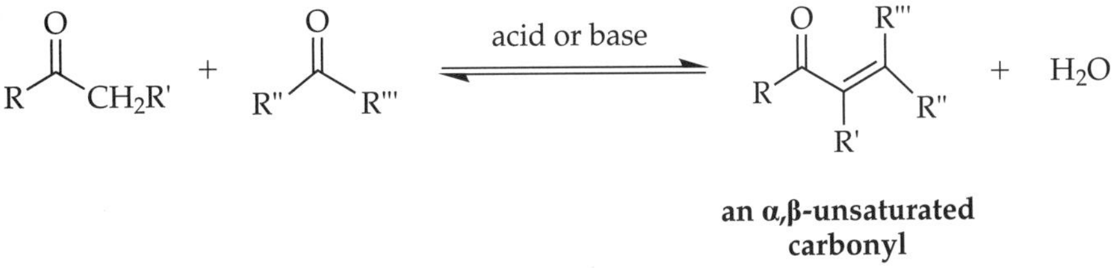

**an α,β-unsaturated
carbonyl**

In this experiment, you will perform a crossed-aldol condensation reaction between acetone and one of three aldehydes: benzaldehyde, *p*-anisaldehyde, or cinnamaldehyde.

Did you know?

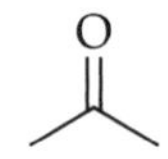

acetone

Acetone is found in very small quantities in normal urine and blood as a result of ketosis. This is a stage of metabolism during which fat is converted into fatty acids and ketones in the liver. Diabetics with severe insulin deficiency, victims of starvation, and people on hunger strikes do not have enough glucose to satisfy the body's need for energy. Consequently, they burn body fat, causing high concentrations of ketones, including acetone, to form and build up in the blood and urine. This condition is known as ketoacidosis. The acetone causes the breath of these individuals to smell fruity, called "acetone breath."

Background Reading:

Before the laboratory period, read about the aldol condensation reaction in your lecture text.

Prelab Checklist:

Write the balanced equation for the reaction of each of the three aldehydes in your prelab write-up. In each case, there are two possible products, not including geometric isomers. Write the structures for both products.

Tabulate the physical properties for the starting materials, solvents, and products. You may not be able to find the physical constants for all of the products. In addition to tabulating the moles of the starting materials, identify the limiting reactant. Indicate the product that is most likely formed based on the stoichiometry of the reactants.

Experimental Procedure

Don't forget to keep your safety glasses on during the entire laboratory period.

CHEMICAL SAFETY NOTE   Sodium hydroxide is corrosive. If any solution containing sodium hydroxide comes in contact with your skin, immediately flush the affected area with water for at least 5 minutes. Sodium hydroxide is particularly harmful to the eyes. If any solution contacts your eyes, proceed immediately to the eyewash station and flush your eyes for at least 5 minutes. In either case, have another student notify the instructor.

EXPERIMENTAL NOTE   The bottles of aldehydes must be kept tightly capped to help avoid air oxidation to the carboxylic acids.

Your instructor will assign you one of the three aldehydes that are in bottles labeled X, Y, and Z. Weigh 200 mg of the aldehyde in a reaction tube/small test tube. Add approximately 1.2 mL of 95% ethanol and mix the solution thoroughly with a stainless steel spatula. Add approximately 0.7 mL of 2N NaOH and stir the solution well. Use a syringe to add 0.05 mL of acetone to the solution and mix thoroughly. If the solution does not cloud after five minutes, warm it very gently in a warm tap water bath (50°C). Let the solution stand for an additional 25 minutes. Stir the solution occasionally during this time to break up any clumps.

Cool the reaction mixture in an ice-water bath. Remove the solvent using the square-tip Pasteur pipette method (Section 7.2.1) and wash the product with ice-cold water. Remove the water and add 1 mL of ice-cold 10% sodium bicarbonate. Stir the mixture well and remove the sodium bicarbonate wash. Repeat the sodium bicarbonate wash once more, removing the final wash by vacuum filtration (Section 7.2.5 and Figure 7.10). Wash the precipitate in the Hirsch funnel with ice-cold water until the wash is neutral (pH paper).

EXPERIMENTAL NOTE   Draw air through the product after the final wash to remove as much water as possible. Water can cause oiling-out problems during recrystallization (Section 7.1.2).

Reserve a small amount of the crude product for TLC analysis. Recrystallize the remaining crude solid from 95% ethanol.

EXPERIMENTAL NOTE   The product from aldehyde X will require more ethanol than the other two products.

Determine the melting point of the recrystallized product. Spot two TLC plates with the starting aldehyde, the crude product, and the recrystallized product. Run one plate in a solvent system of 90:10 hexane:ethyl acetate and the other in 80:20 hexane:ethyl acetate. Visualize the chromatograms by UV light and iodine.

Save On Solvents: Share your TLC Solvent Chambers with a neighbor.

Cleanup

Place the products in an appropriate container. The ethanol solvent and aqueous filtrates should be poured into the appropriate liquid waste containers. Dispose of clogged syringe needles in the container for used syringe needles. After the needle is removed, a clogged syringe should be put in the waste container for chemically contaminated debris.

Discussion

- Analyze the ^1H and $^{13}C[^1H]$ NMR spectra of the product. Discuss how you determined the geometry of the alkene(s) generated in this reaction (Chapter 8). Your instructor will let you know how or where to obtain the spectra.

- Discuss your strategy for determining the aldehyde you were assigned in this experiment.

- Calculate the percent yield for the product.
- Write a rational arrow-pushing mechanism for the formation of the product you obtained.
- Discuss the two TLC plates you developed. Based on this data, discuss whether or not recrystallization of the crude product was warranted.

Questions

1. Explain why 4-hydroxy-4-methyl-2-pentanone is not a major side-product in the aldol reaction you performed.
2. Why is the crude product in this experiment washed with sodium bicarbonate? (Refer to Table 4.1.)
3. Draw all possible geometric isomers of your product.
4. Based on the stoichiometry shown, draw all possible products, including geometric isomers, for the reaction shown below.

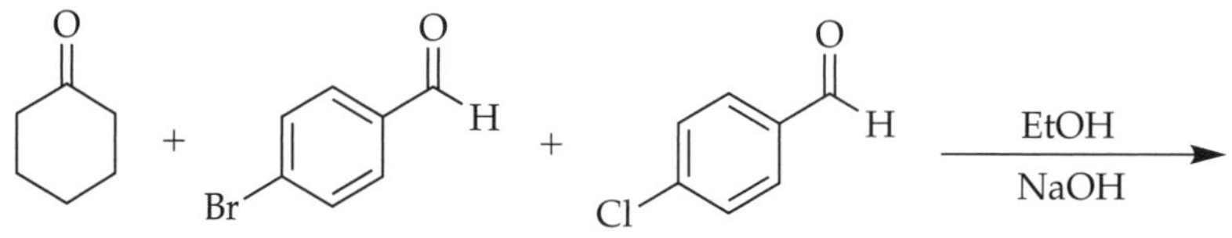

5. Starting with any aldehyde(s) and acetone, outline a strategy for selectively synthesizing the molecule shown below. Explain each step that you would use, including the stoichiometry of the reactants and any reagents and workup conditions. What type of spectral data would you acquire to ensure that you had obtained the desired product?

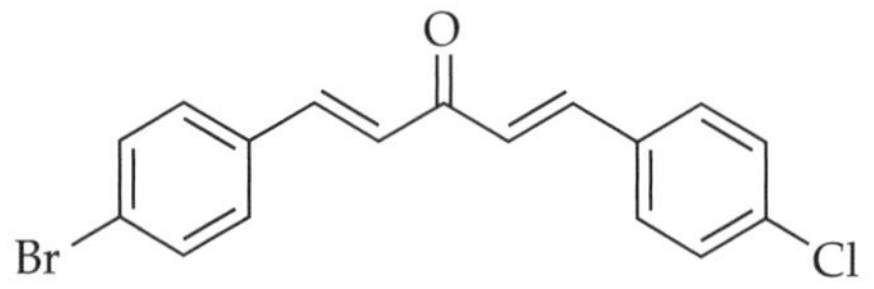

Identifying the Structure of an Aldehyde by Qualitative Analysis

The identification of organic compounds using analytical techniques such as nuclear magnetic resonance spectroscopy and mass spectrometry has become commonplace in the organic chemistry laboratory. But how did organic chemists determine the identity of an unknown compound before the development of these techniques? If the compound has been reported previously, the physical properties of the compound can aid in narrowing down the possibilities (Experiment 4). For example, if the compound is a solid, melting point ranges can be measured with relative ease. However, many compounds have similar melting point ranges so that a unique identification is often not possible without using mixed melting points. In the case of liquids, the boiling point range is more difficult to measure and comparison with literature values is less precise.

There are a plethora of qualitative analytical methods to help identify the functional groups present in a given compound. These tests generally rely on reactions that are specific for a given type of functional group. For example, the Tollen's test for aldehydes makes use of the oxidizing properties of the silver(I) cation. The products are a carboxylic acid and a precipitate of metallic silver, which appears as a mirror on the inside of the reaction vessel. There are many other tests that afford visual confirmation of a reaction to aid in establishing the presence or absence of one or more functional groups. In this way, the number of possible structures for a compound is significantly reduced.

If the compound has been previously reported, with some information about the functional groups in a molecule, chemists can make derivatives of the unknown compound. Derivatives are the products of the reaction of the unknown with known reagents to provide new compounds. The objective is to make derivatives that are solids so that melting point ranges can be obtained. Tables can be found in the literature that list the physical properties of derivatives of many known compounds[1]. With the physical

---

[1] For example, see Furniss, B.S.; Hanneford, A.J.; Smith, P.W.G.; and Tatchell, A.R. *Vogel's Textbook of Practical Organic Chemistry (5th ed)*, Longman Group with John Wiley and Sons, New York, 1989; Shriner, R.L., Hermann, C.K.F., Morrill, T.C., Curtin, D.Y., Fuson, R.C. *The Systematic Identification of Organic Compounds (8th ed)*, John Wiley and Sons, New York, 2003; Cheronis, N.D., and Entriken, J.B. *Identification of Organic Compounds: A Student's Text Using Semimicro Techniques*, Interscience, New York, 1963; Rappoport, Z (ed) *CRC Handbook of Tables for Organic Compound Identification*, Chemical Rubber Company, Boca Raton, FL, 1983.

Figure E24.1 Nitrogenous Derivatives of Aldehydes and Ketones

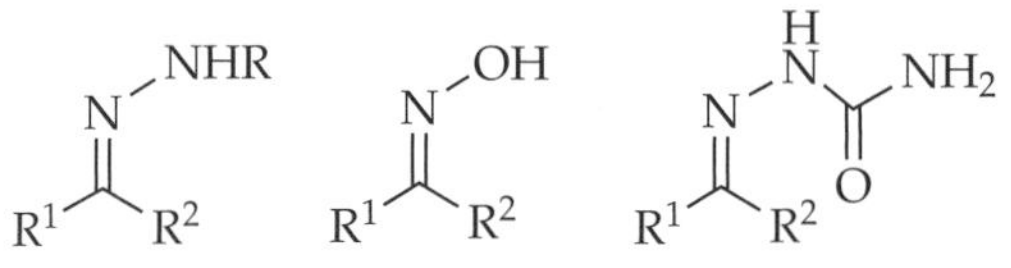

a hydrazone   **an oxime**   **a semicarbazone**

properties of the unknown material and the melting point ranges of one or more derivatives, the identity of the unknown can hopefully be established by comparison with the values in the literature.

In the case of aldehydes and ketones, a variety of derivatives can be made that rely on the transformation of the carbonyl group into a nitrogenous analog. The general reaction is shown in Equation E24.1.

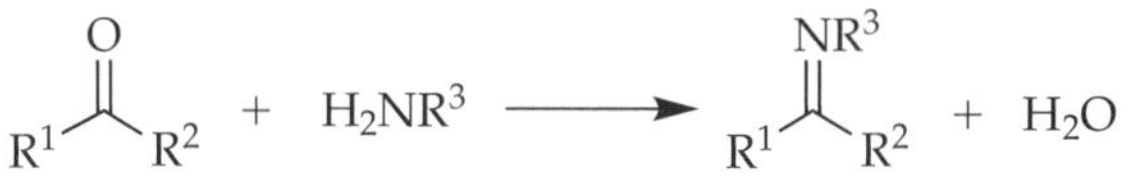

$$ \text{(E24.1)} $$

If a substituted hydrazine is used (R=NHR′), the derivative is a hydrazone. When hydroxylamine is used (R=OH), the product is an oxime. Reaction of the aldehyde or ketone with semicarbazide (R=NH(CO)NH$_2$) gives a semicarbazone (Figure E24.1).

Another class of crystalline aldehyde derivatives is shown in Equation E24.2. The starting diketone is commonly referred to as dimedone and the product is known as a dimethone. Two equivalents of dimedone add to the aldehyde. The reaction does not work with ketones.

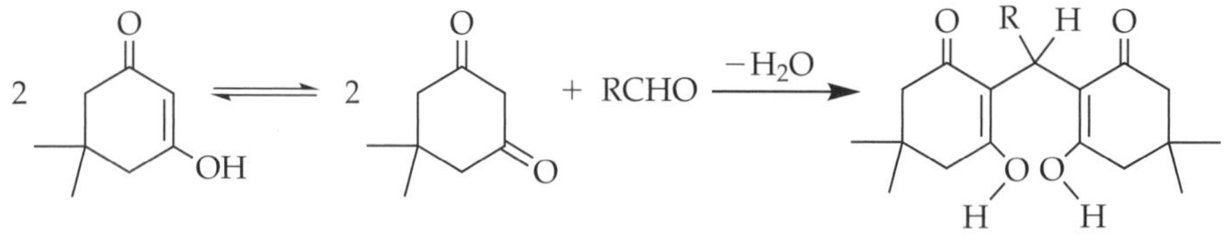

$$ \text{(E24.2)} $$

dimedone   **a dimethone**

The dimethone derivative can be converted into another crystalline derivative, referred to as a dimethone anhydride (Equation E24.3). The ability to obtain two derivatives from the same amount of starting aldehyde is a valuable aspect of dimedone chemistry, especially when only small amounts of an unknown aldehyde are available.

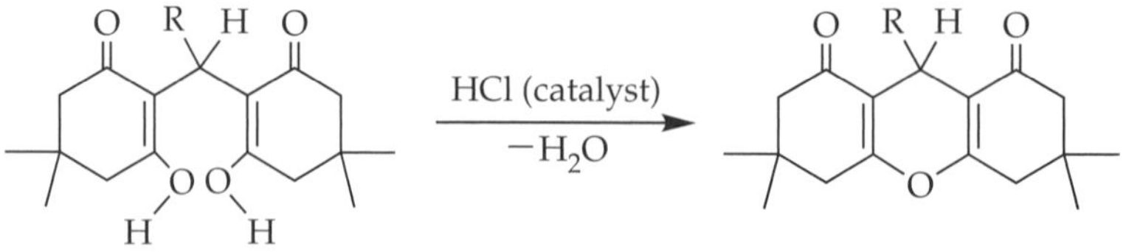

$$ \text{(E24.3)} $$

a dimethone   **a dimethone anhydride**

In this experiment, you will prepare a crystalline dimethone derivative from an unknown aldehyde. After determining the melting point range of this product, you will convert the dimethone into the dimethone anhydride and determine the melting point range of the anhydride derivative. Comparison of the two melting point ranges with the data in Table E24.1 and other physical properties of the unknown will help you establish the identity of the unknown aldehyde.

Background Reading:

Before the laboratory period, read about keto-enol chemistry, the acid-catalyzed aldol condensation, and conjugate addition in your lecture text.

Prelab Checklist:

The prelab write-up should include the structures and physical constants for dimedone and any solvents used in this experiment. You do not need to include the physical constants for the unknown aldehydes.

Table E24.1 Table of Unknown Aldehyde Derivatives

Structure	Name	MP °C Dimethone Derivative	MP °C Dimethone Anhydride Derivative
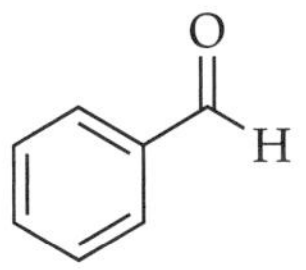	benzaldehyde	194–195	204–206
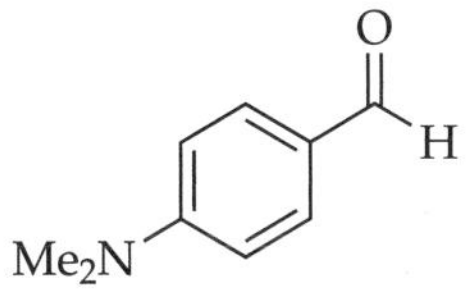	4-dimethylamino-benzaldehyde	194.5–196.5	220–222
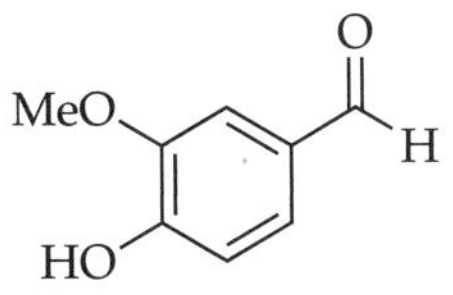	vanillin 4-hydroxy-3-methoxy-benzaldehyde	195.5–196.5	226–228
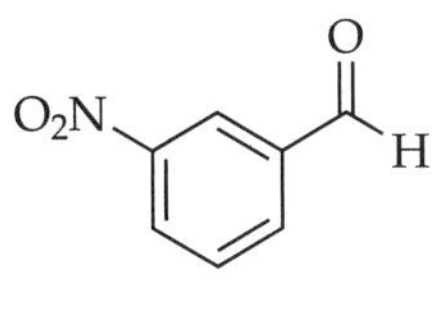	3-nitrobenzaldehyde	197–198	171.5–172.5
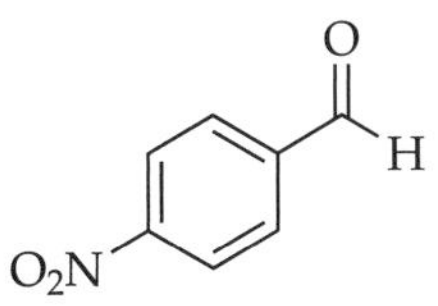	4-nitrobenzaldehyde	188–190	222

(Continued)

Table E24.1 Table of Unknown Aldehyde Derivatives *(Continued)*

Structure	Name	MP°C Dimethone Derivative	MP°C Dimethone Anhydride Derivative
	4-hydroxybenzaldehyde	188–190	246
	2-methoxybenzaldehyde	187–188	190–191
	veratraldehyde 3,4-dimethoxybenzaldehyde	173–174	184–185.5
	cumin aldehyde 4-(1-methylethyl)-benzaldehyde	170–171	172–173
	piperonal	175.5–177	218.5–220
	meta-tolualdehyde 3-methylbenzaldehyde	171–172.5	205–207

Did you know?

vanillin

The bean of the plant *vanilla planifolia* is the sole natural source of 4-hydroxy-3-methoxybenzaldehyde, commonly known as vanillin. The plant is an orchid that blooms only one day a year for just a few hours. During that time, it must be pollinated and the only natural pollinator is the Melipona bee. Because of the critical timing, the flowers are now pollinated by hand. The pollinated flowers mature into small bean pods that grow in bunches like bananas. When these beans are dried and split, they reveal the tiny black grains used in flavorings.

Experimental Procedure

Wear your safety glasses at all times in the lab.

Part I: Preparation of the Dimethone Derivative

Your instructor will assign you an unknown aldehyde.

- *If your unknown aldehyde is a solid,* take the melting point of the compound. Weigh 100 mg of the compound in a large test tube. Add 300 mg of dimedone, 4 mL of a 1:1 ethanol:water mixture, and a boiling stick to the tube.

- *If your unknown aldehyde is a liquid,* weigh 300 mg of dimedone in a large test tube. Add 4 mL of a 1:1 ethanol:water mixture and a boiling stick to the tube. Using a syringe, deliver 0.1 mL of the liquid unknown to the mixture.

Bring the reaction mixture to a gentle reflux on a sand bath. If all of the starting materials have not dissolved, use a clean glass stir rod to *gently* crush any large crystals. Reflux the mixture for 5 minutes. Remove the test tube from the sand bath and let the reaction cool to room temperature.

- *If crystals are present,* cool the tube in an ice-water bath for an additional 10 minutes.

TECHNIQUE TIP   If an oil forms after cooling the reaction mixture to room temperature, swirl the contents of the flask vigorously until crystallization begins. Make sure any oil droplets have dissolved before placing the tube in the ice-water bath.

- *If crystals have not formed,* try one of the following: (1) vigorously swirl the contents of the tube and let it stand for a minute or two or (2) *gently* scratch the inside of the tube with the glass stirring rod (Section 7.1.6).

- *If no crystals are present after these attempts,* add distilled water one drop at a time. After each drop, swirl the mixture to re-dissolve any solids that form. Stop adding water once crystals appear or a cloudiness persists (Section 7.1.3.a). Then, cool the mixture for 10 minutes in an ice-water bath.

Isolate the crystals on a Hirsch funnel, using a spatula to remove any remaining crystals. Use an ice-cold 1:1 ethanol:water mixture to wash the crystals. Dry the crystals by drawing a vacuum through the filtered material.

Determine the melting-point range of the product. Check this range with the data for the dimethone derivatives in Table E24.1. If you can start to narrow down the possible structures at this point, record the possibilities in your laboratory notebook.

Part II: Preparation of the Dimethone Anhydride Derivative

Weigh 50 mg of the dimethone derivative prepared in Part I in a reaction tube/small test tube. Add 2 mL of a 4:1 ethanol:water mixture and a boiling stick. Gently heat this solution on the sand bath until all of the starting material dissolves. Remove the tube from the heat and add 1 drop of concentrated hydrochloric acid.

CHEMICAL SAFETY NOTE Hydrochloric acid is corrosive. If any solution containing hydrochloric acid comes in contact with your skin, immediately flush the affected area with water for at least five minutes and have another student notify the instructor.

Continue heating the solution at a gentle reflux for 5 minutes. Remove the test tube from the sand bath and allow it to cool to room temperature.

- *If crystals are present*, cool the tube in an ice-water bath for an additional 10 minutes.

- *If crystals have not formed*, follow the suggestions outlined in Part I. Then, cool the mixture for 10 minutes in an ice-water bath.

Isolate the crystals on a Hirsch funnel, using a spatula to remove any remaining crystals. Wash the crystals with a small amount of an ice-cold 4:1 ethanol:water mixture. Dry the crystals by drawing a vacuum through the filtered material.

Determine the melting-point range of the product. Check this range with the data for dimethone anhydride derivatives given in Table E24.1. Use this information along with the melting-point range of the dimethone derivative from Part I to establish the identity of your aldehyde.

Cleanup

Place the products in appropriate containers. Place any remaining solvents in the appropriate liquid waste containers. Dispose of all contaminated Pasteur pipettes, used weigh paper, and melting point capillaries in the container for contaminated laboratory debris. Dispose of any clogged syringe needles in the container for used syringe needles.

Discussion

- Briefly describe how you determined the structure of your unknown aldehyde. Was it necessary to make two derivatives?

- Look up the physical properties of your unknown in the literature. Does this information confirm your answer? If your unknown was a solid, is the literature melting point consistent with your experimental measurement? If not, reexamine the data and explain any discrepancies.

- The preparation of derivatives, such as the dimethone and its anhydride, are no longer routinely used by organic chemists. NMR spectroscopy and mass spectrometry are the main tools used for structure determination in the modern organic chemistry laboratory. Sketch the

^{1}H and ^{13}C[^{1}H] NMR spectra of the aldehyde you were assigned in this experiment. For the ^{1}H NMR spectrum be sure to show any multiplets you would expect to see. For both the ^{1}H and ^{13}C[^{1}H] NMR spectra, the chemical shifts should be reasonable and you should assign all of the resonances (Chapter 8).

- Would the NMR spectra you sketched be adequate to definitively assign the correct structure to your aldehyde? Give an example of another compound that might have similar spectra and explain how you would tell the difference between your unknown and the alternate compound. You may use other analytical techniques to distinguish the two.

- Provide one example of a qualitative analytical method that is commonly used today. Your example can include methods used in commercial products. Briefly explain the chemistry behind the technique. Provide the complete citation for your reference including, for example, the exact book, author, article, or Internet address.

Questions

1. Using an acid catalyst, write a logical arrow-pushing mechanism for the reaction shown in Equation E24.2.

2. Write a logical arrow-pushing mechanism for the reaction shown in Equation E24.3.

3. If the melting point ranges of your dimethone and dimethone anhydride derivatives were the same, what common analytical technique(s), not including NMR or mass spectrometry, could you use to establish if the compounds were the same or different?

4. A student accidentally used the 4:1 mixture of ethanol:water intended for Part II in Part I. What effect(s) might this mistake have on the outcome of the reaction in Part I?

5. When 2-hydroxybenzaldehyde is reacted with two equivalents of dimedone, a product different from the dimethone structure shown in Equation E24.2 is obtained. The new compound consists of two dimedone units linked together by the aldehyde. The molecular formula of $C_{23}H_{26}O_4$ suggest that it has lost water, but it is not a dimethone anhydride. Propose a structure for this molecule. Hint: Consider what the 2-hydroxy group of the starting aldehyde might do once it is part of the dimethone structure.

The Benzoin Reaction[1]

In the 1870s, Christiaan Eijkman, a Dutch physician, was sent to the East Indies to investigate a disease known as beriberi, which was thought to be caused by a microorganism. The disease was common in Asia among sailors and prisoners, but was not found among those from Europe[2]. Symptoms of the disease included mental confusion, muscle wasting, high blood pressure, difficulty walking, and heart problems. Eijkman observed that chickens suffered from symptoms similar to beriberi when they were fed polished white rice, but that the addition of rice polishings to their diet alleviated these symptoms. The disease was cured in humans by adding green peas, green beans, meat, and unpolished rice to the diet. In 1926, 40 mg of a pure crystalline material was isolated from rice polishings. Eijkman confirmed that this compound cured birds of the beriberi-like symptoms. Beriberi was therefore one of the first diseases found to be caused by a dietary deficiency and Eijkman was a co-recipient of the Nobel Prize in physiology or medicine in 1929 for his work[3].

Beginning in 1935, groups of chemists in the United States, Germany, and England worked feverishly to determine the structure and synthesize the material. An American, R.R. Williams, was the first to publish the details of the complete synthesis in August 1936[4]. Williams and Cline established that the synthetic material was identical to the crystals isolated from natural sources. Thiamine was the first B vitamin discovered and consequently it is known as vitamin B_1. Vitamins are molecules that are not synthesized in the body and must be present in the diet because they function with enzymes to catalyze chemical reactions in the body.

Thiamine (vitamin B_1) is a water-soluble vitamin found naturally in whole grains, egg yolks, fresh legumes, and meat. In addition, it is added to flour, bread, breakfast cereals, pasta, and milk to improve their nutritive value. It is not stored in the body and, because of its water solubility, it is readily excreted. Humans need less than 1 mg a day to prevent beriberi. The amount required depends on lifestyle, age, and disease. Beriberi results from severe deficiency and the disease is still prevalent today in areas where famine exists, as evidenced by the resulting swollen abdomens of children. Symptoms of mild deficiency include reduced stamina, depression, and irritability. On the other hand, very large doses of vitamin B_1 may contribute to the imbalance of B vitamins for which an overdose of one may lead to a deficiency of others. The recommended daily allowance (RDA) of vitamin B_1 for adults is 1 to 1.5 mg.

---

[1] Ide, W.S., and Buck, J.S. *Organic Reactions* **1948**, *4*, 269.

[2] Sebrell, W.H., Jr., and Harris, R.S., eds *The Vitamins* Vol. V, 97, **1972**.

[3] http://www.almaz.com/nobel/medicine/.

[4] Williams, R.R. and Cline, J.K. *J.Am.Chem.Soc.* **1936,** *58*, 1504.

Thiamine is stable in acid, but decomposes to thiazole and a pyrimidine when boiled or stored in water. In strong base, the thiazole ring opens. Thiamine is sensitive to alcohol, tannins in coffee and black tea, and to sulfites. Baking soda also destroys thiamine. As a result, much of the thiamine is lost from foods during cooking. Overconsumption of alcohol results in depletion of thiamine available for metabolic processes and is one of the reasons for the symptoms known as a hangover. Consequently, some sources recommend taking B$_1$ pills both before and after heavy alcohol consumption.

Some enzymes require small molecules called **cofactors** in order to catalyze reactions. The enzyme functions to bind, but not covalently, the cofactor (the catalyst) and the substrate (the starting material). Thiamine pyrophosphate (TPP) is a cofactor that is essential for the metabolism of carbohydrates and alcohol in the body and for the polarization-depolarization of membranes in nerve transmission. The hydrogen on the carbon atom between the nitrogen and the sulfur on the thiazole ring is relatively acidic (pK$_a$ 17–19).

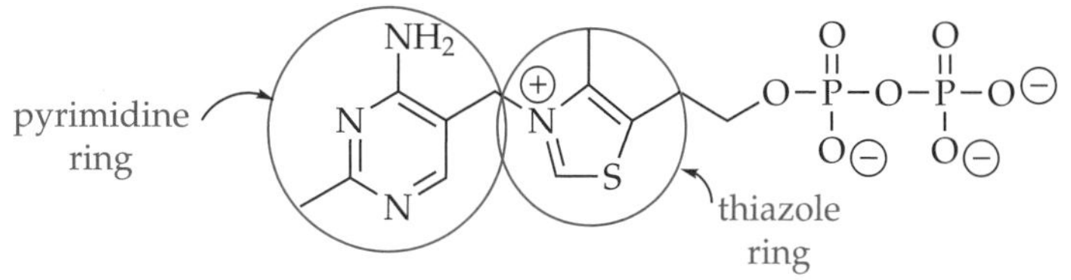

thiamine pyrophosphate

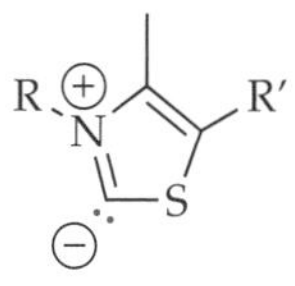

The ylide that results from the deprotonation of the thiazole ring acts as a nucleophile toward carbonyl groups to form a tetrahedral intermediate. The loss of carbon dioxide (from α-ketoacids) or a proton (from aldehydes) results in a stabilized carbanion (Figure E25.1). The thiazolium moiety acts as an electron sink to stabilize the negative charge on the carbon in the intermediates involved in the reactions. This enamine behaves like a carbanion that can react with a proton, a carbonyl group, or another good electrophile.

TPP is an important cofactor for a variety of enzymes called decarboxylases because they are involved in decarboxylation reactions. For example, pyruvate decarboxylase requires TPP for the decarboxylation of pyruvate to acetaldehyde (Equation E25.1). The tetrahedral intermediate formed when the thiazolium moiety adds to the ketone carbonyl loses carbon dioxide to form the resonance stabilized carbanion (enamine). This carbanion then reacts with a proton before collapsing to eliminate the TPP catalyst.

An ylide is a compound having opposite charges on adjacent atoms.

an ylide

Figure E25.1   The Stabilized Carbanion

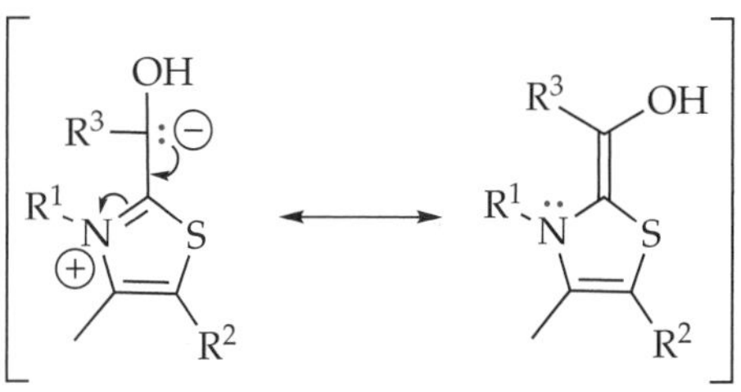

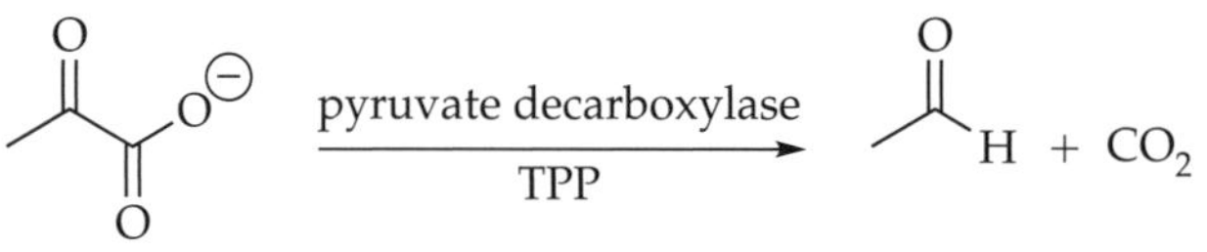

$$\text{pyruvate} \xrightarrow[\text{TPP}]{\text{pyruvate decarboxylase}} \text{acetaldehyde} + CO_2 \qquad (E25.1)$$

Another enzyme, acetolactate synthase, requires thiamine pyrophosphate to convert two molecules of pyruvate to acetolactate (Equation E25.2). This is the first step in the biosynthesis of the amino acids valine and leucine.

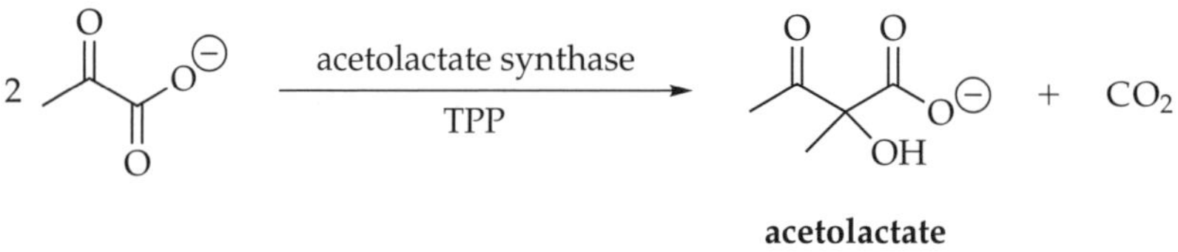

$$2 \; \text{pyruvate} \xrightarrow[\text{TPP}]{\text{acetolactate synthase}} \text{acetolactate} + CO_2 \qquad (E25.2)$$

acetolactate

TPP is also essential for the decarboxylation of α-ketoglutaric acid to succinic acid in the citric acid cycle (Equation E25.3).

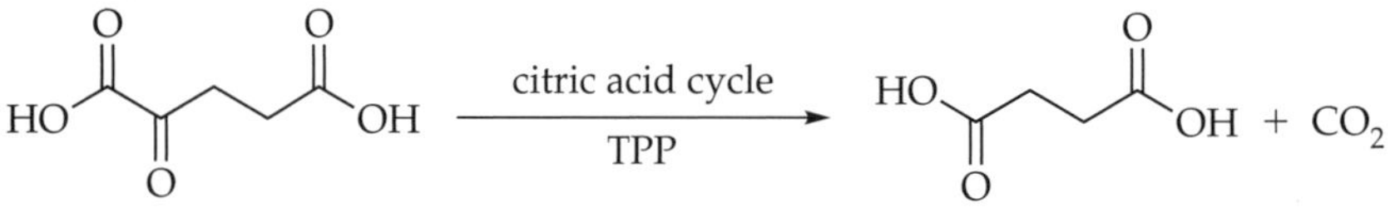

$$\text{α-ketoglutaric acid} \xrightarrow[\text{TPP}]{\text{citric acid cycle}} \text{succinic acid} + CO_2 \qquad (E25.3)$$

In addition to the decarboxylases, transketolase also requires the coenzyme thiamine pyrophosphate (Equation E25.4). This enzyme transfers the two-carbon (hydroxymethyl)carbonyl unit from one carbohydrate to another.

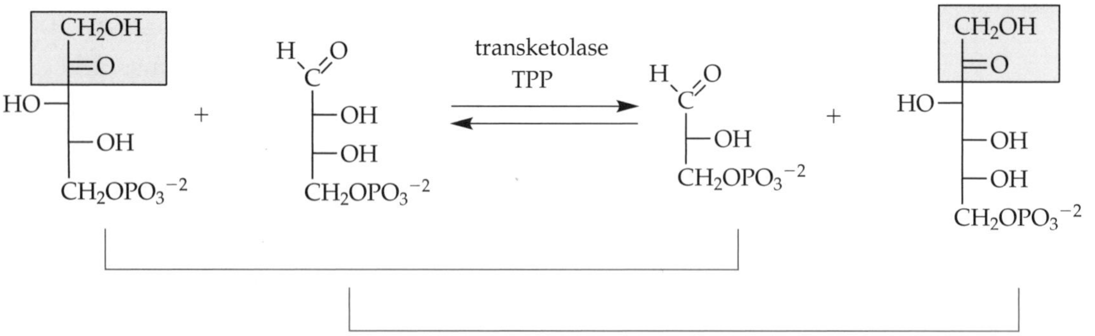

$$(E25.4)$$

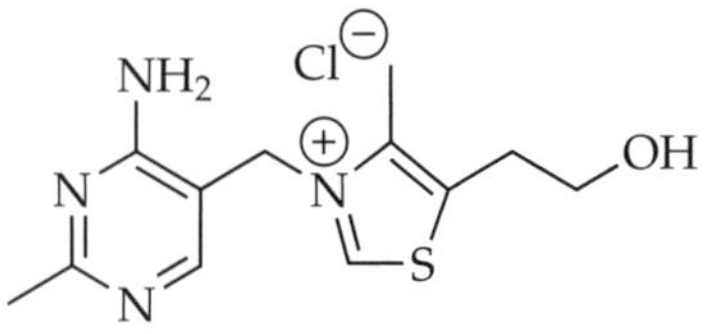

thiamine chloride

In the laboratory, thiamine chloride is used as a model system to catalyze reactions in the absence of an enzyme. In this experiment, you will use the hydrochloride salt of thiamine chloride as a catalyst for the synthesis of an α-hydroxyketone in the benzoin reaction (Equation E25.5).

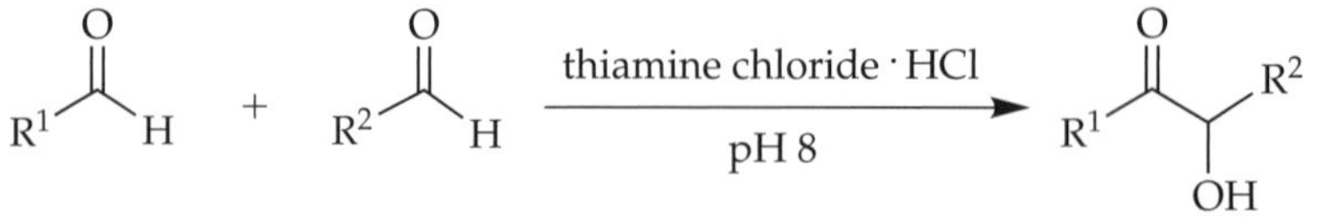

$$R^1\text{CHO} + R^2\text{CHO} \xrightarrow[\text{pH 8}]{\text{thiamine chloride}\cdot\text{HCl}} R^1\text{C(O)CH(OH)}R^2 \qquad (E25.5)$$

Because it is a catalyst, the rate of the reaction is proportional to the concentration of the thiamine. Aqueous solutions of thiamine chloride hydrochloride are acidic so that the amount of thiamine used alters the pH of the solution. The control of pH is very critical for the reaction. The hydrogen atom on the two position of the thiazolium ring is relatively acidic and ionizes to give an ylide that is the catalytically active form. Although thiamine is stable in acid, at a pH less than 8, the thiamine is not sufficiently ionized and the reaction will not proceed. A pH greater than 10 interferes with the reaction by hydrolyzing the thiazole ring.

Did you know?

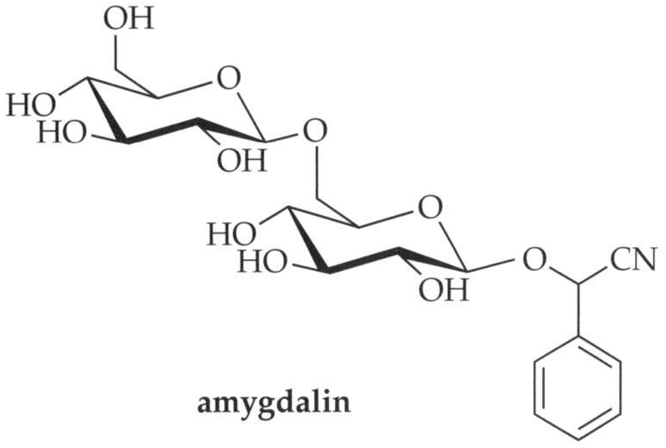

amygdalin

Amygdalin, also known as laetrile, is a toxic glycoside found in the seeds of *Rosaceae* such as apricots, almonds, and peaches. Enzymes catalyze its decomposition to form benzaldehyde, two molecules of glucose, and hydrocyanic acid that is highly poisonous. In 1892, laetrile was tested as an anti-cancer treatment in Germany, but ruled ineffective and toxic. In spite of this, its use was promoted after 1950 in the United States and it wasn't until the 1980's that tests in animals and humans confirmed its toxicity and ineffectiveness and its use is not approved by the FDA.

Background Reading:

Before the laboratory period, read about the benzoin condensation in your lecture text, in reference 1, or on the Internet at http://www.organic-chemistry.org/namedreactions/benzoin-condensation.shtm.

Prelab Checklist:

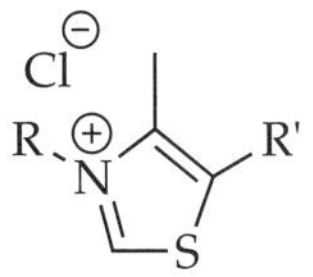

Include the structures and the physical properties of the starting materials and product in your prelab write-up. Write the mechanism of the reaction using curved arrows to show the movement of the electrons. Use the structure to the left to represent thiamine chloride. Predict the relative R_f values of the starting material and the product.

Experimental Procedure

Don't forget to keep your safety glasses on during the entire laboratory period.

Weigh 60 mg of thiamine chloride hydrochloride in a reaction tube/small test tube. Add up to three drops of water dropwise with stirring until the thiamine dissolves. Then, add 0.30 mL of 95% ethanol and stir the solution well. Cool the solution in an ice-water bath and add ice-cold 2N sodium hydroxide drop-by-drop, stirring well between additions until the solution is yellow. If the yellow color fades after a few minutes, add drops of 2N sodium hydroxide until the yellow color persists. Check to confirm that the pH is about 8.

CHEMICAL SAFETY NOTE   Sodium hydroxide is corrosive. If any solution containing sodium hydroxide comes in contact with your skin, flush the affected area immediately with water for at least 5 minutes. Sodium hydroxide is particularly harmful to the eyes. If any solution contacts your eyes, proceed immediately to the eyewash station and flush your eyes for at least 5 minutes. In either case, have another student notify the instructor immediately.

Weigh the reaction tube/small test tube containing the thiamine chloride solution and add about 1.5 mmol of benzaldehyde. Reweigh the tube to determine the exact amount of benzaldehyde added. Stir the solution well using the stainless steel spatula. If the yellow color fades, add 2N NaOH dropwise with stirring between drops until a yellow color persists and the pH is 8. If the solution is cloudy, add 95% ethanol drop by drop with stirring until the solution is homogeneous. Spot a solution of benzaldehyde in lane 1 and the reaction mixture in lane 2 of a TLC plate, leaving room for one more lane.

EXPERIMENTAL NOTE   If the solution turns orange, it is an indication that the pH is too high. Discard the solution and start again.

Connect the air condenser to the reaction tube/small test tube, add a boiling chip, and heat the solution in a water bath kept between 60°C and 80°C on the sand bath for 60 minutes. Monitor the temperature of the water bath with a digital thermometer.

After 15 minutes of heating, spot the reaction mixture in lane 3 of the TLC plate. Develop the plate in 90:10 hexane:ethyl acetate and visualize it with UV light and iodine. If there is no evidence of product after 15 minutes, you should start over.

Save On Solvents: Share your TLC solvent chamber with a neighbor.

Follow the reaction by TLC using a second plate to spot the reaction mixture after 30, 45, and 60 minutes. Develop the plate in 90:10 hexane:ethyl acetate and visualize it with UV light and iodine.

Cool the reaction mixture in an ice-water bath. If no precipitate appears upon cooling, scratch the inside of the reaction tube/small test tube *gently* with a glass stirring rod. The addition of a drop of water may also help initiate crystallization. After crystallization has ceased, remove the solvent using the square-tip Pasteur pipette method (Section 7.2.1). Place the reaction tube/small test tube in an ice-water bath and wash the product with a 50-50 mixture of ice-cold 95% ethanol and water. Recrystallize the product from 95% ethanol.

Dry the crystals and determine the weight, percent yield, and melting point of the product.

Cleanup

The product should be placed in a suitable container. It can be used as a starting material in Experiment 13. The filtrate from the reaction and the ethanol/water washes, the solvent from the recrystallization, and the TLC solvents should be poured into the appropriate waste containers. Dispose of the contaminated Pasteur pipettes, used weigh paper, TLC plates, and melting point and TLC capillaries in the container for contaminated laboratory debris.

Discussion

- Discuss the rationale behind each step that you performed prior to the addition of the benzaldehyde. What chemical species do you think is responsible for the yellow color of the solution?

- Discuss the details associated with the progress of the reaction as a function of time.

- Discuss the yield and purity of your product. Explain what happened to the thiamine.

Questions

1. Draw all reasonable structures, including their resonance forms, to indicate the possible position of the proton of HCl in thiamine chloride hydrochloride. Explain which of the structures you have drawn is most reasonable.

2. Write an equation showing the formation of the ylide from the ionization of the hydrogen on the two position of the thiazolium cation. Include all resonance forms for the thiazolium cation and the ylide. Does resonance stabilization explain the acidity of the hydrogen on the two position of the thiazolium salt? Explain your answer.

3. Examine the mechanism and the intermediates in the reaction to explain why the thiamine-catalyzed synthesis of benzoin is sensitive to electron-withdrawing and electron-donating substituents in the four position (the para position) of the benzene ring of the aldehyde. For each of the steps in the mechanism, indicate whether electron-donating or electron-withdrawing substituents stabilize the intermediate.

4. Identify the chiral center in the product from this experiment. Is the product optically active? Explain your answer.

5. In nature, thiamine pyrophosphate (vitamin B_1) is a cofactor used by transketolase, as shown in Equation E25.4. Write the mechanism for this reaction using curved arrows to show the movement of the electrons.

The Diels-Alder Reaction

The Diels-Alder reaction is an important tool in synthetic chemistry. The cycloaddition of a conjugated diene and an alkene, called a dienophile, allows control over the regio- and stereochemical placement of the substituents on the resulting cyclohexene ring.

Otto Paul Hermann Diels and Kurt Alder published their first paper on this reaction in 1928[1]. Twenty-two years later, the reaction was so important in organic synthesis that these two men shared the Nobel Prize in chemistry[2]. In this experiment you will be examining the reaction of maleic anhydride, a classic Diels-Alder dienophile, with one of three conjugated monoterpenes, the diene. The reaction of one of these terpenes, α-phellandrene, with maleic anhydride was reported in the original 1928 paper (Equation E26.1).

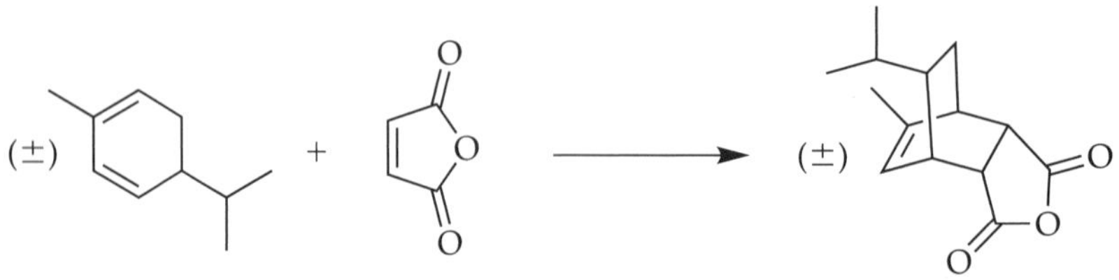

$$(\pm) \quad + \quad \longrightarrow \quad (\pm) \qquad \text{(E26.1)}$$

α-phellandrene   maleic anhydride

Your instructor will assign you one of the monoterpenes. The structures of the unknown monoterpenes are shown in Figure E26.1. Because not all of the Diels-Alder adducts from the reaction with maleic anhydride are readily crystallized, you will make a derivative of your product by treating it with aqueous base. You will identify the monoterpene you were assigned by examining the NMR spectral data for the derivative. It will be necessary to consider the differences and similarities among the spectra of the possible derivatives from the three monoterpenes.

Figure E26.1   The Structures of the Unknown Monoterpenes

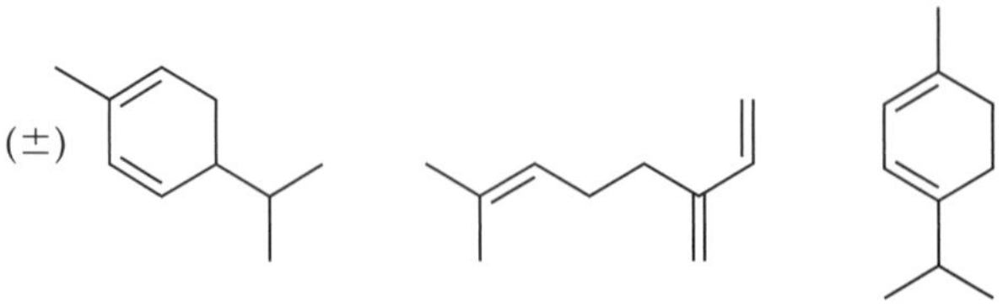

α-phellandrene   β-myrcene   α-terpinene

---

[1] Diels, O., Alder, K. *Ann.* **1928**, *460*, 98.

[2] See http://nobelprize.org/chemistry/laureates/1950/index.html.

344

Did you know?

Chemicals used to elicit behavioral or physiological effects on individuals of the same species are called pheromones. When different species are involved, they are called allelochemicals and can produce a variety of effects. A compound that is formed by one species that is of benefit to itself and to another species is called a synomone. The volatile compounds that are produced by plants to attract pollinators are synomones. Allomones help the emitter to the detriment of the receiver, for example, a chemical emitted for defense against a predator. Myrcene, α-terpinene, and α-phellandrene are all allomones emitted by some plant species to repel insects. Kairomones benefit the receiver, but not the emitter. When it is damaged by the Western pine beetle, the Ponderosa Pine emits myrcene. This attracts more beetles to the tree, causing further damage.

Background Reading:

Before the laboratory period, read about the Diels-Alder reaction, terpene chemistry, and anhydride reactions in your lecture text.

Prelab Checklist:

Include the structures and physical constants for maleic anhydride, β-myrcene (listed as myrcene), α-phellandrene, α-terpinene, and the solvents used in this experiment in your prelab write-up. Write a balanced equation for the Diels-Alder reaction between each diene and maleic anhydride, being sure to include your predictions about the stereochemical outcome of each reaction.

Experimental Procedure

Don't forget to keep your safety glasses on during the entire laboratory period.

Add 1 mmol of maleic anhydride, a stir bar, and 0.5 mL of toluene to a reaction tube/small test tube. Cap the tube with a rubber septum and insert a needle in the septum. Start heating and stirring this mixture on a sand bath. The sand bath temperature should eventually be between 120 °C and 140 °C in order to get the reaction mixture to reflux. While the mixture is heating, remove the septum briefly and use a syringe to add 0.34 mL of the monoterpene you were assigned. After the addition is complete, immediately sample the reaction mixture using a TLC capillary attached to a boiling stick and then replace the septum. Spot the sample in lane 3 of a TLC plate and call this $t = 0$. Spot a solution of maleic anhydride in lane 1 and a solution of the unknown monoterpene in lane 2. On a second TLC plate, spot samples of the mixture at reaction times of 30 minutes and 1 hour. Develop the plates in 60:40 hexane:ethyl acetate and visualize them with UV light and iodine.

After refluxing the reaction mixture for 1 hour, remove it from the sand bath and allow it to cool to room temperature. Add 2 mL of 3N NaOH to a centrifuge tube and cool this solution in an ice-water bath. Add the reaction mixture to this cold solution. Cap the tube and shake it for several seconds. Then add 1.5 mL of ether and shake the contents of the tube again. If separation of the layers does not occur readily, centrifuge the contents of the centrifuge tube. After the layers separate, remove the ether layer and place it in a large test tube.

SAFETY NOTE   Be sure to balance your centrifuge tube with another centrifuge tube that has approximately the same amount of liquid.

Repeat the extraction with ether twice more, combining the ether layers. Spot some of the ether solution on a third TLC plate along with solutions of the monoterpene and maleic anhydride. Develop the plate in 60:40 hexane:ethyl acetate and visualize it with UV light and iodine.

Transfer the aqueous solution from the centrifuge tube to a 10-mL Erlenmeyer flask and cool it in an ice-water bath. Check the pH of the solution and then slowly add 1 mL of 3N H_2SO_4 to the solution while gently swirling the contents of the flask. After the addition is complete, check the pH of the solution again to make sure it is acidic. Scratch the inside walls of the Erlenmeyer flask with a glass stir rod until crystallization begins. Once crystallization begins, allow the flask to remain in the ice-water bath for an additional 10–15 minutes.

Isolate the solid product on a Hirsch funnel and wash it with ice-cold water. Dry the crystals to a constant weight. Determine the final weight and melting point of the derivative. Dissolve a small amount of the solid in ethyl acetate and spot it on a TLC plate. Develop the plate in 98:2 ethyl acetate:acetic acid and visualize it with UV light and iodine.

Cleanup

Place the solid product in a suitable container. Place any remaining solvents, TLC chamber solvents, and aqueous solutions in the appropriate liquid waste containers. Dispose of contaminated Pasteur pipettes, used weigh paper, TLC plates, melting point and TLC capillaries in the container for contaminated laboratory debris. Dispose of any clogged syringe needles in the container for used syringe needles.

Discussion

• One of the goals in this experiment is to determine which monoterpene you were assigned based on an analysis of the NMR spectra of the derivative obtained from the Diels-Alder product. Before attempting to analyze the spectra, make a table listing the expected resonances for the derivatives of all three unknowns. For each derivative, indicate the number of different alkene carbons and hydrogens you would expect. Also, predict how many different methyl groups and/or isopropyl groups you would expect to observe. Note any other resonances that you think would stand out in the spectra of these compounds. Your table should include the approximate chemical shift information for these groups (Chapter 8).

• After making the table of predictions, examine the ^1H and $^{13}C[^1H]$ NMR spectra of the derivative of your Diels-Alder product to determine which monoterpene you were assigned (Chapter 8). Your instructor will let you know how or where to obtain the spectra.

• Calculate the percent yield of the derivative.

- Discuss the results of the TLC analysis of the ether extracts obtained after the crude Diels-Alder product was treated with 3N NaOH.

- Provide a rational arrow-pushing mechanism for the reaction of the Diels-Alder product with aqueous base. Explain how this reaction allows the Diels-Alder adduct to be separated from the other reagents and impurities present at the end of the reaction (Section 7.6.1).

Questions

1. The reaction in Equation E26.1 shows the formation of only one pair of enantiomers. Draw the other possible products from the Diels-Alder reaction of α-phellandrene and maleic anhydride and explain why they are either not formed or are formed in low yield.

2. The anhydrides formed in this experiment are reacted with aqueous NaOH followed by acidification. Reaction with $NaOCH_3$ followed by an acidic workup also produces a new product. Write the structure of this product(s) and explain why it would not be a good derivative to use to characterize the Diels-Alder product.

3. The solvent used in this experiment is toluene. Draw one of the Diels-Alder products from the reaction of toluene with maleic anhydride. Why is this product not formed?

4. Write the product of the reaction between myrcene and dimethyl acetylenedicarboxylate (Equation PE26.4). If this product is heated in the presence of palladium on carbon, a new product is formed. Propose a structure for the product. Hint: Refer to Experiment 17.

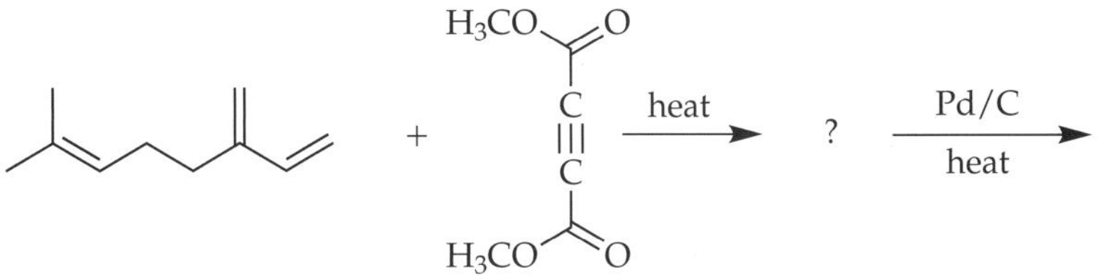

5. Propose a synthesis of the molecule below starting with maleic anhydride and one of the terpenes used in this experiment. You may use any other reagents to complete the synthesis.

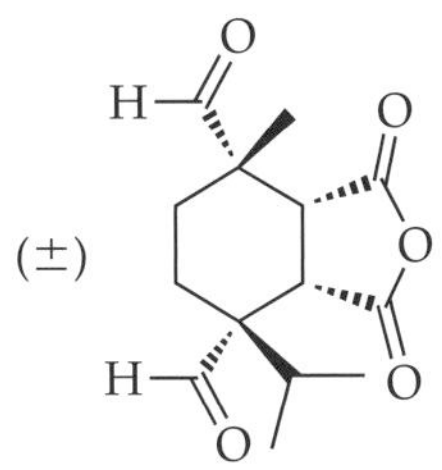

Synthesis of an α,β-Unsaturated Carboxylic Acid Derivative

α, β-Unsaturated carboxylic acids and their derivatives are widely distributed in nature. Some examples are presented in Figure E27.1.

Figure E27.1 Some Naturally Occurring α,β-Unsaturated Carboxylic Acid Derivatives

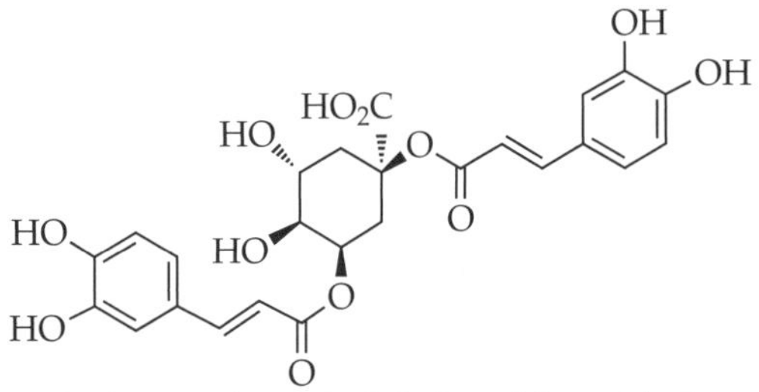

Cynarin is an active principle found in artichokes.

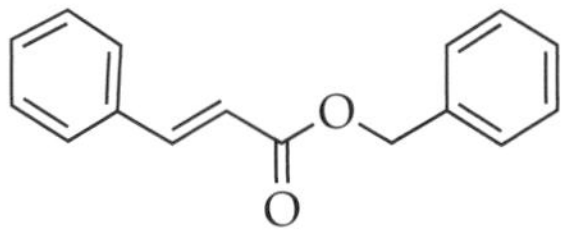

Benzyl cinnamate is found in Peru balsam and is used in the perfume industry and as an artificial flavor. It is characterized by a sweet, balsamic fragrance.

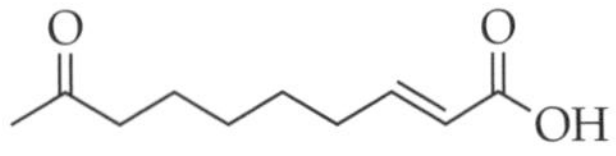

Queen substance is produced in the mandibular gland of queen honey bees. It is used to attract male bees for the purpose of mating and also inhibits ovary development in worker bees.

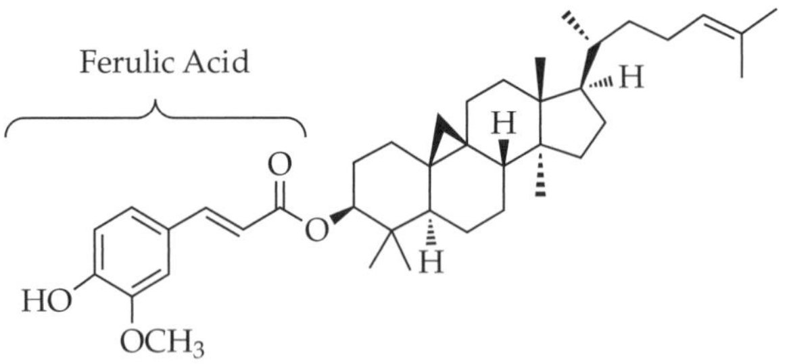

Ferulic Acid

Oryzanol A is found in rice bran and corn and barley oils. It has been examined for antiulcer effects.

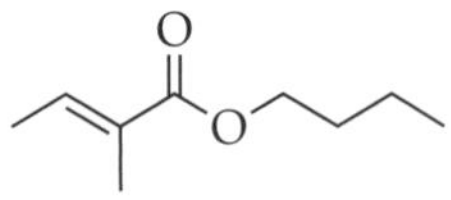

Butyl tiglate is found in the oil of Roman camomile.

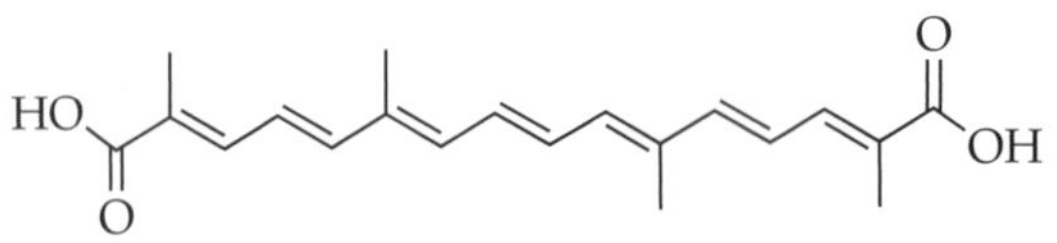

Crocetin is one of the yellow-red pigments isolated from saffron.

Figure E27.2   Retrosynthesis of an α,β-Unsaturated Carboxylic Acid

$$X = OR, NR_2$$

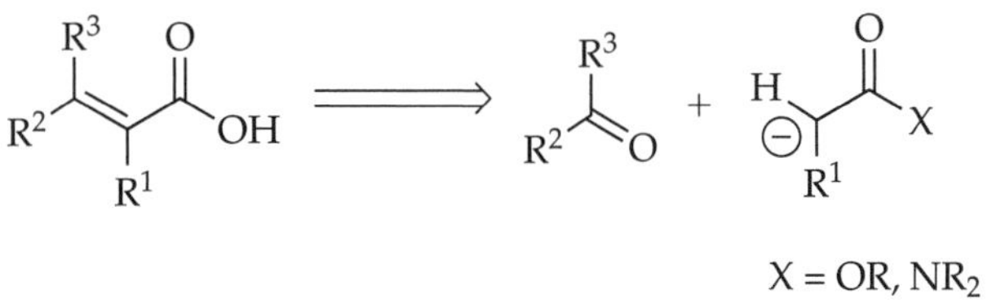

One of the most versatile methods of synthesizing these compounds involves constructing the double bond by condensing an enolate with an aldehyde or ketone (Figure E27.2).

Enolates of carboxylic acids cannot be used because the carboxy proton is more acidic than the α-hydrogen. The dianion that would result is a high energy species (Equation E27.1). Consequently, carboxylic acid derivatives such as esters and amides are used (X = OR or NR$_2$ in Figure E27.2).

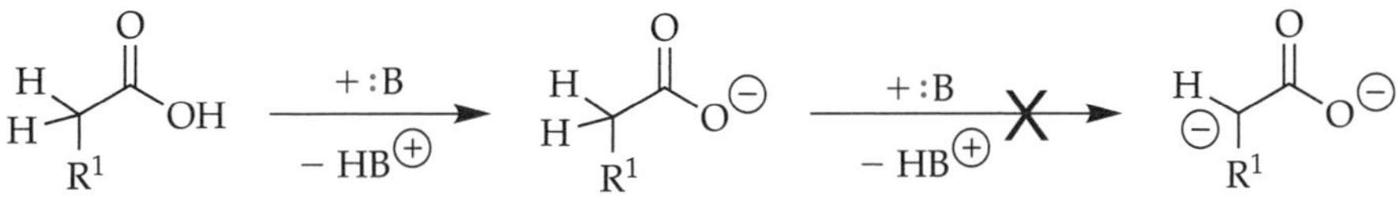

In 1868, William Henry Perkin reported the Perkin reaction[1,2] that utilizes the reaction of acid anhydrides with non-enolizable aldehydes and ketones to synthesize α, β-unsaturated carboxylic acids (Equation E27.2)

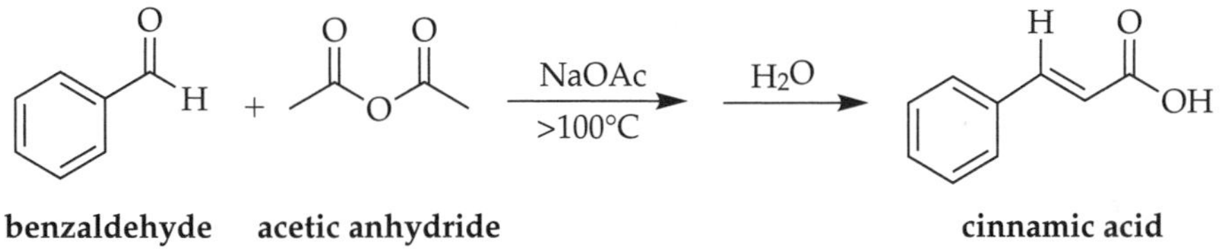

benzaldehyde   acetic anhydride   **cinnamic acid**

Carboxylic acids with α-hydrogens can be used as the source of the enolate in a modification of the Perkin reaction. If a carboxylic acid is reacted with acetic anhydride in the presence of a base such as sodium acetate or a tertiary amine, an equilibrium is established between the acetic anhydride and a mixed anhydride (Equation E27.3).

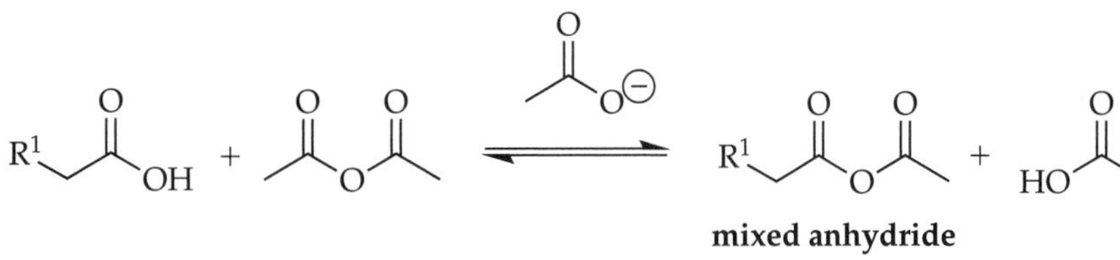

mixed anhydride

[1] Perkin, William H. *J. Chem. Soc.* **1868**, *21*, 53.

[2] Perkin, William H. *J. Chem. Soc.* **1868**, *21*, 181.

Figure E27.3 **Proposed Mechanism of the Modified Perkin Reaction**

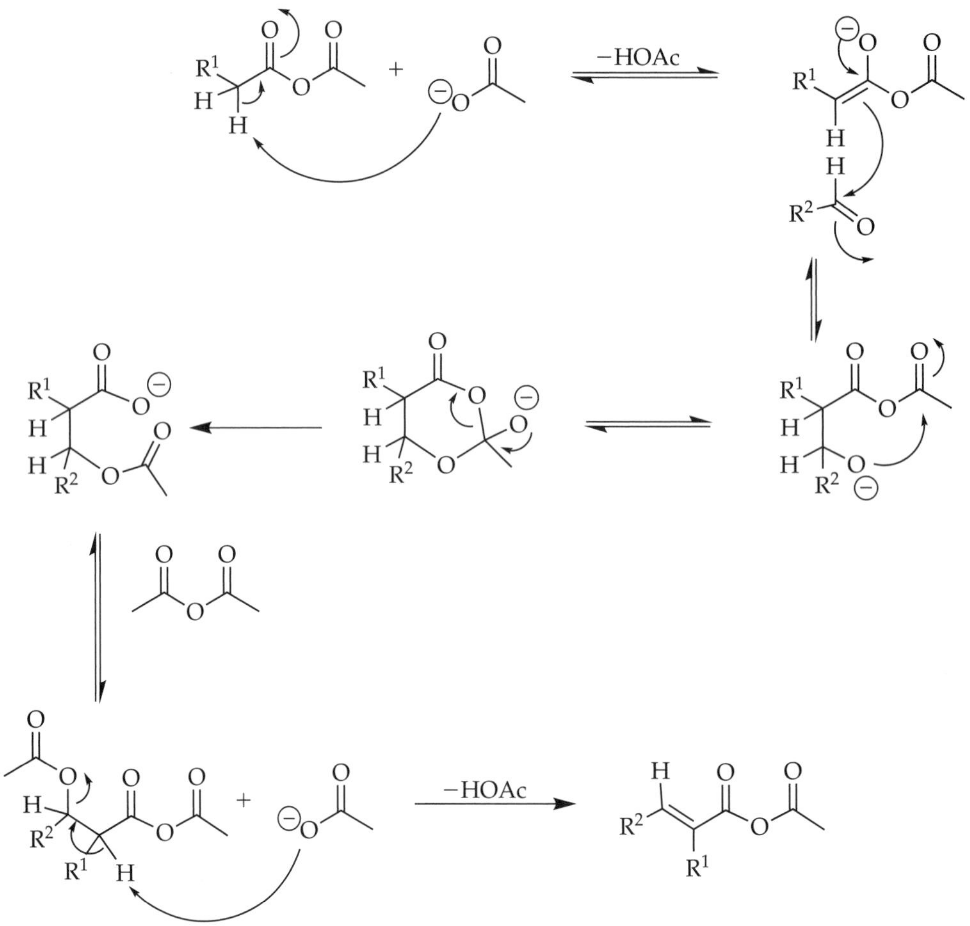

In addition to protecting the carboxy group, the acyl group undergoes an intramolecular reaction with the alkoxide formed from the aldol-type reaction between the enolate from the mixed anhydride and the non-enolizable aldehyde or ketone (Figure E27.3). The resulting acetoxy group is an excellent leaving group that also acts as a base in the formation of the double bond.

A specific example is shown in Equation E27.4. In this reaction, the major product arises from enolization at the more acidic of the two types of enolizable hydrogen atoms found in the mixed anhydride. This is expected under the thermodynamic conditions employed in these reactions.

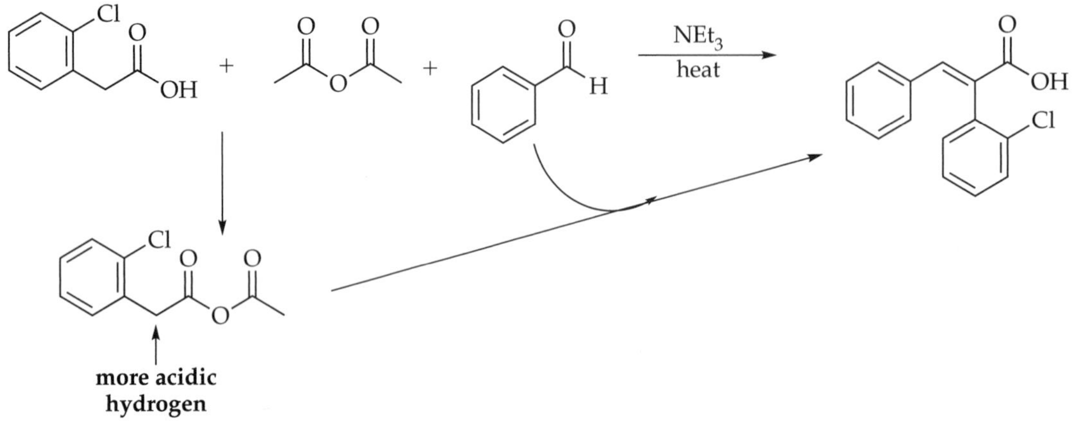

(E27.4)

In the example shown in Equation E27.5, the intermediate mixed anhydride has four types of acidic hydrogen atoms. Deprotonation of the amide N-H, the most acidic of the four types, leads to a cyclic ester known as an azlactone. Enolization of the azlactone followed by reaction with the non-enolizable aldehyde ultimately leads to the new double bond (Figure E27.3).

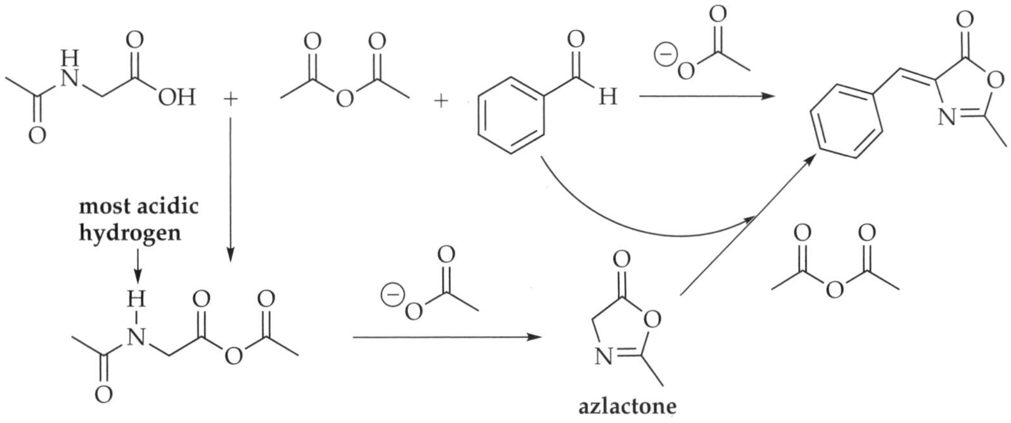

(E27.5)

most acidic hydrogen

azlactone

Hydrolytic ring opening of these cyclic esters followed by catalytic hydrogenation is known as the Erlenmeyer-Plochl azlactone/amino acid synthesis[3,4]. An example of this synthetic method, which provides an important synthetic route to natural and unnatural amino acids, is shown in Equation E27.6.

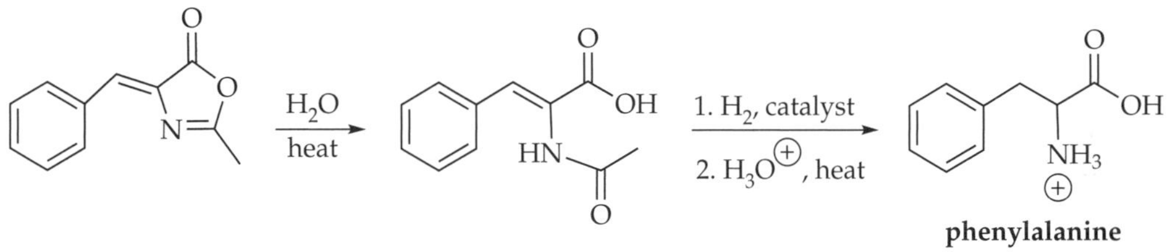

(E27.6)

phenylalanine

In this experiment, you will be performing a modified Perkin reaction using the starting materials shown in Equation E27.7.[5] Your goal is to isolate and purify the product from this reaction and determine its structure from the spectral data.

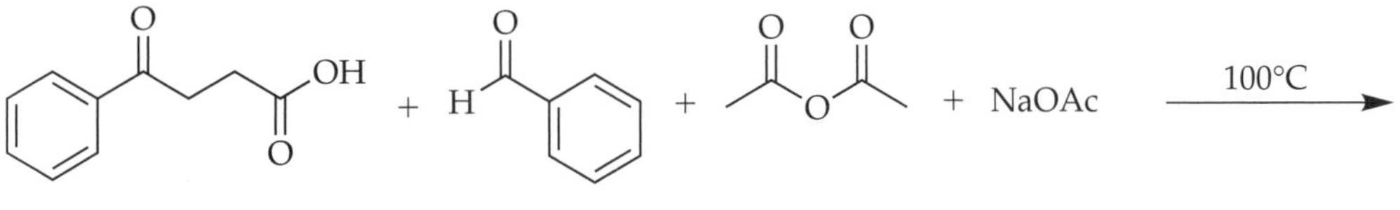

? (E27.7)

3-benzoylpropionic acid

---

[3] Erlenmeyer, E. *Ann.* **1893**, *275*, 1.

[4] Plochl, J. *Ber.* **1884**, *17*, 1616.

[5] Filler, R., Piasek, E. J., Leipold, H. A., *Org. Syn.* Coll. Vol. 5, 80.

Background Reading:

Before the laboratory period, read about enolate and anhydride chemistry in your lecture text.

Prelab Checklist:

Include the structures and all relevant physical constants for the starting materials and solvents used in this reaction in your prelab write-up.

Experimental Procedure

Don't forget to keep your safety glasses on during the entire laboratory period.

Attach a gas trap to the condenser as shown in Section 7.4.1.a and Figure 7.16. Weigh approximately 1 mmol of benzaldehyde directly into a 5–mL long-neck round-bottom flask. Add 1 mmol of 3-benzoylpropionic acid and 1 mmol of sodium acetate. Introduce a stir bar and cap the flask. Take the reaction flask to the fume hood and add 0.5 mL of acetic anhydride using a syringe. Re-cap the flask. Take the mixture back to the laboratory bench and attach it to the condenser and gas trap.

CHEMICAL SAFETY NOTE   Acetic anhydride has a strong odor resembling acetic acid and it is a lachrymator, which is a substance that irritates the eyes, nose, mouth, and lungs. It is also a skin irritant and contact with skin should be avoided. If some gets on your skin, flush the affected area with water and have another student notify your instructor.

While stirring, heat the reaction mixture in a sand bath, keeping the temperature of the bath between 95–105°C for 1 hour.

Technique Tip   To help maintain the sand bath temperature, it is convenient to turn the control off as the desired temperature is approached. When the temperature starts to decrease, turn the control back on until the temperature rises again. A low setting on the control should allow you to obtain the temperature necessary for the reaction.

After 1 hour, remove the flask from the sand bath and allow the mixture to cool to room temperature. Add 1 mL of 80% ethanol to the mixture and stir it thoroughly for 1 minute. Filter the contents of the flask using a Hirsch funnel and wash the solid once with 1 mL of ice-cold 80% ethanol and three times with 1 mL portions of hot water. Apply a vacuum to the filter flask to help dry the crude solid. Clean the Hirsch funnel for the next step.

Recrystallize the solid from 1–2 mL of absolute ethanol. Collect the final product on the Hirsch funnel and wash it with 0.5 mL of ice-cold absolute ethanol. Determine the weight and melting point of the product.

Cleanup

Place your product in a suitable container. Pour the original filtrate and the washes into the appropriate liquid waste containers. Contaminated Pasteur pipettes, melting point capillaries, and used weigh paper should be disposed of in the container for contaminated laboratory debris. Clogged syringe needles must be placed in the container for used syringe needles.

Discussion

- Make reasonable assignments in the ^{1}H and ^{13}C[^{1}H] NMR spectra of 3-benzoylpropionic acid and the product. In the EI mass spectrum of the product, assign the ions at m/z 248, 105, and 77 (Chapter 8). Your instructor will let you know how or where to obtain the spectra.

- Discuss how you used all of the spectral data to deduce the structure of the product.

- Write a rational arrow-pushing mechanism leading to the product of this reaction.

- Calculate the percent yield of the product.

- Based on your mechanism, discuss whether or not the acetic anhydride was consumed in this reaction. If it was not consumed, how did you separate it from your final product? If it was consumed, write the structures of the by-products and explain how they were removed from the final product.

Questions

1. Explain why only non-enolizable aldehydes and ketones can be used in the Perkin reaction.

2. In the Perkin reaction shown in Equation E27.2, a small amount of styrene is formed. Propose a mechanism for the formation of this minor product. Hint: Examine the mechanism outlined in Figure E27.3 and determine if there is some point at which the alkoxide could react differently than is shown in the figure.

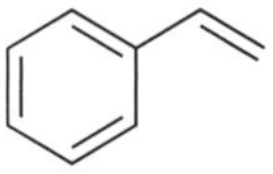

styrene

3. Design a synthesis of the amino acid below using the Erlenmeyer-Plochl synthesis (Equations E27.5 and E27.6). Show the structures of all isolable intermediates.

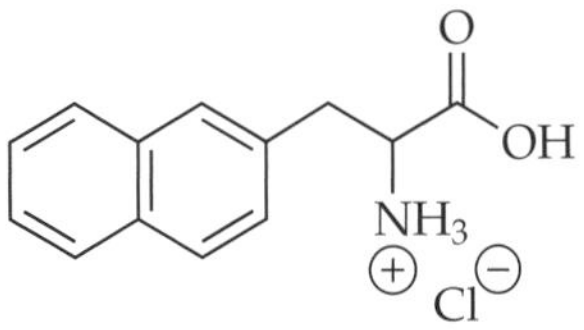

4. The reaction shown in Equation PE 27.4 is known as the Stobbe condensation. Propose a mechanism for the reaction, accounting for the fact that the product has one ester group and one carboxylate function.

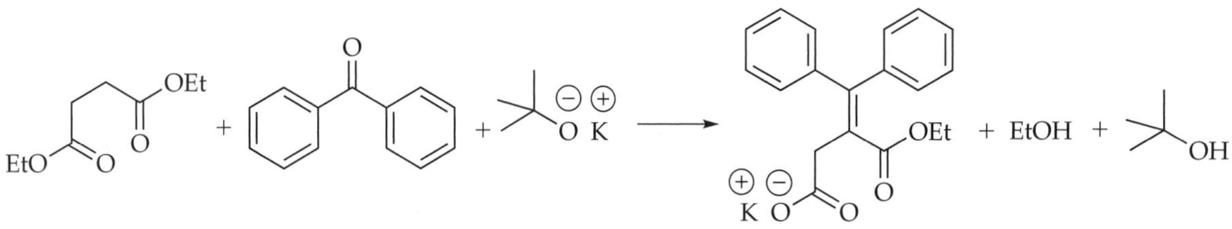

(PE27.4)

5. The reaction of the product you obtained in this experiment with methanol and triethylamine is shown in Equation PE27.5.[6] Write a mechanism for this reaction.

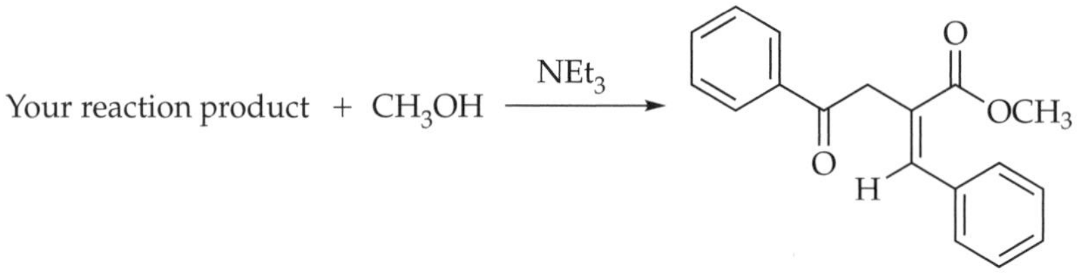

Your reaction product + CH₃OH $\xrightarrow{\text{NEt}_3}$ (PE27.5)

[6] Reddy, G. S.; Neelakantan, P.; Iyengar, D. S. *Indian J Chem Sect B: Org Chem* **2000**, *39B*, (12), 894.

The Reaction of 2-Acetylphenyl Benzoate with Potassium Hydroxide. Part I

The hydroxide ion plays an important role in many types of organic reactions, serving as a base or a nucleophile. In the molecule shown in Equation E28.1, there are three sites where the hydroxide ion can react: the carbonyl carbon of the ester, the carbonyl carbon of the ketone group, and the α-hydrogens of the ketone. In this experiment, you will be reacting 2-acetylphenyl benzoate with an excess of potassium hydroxide in the polar, aprotic solvent dioxolane. These conditions are significantly different from reactions where the hydroxide ion is dissolved in water. The goal of this experiment is to investigate the consequences of the reaction conditions by isolating the product and determining its structure using ^{1}H and ^{13}C[^{1}H] NMR spectroscopy and mass spectrometry.

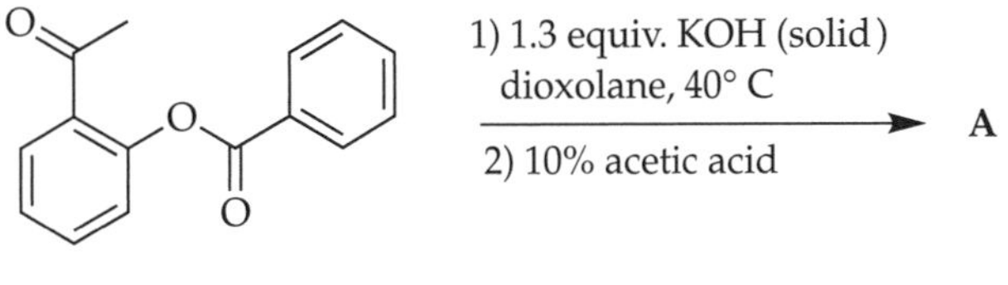

1) 1.3 equiv. KOH (solid)
   dioxolane, 40° C
   ———————————————→ A
2) 10% acetic acid

2-acetylphenyl benzoate

(E28.1)

Background Reading:

Before the laboratory period, read about enolate, ester, and β-dicarbonyl chemistry in your lecture text.

Prelab Checklist:

The prelab write-up should include the molecular weight for the starting ester and potassium hydroxide as well as the physical constants for dioxolane.

Write the structures of the products that might result from this reaction. Develop at least two reaction pathways that lead to different products under the conditions shown in Equation E28.1. The pathways should include mechanisms that indicate detailed arrow pushing. This experiment makes use of an excess of solid potassium hydroxide so that the reaction goes to completion. For the purpose of writing reaction mechanisms, use one equivalent of hydroxide.

Experimental Procedure

Don't forget to keep your safety glasses on during the entire laboratory period.

Weigh approximately 0.13 g of potassium hydroxide pellets or flakes on a piece of weighing paper. Fold the weighing paper over the pellet and crush it as demonstrated by your instructor. Transfer this material to a 5–mL long-neck round-bottom flask and then grind the potassium hydroxide *carefully* and thoroughly with a glass stirring rod. Add a magnetic stir bar and 1.0 mL of dioxolane to the reaction flask. The potassium hydroxide has limited solubility in dioxolane and some of it will not dissolve. Put the reaction flask in a warm water bath (35°– 40°C, that is, hot tap water) on top of the stir plate.

Based on the amount of potassium hydroxide used, calculate the weight of 2-acetylphenyl benzoate required to satisfy the stoichiometry shown in Equation E28.1. Weigh the ester in a reaction tube/small test tube and then dissolve it in 1.0 mL of dioxolane. Add the dioxolane solution of the benzoate ester to the reaction mixture in the hot water bath.

Maintain vigorous stirring for 15 minutes. Keep the water bath temperature in the 35°C to 40°C range by adding more warm water if necessary.

After 15 minutes, remove the long-neck round-bottom flask from the water bath, attach an inverted distillation head to it, and cap the top.

Stir the solution on a stir plate. Attach the arm of the distillation head to a vacuum source that is equipped with a cold-temperature trap and slowly apply a vacuum. The system must be stirred to prevent the solution from bumping. Adjust the vacuum if bumping becomes a problem. Continue applying the vacuum until the flask is no longer cool to the touch, indicating that most of the dioxolane is removed.

After evaporation is complete, remove the distillation head and add 10% acetic acid to the reaction mixture until the pH is approximately 4–5. Stir the mixture vigorously during this process.

Filter the contents of the round-bottom flask using a Hirsch funnel and wash the product with 1–2 mL of distilled water. Empty the filtrate into the appropriate liquid waste container. Place the Hirsch funnel back on the filter flask and apply a vacuum to the system to aid in drying the solid. Weigh the crude product.

Recrystallize the crude product in a 25–mL Erlenmeyer flask using 95% ethanol. Begin with 2 mL of ethanol and bring the mixture to a gentle reflux. Continue adding hot ethanol in small portions until all of the solid has dissolved. Remove the flask from the heat, let it cool to room temperature, and then place it in an ice-water bath.

Isolate the crystals by vacuum filtration and wash them with cold 95% ethanol. Weigh the dried product and measure a melting point. Dissolve a few crystals in a minimum amount of ethyl acetate and spot some of this solution on a TLC plate. Spot a solution of 2-acetylphenyl benzoate on the same plate and develop the plate in 80:20 hexane:ethyl acetate. Visualize the plate with UV light and iodine.

Save on Solvents: Share your TLC solvent chamber with a neighbor.

Cleanup

Put the product in an appropriate container and label it with your name, the name of the experiment, and the date. It will be used in Experiment 29. Pour the mother liquor, the TLC chamber solvent, and the 10% acetic acid washes in the appropriate liquid waste containers. Dispose of the contaminated Pasteur pipettes, TLC plates and capillaries, used weighing paper, and melting point capillaries in the container for contaminated laboratory debris.

Discussion

- Determine the structure of the product from the ^1H and $^{13}C[^1H]$ NMR and the EI mass spectra. Compare the spectra of the product with those of the starting material and explain how you used all of this information to deduce the structure of the product. Assign the signals in the NMR spectra. In the EI mass spectra, identify the molecular ions (Chapter 8). Your instructor will let you know how or where to obtain the spectra.

- Calculate the percent yield of the product.

- Discuss the role of the acetic acid in this reaction. In particular, what was the significance of adjusting the pH of the solution to between 4 and 5?

- Comment on whether or not the relative R_f values of the starting material and product are what you would expect.

- Knowing the structure of the product, discuss what might have happened if aqueous potassium hydroxide had been used.

Questions

1. Draw the possible products from the following reaction.

2. In the introduction to this experiment, the text stated that the hydroxide ion could react with the carbonyl carbon of the ketone. Draw the structure of the product of this reaction.

3. Write a detailed arrow-pushing mechanism for an aldol-type condensation reaction (Experiment 23) between two equivalents of acetone using 0.1 equivalents of KOH in dioxolane. Does the reaction go to completion under these conditions?

4. Using equilibrium arrows, indicate the direction of the equilibrium in the following reactions (Chapter 4.1.1.b).

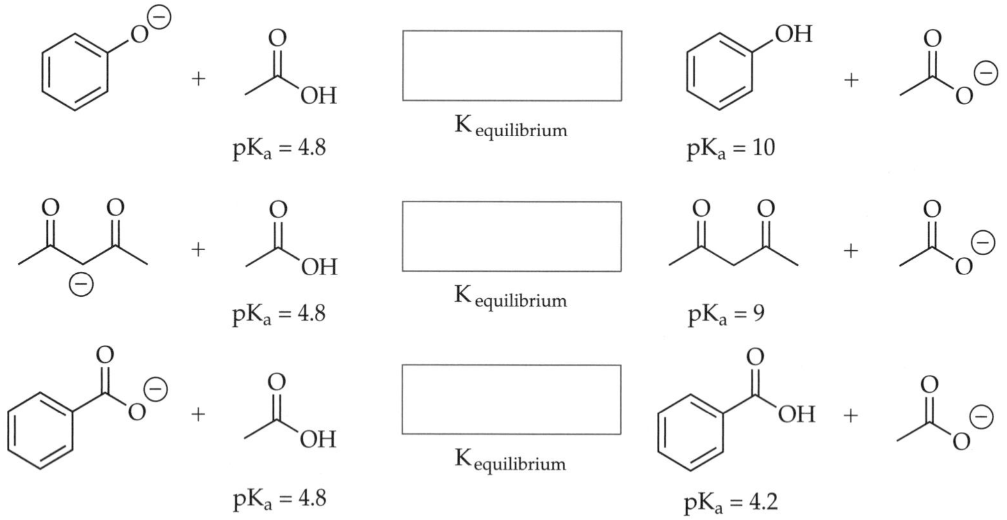

pK$_a$ = 4.8   K$_{equilibrium}$ pK$_a$ = 10

pK$_a$ = 4.8   K$_{equilibrium}$ pK$_a$ = 9

pK$_a$ = 4.8   K$_{equilibrium}$ pK$_a$ = 4.2

5. Propose a synthesis of 2-acetylphenyl benzoate starting with 2-hydroxyacetophenone.

The Reaction of 2-Acetylphenyl Benzoate with Potassium Hydroxide. Part II

In this experiment, the product obtained from Experiment 28 is refluxed with an acid catalyst (Equation E29.1).

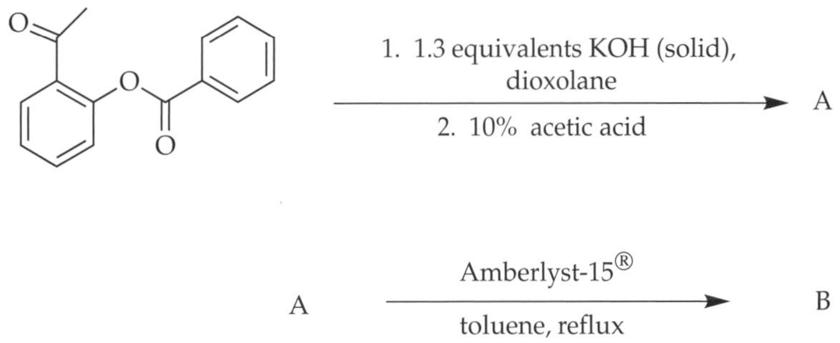

$$
\begin{array}{c}
\text{[2-acetylphenyl benzoate structure]} \xrightarrow[\text{2. 10\% acetic acid}]{\substack{\text{1. 1.3 equivalents KOH (solid),} \\ \text{dioxolane}}} \text{A}
\end{array}
$$

(E29.1)

$$
\text{A} \xrightarrow[\text{toluene, reflux}]{\text{Amberlyst-15}^{\circledR}} \text{B}
$$

Unlike a traditional acid catalyst such as sulfuric or hydrochloric acid, the catalyst is attached to an insoluble resin, Amberlyst-15® (A-15). The resin consists of a polystyrene framework in which some of the phenyl groups have been sulfonated to give arylsulfonic acid units (Figure E29.1). The spherical beads of the resin have pores of sufficient diameter and length to allow diffusion of most small organic molecules into the structure. Acid-catalyzed reactions can occur within these pores.

The A-15 is completely insoluble in the reaction medium so that it is easily recovered for reuse by filtering the reaction mixture. The insolubility of the catalyst also eliminates the need for the aqueous neutralization step required when a soluble acid catalyst such as sulfuric or hydrochloric acid is used. Because the catalyst is recycled and an additional workup step is eliminated, the reaction is environmentally friendly.

359

Figure E29.1   The Molecular Structure of the Amberlyst®-15 Resin

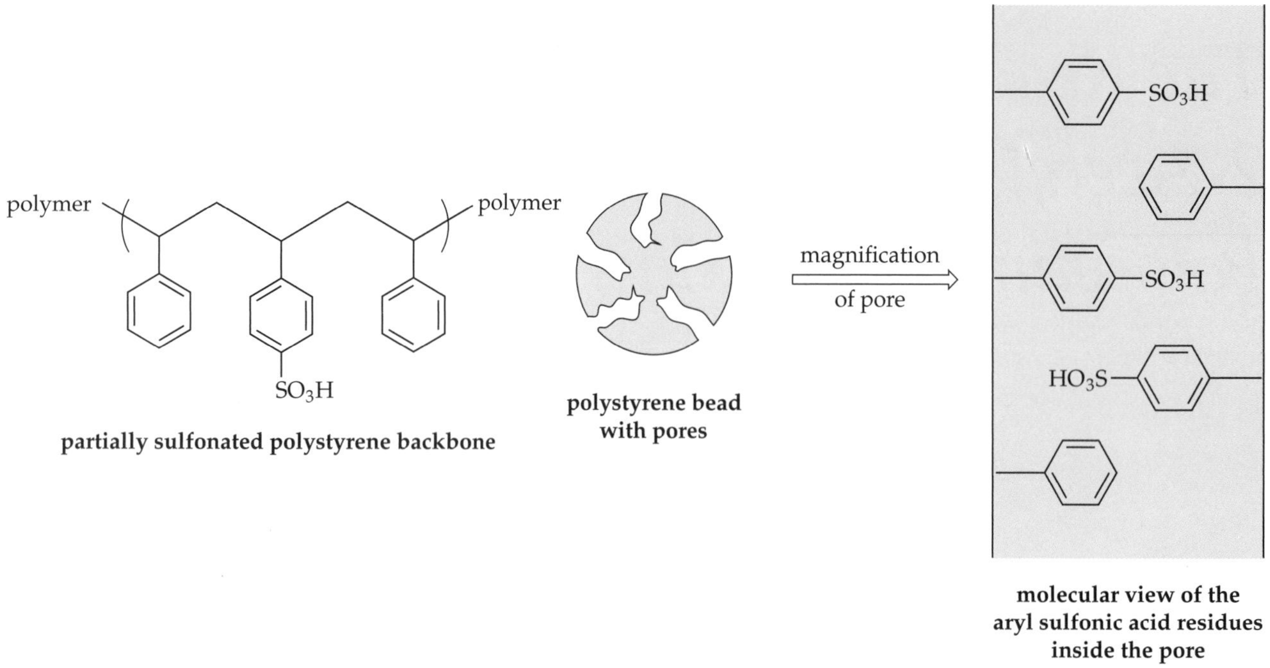

polymer — partially sulfonated polystyrene backbone

polystyrene bead with pores

magnification of pore

molecular view of the aryl sulfonic acid residues inside the pore

Background Reading:

Before the laboratory period, read about acid-catalyzed carbonyl addition reactions in your lecture text.

Prelab Checklist:

The prelab write-up should include the structure of the starting material for the reaction along with its molecular weight. Include the relevant physical properties of the solvent.

Experimental Procedure

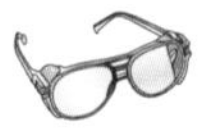

Please make sure you are wearing your safety glasses during the entire laboratory period.

Weigh approximately 100–120 mg of dry Amberlyst-15® beads in a 5–mL long-neck round-bottom flask.

Technique Tip   This is best accomplished using a small funnel to guide the Amberlyst-15® beads into the reaction flask.

Weigh 200 mg of the recrystallized product from Experiment 28 and add this to the reaction flask along with a stir bar. Add 1 mL of toluene, making sure that the majority of Amberlyst-15® is below the solvent level. Spot the reaction mixture in lane 1 of a TLC plate that can accommodate three lanes and label it $t = 0$.

EXPERIMENTAL NOTE   If you have more than 200 mg of product, give the extra material to your instructor for use by others in your laboratory section who may not have enough.

Attach an air condenser. Stir the mixture while heating to a gentle reflux. Monitor the reaction by TLC every 10 min for 1 hour or until TLC indicates that the reaction is complete.

Technique Tip   Use a second plate that can accommodate four lanes.

Develop the plates in an 80:20 hexane:ethyl acetate solvent system and visualize them using UV light and iodine.

Save on Solvents: Share your TLC solvent chamber with a neighbor.

When the reaction is complete, use a Pasteur pipette to separate the hot reaction mixture from the Amberlyst-15® beads. To remove any remaining crushed Amberlyst-15® beads, filter the solution into a large test tube using the glass-fiber Pasteur pipette filtration method (Section 7.2.2). Rinse the reaction flask with 0.5 mL of toluene and pass this solution through the pipette filter into the same large test tube.

Add hexane to the test tube in 0.25 mL portions, swirling the contents of the tube after each addition until the solution remains cloudy. Then, place it in an ice-water bath for 15–20 minutes. If crystallization does not take place after this period of time, warm the solution to room temperature, add a few more drops of hexane, and return it to the ice-water bath for a few minutes.

After crystallization has ceased, remove the mother liquor and rinse the crystals with 1 mL of cold hexane. Repeat the rinse once more if the filtrate remains yellow. Dry the crystals and determine the weight and melting point of the product.

Cleanup

Place the product in a suitable container. Pour the mother liquor and TLC chamber solvent into the appropriate liquid waste containers. Put the used A-15 catalyst in a container for recycling. Dispose of contaminated Pasteur pipettes, TLC plates and capillaries, used weighing paper, and melting-point capillaries in the container for contaminated laboratory debris.

Discussion

- Determine the structure of the product from the ^{1}H and ^{13}C[^{1}H] NMR and EI mass spectra. Assign the signals in the NMR spectra, but you do not need to make any assignments on the mass spectum other than for the molecular ion (Chapter 8). Your instructor will let you know how or where to obtain the spectra.

- Calculate the percent yield of the product.

- Discuss the detail(s) of the spectral data that were key to helping you deduce the structure of the product.

- Write a detailed arrow-pushing mechanism leading to the product.
- Comment on whether or not the relative R_f values of the starting material and product are what you would expect.

Questions

1. Amberlyst-15® (A-15) is available in both dry and wet forms. The wet form is prepared by suspending the A-15 beads in water and filtering them. Explain why the reaction in this experiment does not proceed if wet A-15 beads are used.

2. What method of crystallization was used in this experiment (Section 7.1)?

3. Explain what the sulfonic acid residues of the A-15 would look like if the beads were suspended in aqueous ammonium hydroxide and then washed with water.

4. Explain why the reaction from this experiment will not proceed if a basic catalyst, such as sodium hydroxide, is used.

5. There are two carbonyls in the starting material for this experiment. The product results from reaction at one of these. What product would be obtained if the reaction took place at the other carbonyl? Explain why this does not happen.

Synthesis of 1,2,3,4-Tetrahydro-β-Carboline[1]

The discovery and isolation of plant-derived organic compounds containing nitrogen (Figure E30.1) began in the early 1800s when a young German apothecary, Friedrich Wilhelm Adam Sertürner, isolated morphine from opium. Shortly thereafter, two French pharmacists, Pierre-Joseph Pelletier and Joseph-Bienaimé Caventou, extracted and isolated emetine from the root of ipecac and brucine and strychnine from ignatia beans. In 1820, Pelletier and Caventou isolated two alkaloids from the bark of the cinchona tree. This discovery afforded the world the potent anti-malarial agent quinine and provided a cure for this dreaded disease. It also marked the beginning of the first major deforestation of the South American rain forests.

Because these nitrogen compounds, called alkaloids, were the principle components responsible for the pharmacological effect rendered by some herbs and plants, they attracted the interest of synthetic organic chemists.

A key step in the synthesis of many alkaloids involves the condensation of an amine with an aldehyde or ketone to give an imine or immonium ion (Equation E30.1).

Figure E30.1 Some of the First Alkaloids Isolated

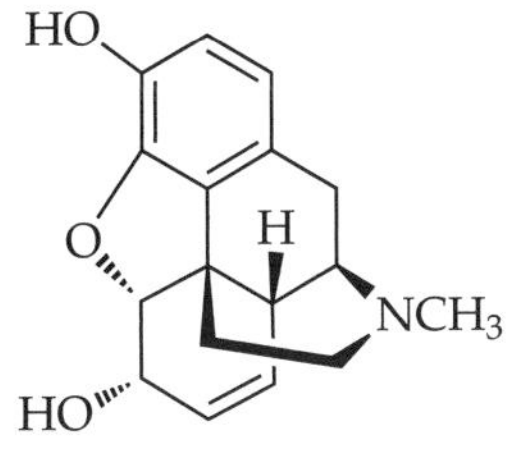

morphine

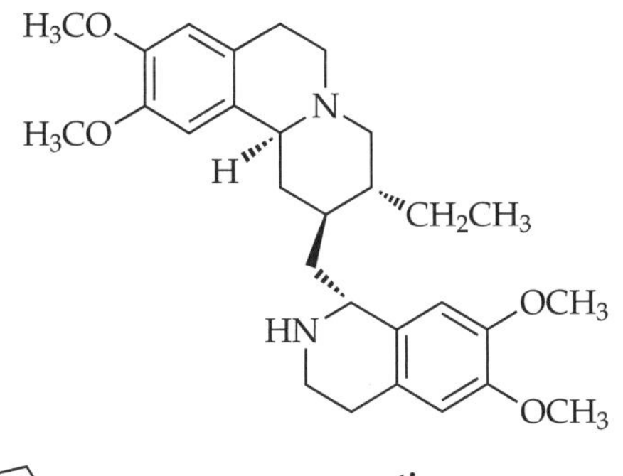

emetine

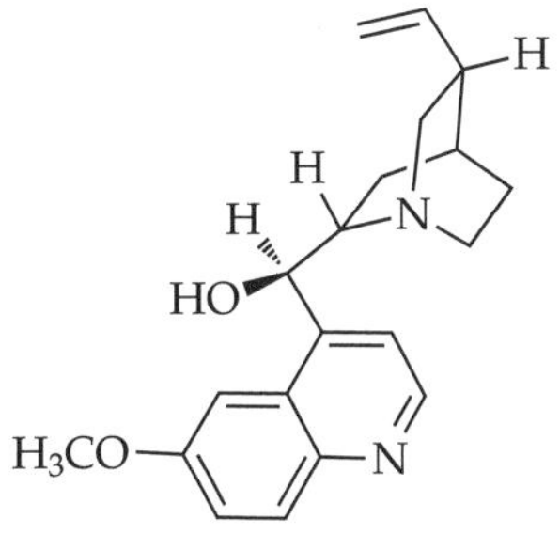

quinine

[1]Ho, B.T., Walker, K.E., *Org. Syn. Coll. Vol. 6* **1988** 965.

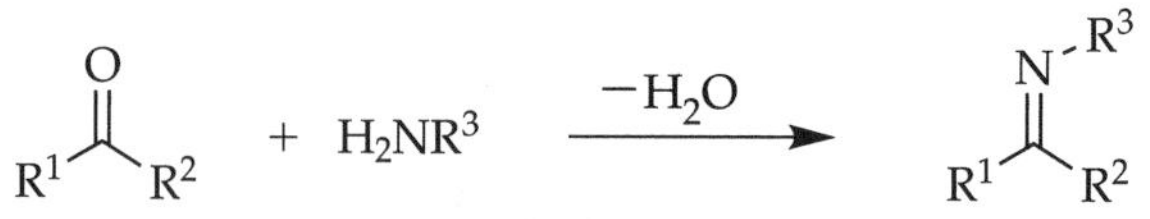

$$R^1 \overset{O}{\underset{}{\Vert}} R^2 \quad + \quad H_2NR^3 \quad \xrightarrow{\,-H_2O\,} \quad R^1 \overset{N\text{-}R^3}{\underset{}{\Vert}} R^2 \qquad \text{(E30.1)}$$

Like their carbonyl precursors, these compounds can act as electrophiles in carbon-carbon bond-forming reactions. Many types of nucleophiles can be used. In this experiment, the double bond of a heterocycle serves in this role. In addition to an interesting carbon-carbon bond-forming reaction, an unusual decarboxylation reaction also occurs.

Both of the starting materials for this reaction, tryptamine and glyoxylic acid, are naturally occurring compounds found in many species of plants. The overall reaction is shown in Equation E30.2.

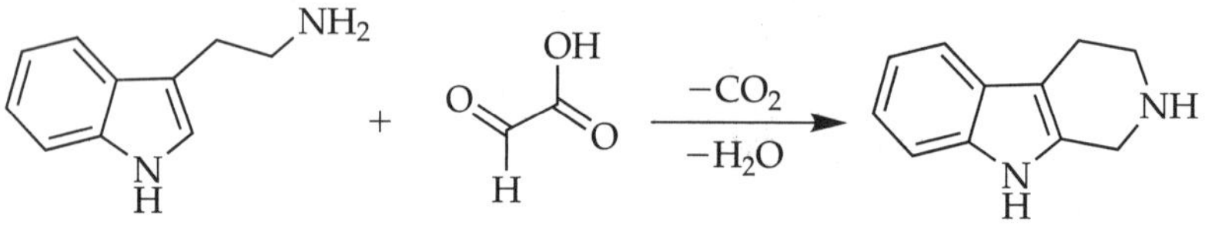

| tryptamine | glyoxylic acid | 1,2,3,4-tetrahydro-β-carboline |

(E30.2)

Background Reading:

Before the laboratory period, read about imine chemistry, electrophilic aromatic substitution, and heterocycle chemistry in your lecture text.

Prelab Checklist:

Include the structures and physical properties for all starting materials and solvents used in this reaction in your prelab write-up. Because some reference sources use a different name for the product, use the molecular formula to find the required information.

Experimental Procedure

Don't forget to keep your safety glasses on during the entire laboratory period.

Add 1.5 mmol of tryptamine to a 10-mL Erlenmeyer flask and crush any chunks into a powder using a glass stirring rod. Add a stir bar and 2 mL of distilled water to the flask and stir for several minutes. Next, add 1.65 mmol of glyoxylic acid monohydrate and continue stirring for 30 minutes. Isolate the resulting solid on a Hirsch funnel and wash it with three 1-mL portions of water. Weigh the solid and then transfer it to a short-neck 5-mL round-bottom flask.

TECHNIQUE TIP   The Hirsch funnel is used three times during the laboratory period and it should be cleaned after each use.

Prepare a gas trap as shown in Section 7.4.1.a and Figure 7.16 and attach it to a condenser. Add 2 mL of 3M HCl and a stir bar to the reaction flask.

Attach the gas trap and condenser to the reaction flask and reflux the reaction for 30 minutes. Allow the mixture to cool to room temperature, then isolate the solid on a Hirsch funnel and wash the solid with 1 mL of water.

Transfer the solid to a clean 10-mL Erlenmeyer flask along with 2 mL of distilled water and a boiling stick. Heat the mixture until all of the solid dissolves. While the solution is still warm, swirl the contents of the flask while adding 1M potassium hydroxide dropwise until the pH is 12.

Cool the mixture to room temperature. Isolate the solid on a Hirsch funnel and wash it with three 1-mL portions of water. Dry the solid and determine the weight and melting point. Calculate the percent yield of the product.

Cleanup

Put the product in a suitable container. Dispose of all solvents in the appropriate liquid waste containers. Contaminated Pasteur pipettes, weigh paper, and melting point capillaries should be disposed of in the container for contaminated laboratory debris.

Discussion

- Analyze the ^{1}H and ^{13}C[^{1}H] NMR spectra of tryptamine and 1,2,3,4-tetrahydro-β-carboline. In the EI mass spectrum of the product, assign the ions at m/z 172, 143, and 115 (Chapter 8). Your instructor will let you know how or where to obtain the spectra.

- The synthesis of the β-carboline shown in Equation 31.2 involves three discrete steps. Write a rational arrow-pushing mechanism for each step.

- Based on the conditions used in the second step, where 3M HCl was added and the solution was refluxed, and on the arrow-pushing mechanism, draw a reaction-coordinate (energy) diagram for this step.

Questions

1. In the first step of this reaction, there are two possible reaction pathways. One of these would occur rapidly leading to an ionic product. The other would occur more slowly leading to the imine. Draw the structure of the ionic species.

2. In the carbon-carbon bond-forming step of the reaction, there are two possible sites of attack on the benzopyrrole ring. One leads to the product and the other leads to a five-membered ring. Write the initial step leading to each of these products and examine the resulting intermediates. Using what you have learned about electrophilic aromatic substitution reactions, rationalize the observed position of attack on the imine that results in the product.

3. What is the role of the HCl in the decarboxylation step of this synthesis?

4. Draw the expected product from the following reaction.

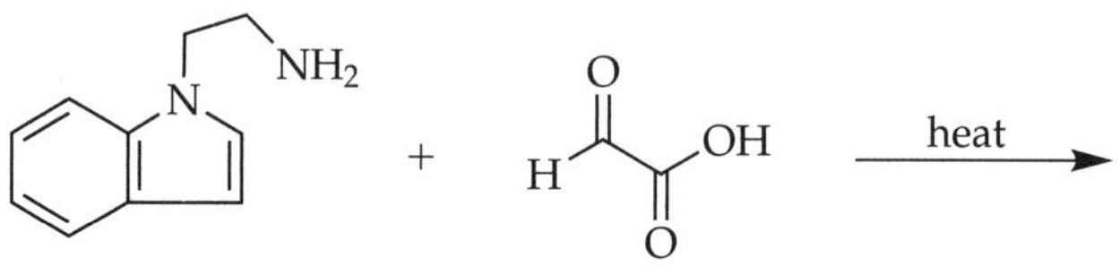

Base-Catalyzed Hydrolysis of Nicotinonitrile Using an Anion-Exchange Resin

The hydrolysis of the nitrile group to a carboxylic acid is a common reaction in organic chemistry. The utility of this method arises from the relative ease with which a cyano group can be introduced into a carbon framework. Not only is the cyanide ion an excellent nucleophile in S_N2 reactions, but it can also be added to aldehydes, ketones, and imines. The nitrile function can also be substituted on aromatic rings using diazonium ion chemistry.

Hydrolysis of the nitrile function is catalyzed by either acids or bases. In both reactions, the corresponding amide is a common intermediate. The amide is then further hydrolyzed to the carboxylic acid (Equation E31.1).

$$RCN \ + \ H_2O \ \longrightarrow \ \underset{R}{\overset{O}{\|}}{NH_2} \ \xrightarrow{H_2O} \ \underset{R}{\overset{O}{\|}}{OH} \qquad (E31.1)$$

In this experiment, you will determine if the product of the base-catalyzed hydrolysis of 3-cyanopyridine, commonly known as nicotinonitrile, is the amide, the carboxylic acid, or a mixture of the two. Hydroxide ion supported on an anion-exchange resin[1] is used as the base. These resins are typically composed of a polystyrene backbone cross-linked with divinylbenzene to give a three-dimensional polymer that is usually processed as spherical beads. The polymer is modified by linking tetraalkylammonium ions covalently to the para position of the phenyl residues. In the experiment, the chloride salt of the Amberlite® IRA-400 is transformed into the hydroxide form by washing it with aqueous sodium hydroxide. Consequently, it is called an anion-exchange resin (Figure E31.1).

By using polymer-supported hydroxide, the reaction mixture can be easily filtered away from the catalyst therefore avoiding the neutralization step required when a soluble base is used. This minimizes the amount of solvent used as well as the amount of waste generated.

---

[1] Galat, A. *J. Am. Chem. Soc.* **1948**, *70*, 3945.

Figure E31.1   Preparation of the Hydroxide Form of Amberlite® IRA-400

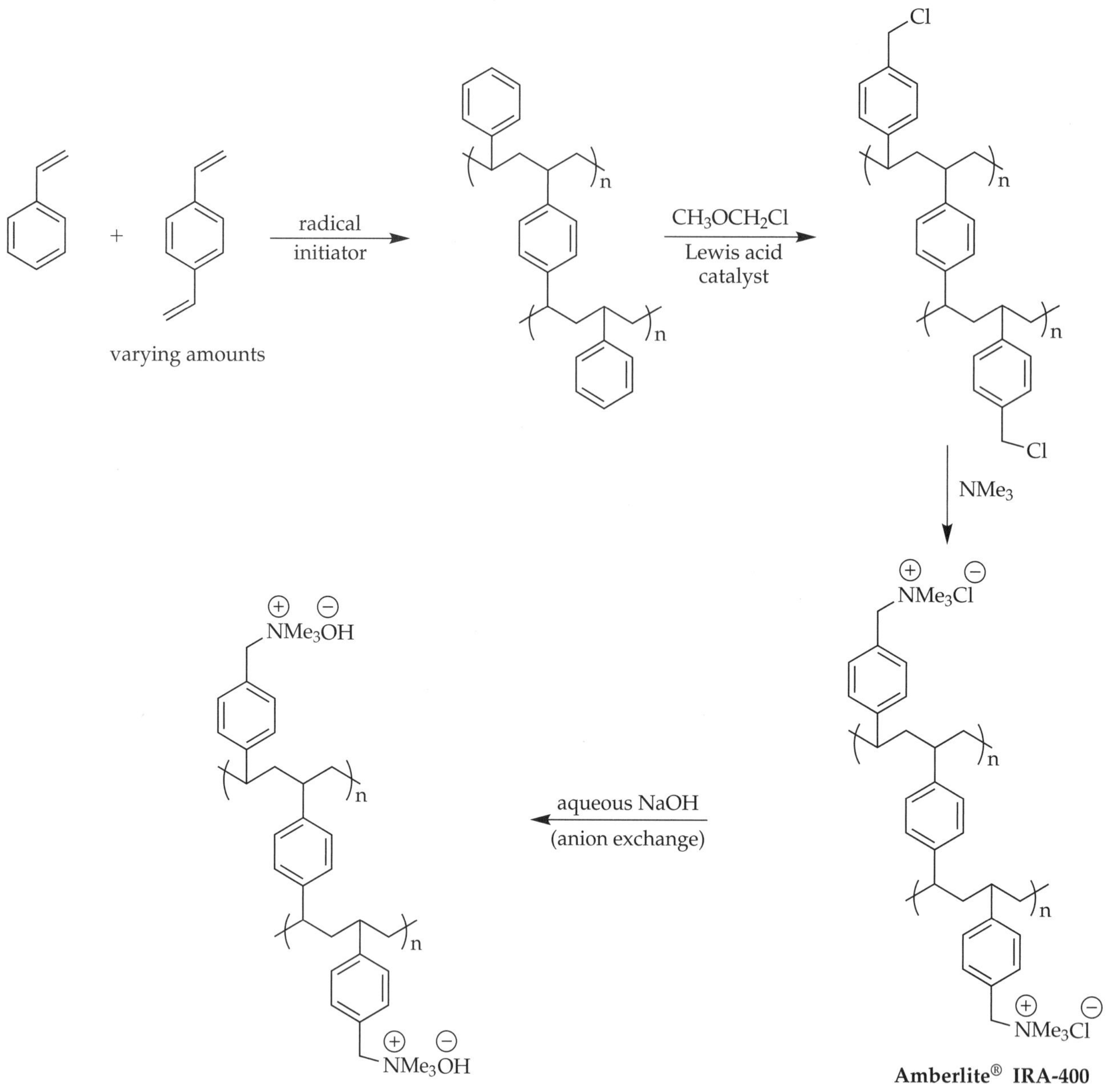

Did you know?

Nicotinic acid and nicotinamide are collectively referred to as Niacin or vitamin B_3. Even though niacin can be synthesized in the liver from the amino acid tryptophan, the synthesis is not very efficient and therefore niacin must be obtained from the diet. The vitamin is involved in many biological processes including the production of energy and the synthesis of fatty acids, cholesterol, and steroids. Nicotinamide adenine dinucleotide is formed from the metabolism of niacin. A deficiency of vitamin B_3 causes pellagra, a disease that leads to dermatitis, diarrhea, dementia, and also death.

Background Reading:

Before the laboratory period, read about the hydrolysis of nitriles in your lecture text.

Prelab Checklist:

In your prelab write-up, write the balanced equation for the hydrolysis of 3-cyanopyridine leading to both the amide and the carboxylic acid. Include the structures and physical constants for all starting materials, possible products, and solvents.

Experimental Procedure

Don't forget to keep your safety glasses on during the entire laboratory period.

Add 400 mg of Amberlite® IRA-400 ion-exchange beads to a large test tube followed by a stir bar and 2 mL of 5% NaOH.

EXPERIMENTAL NOTE   Use the solution of sodium hydroxide to rinse down any beads sticking to the sides of the test tube.

CHEMICAL SAFETY NOTE   If any solution containing sodium hydroxide comes in contact with your skin, flush the affected area immediately with water for at least 5 minutes. Sodium hydroxide is particularly harmful to the eyes. If any solution contacts your eyes, proceed immediately to the eyewash station and flush your eyes for at least 5 minutes. In either case, have another student notify the instructor.

Using a stir plate, stir the mixture vigorously for 10 minutes. Isolate the beads on a Hirsch funnel and wash them with five 2–mL portions of distilled water.

EXPERIMENTAL NOTE   These washes can be disposed of down the drain.

Scrape the beads off the funnel and place them in a clean, large test tube. Add 1 mL of distilled water followed by 210 mg of 3-cyanopyridine. Use another 0.5 mL of water to wash down the sides of the test tube.

TECHNIQUE TIP   Do not use more water than is specified because it is difficult to remove at the end of the reaction.

Add a boiling stick and reflux the mixture on the sand bath.

Take samples for TLC analysis every 5 minutes until the reaction is complete or until 45 minutes have passed, whichever comes first. Spot up to 4 lanes on each TLC plate. Spot one of the lanes with a solution of 3-cyanopyridine. Develop the plates in 80:20 ethyl acetate:hexane and visualize them using UV light and iodine.

TECHNIQUE TIP   To apply the aqueous reaction mixture to the TLC plate, use a Pasteur pipette to remove no more than *one* drop of the reaction mixture and put it in a reaction tube/small test tube containing 1 mL of ethyl acetate. Mix this solution and then spot it on the TLC plate.

Save On Solvents: Share your TLC solvent chamber with a neighbor.

Filter off the beads and wash them with two 1–mL portions of hot water.

Transfer the filtrate to a 50–mL beaker and add a boiling stick. Blow a stream of air over the solution while heating it on a sand bath at a temperature between 60–80°C to remove the water. Once a white precipitate forms, remove the beaker from the heat and continue blowing a stream of air over it until the precipitate is dry.

Scrape out the crude product and record its weight and melting point. Recrystallize the solid from acetone. Allow the solution to cool to room temperature and then place it in an ice-water bath. If no crystals form, *gently* swirl the solution and scratch the inside of the container with a glass stirring rod. Remove the acetone and wash the crystals with cold acetone. Dry the product and record the weight and melting point.

Cleanup

Put the product in a suitable container. All aqueous solutions from this reaction can be disposed of down the drain, providing they do not contain any 3-cyanopyridine. The aqueous solutions of 3-cyanopyridine and the organic solvents should be disposed of in the appropriate liquid waste containers. Dispose of contaminated Pasteur pipettes, used TLC plates, and melting point and TLC capillaries in the container for contaminated laboratory debris.

Discussion

- Analyze the ^1H and $^{13}C[^1H]$ NMR spectra of nicotinonitrile and the product of the reaction (Chapter 8). Your instructor will let you know how or where to obtain the spectra.

- Discuss your strategy for determining the structure of the product.

- Calculate the percent yield of the product.

- In reading the experiment to prepare the prelab write-up, did you find any clue as to which product you would obtain? Explain.

- Write a rational arrow-pushing mechanism for the formation of the product.

- What were the differences/similarities between the melting points of the crude and recrystallized products? What information could you deduce from these melting points? Was it necessary to take a melting point of the crude product?

Questions

1. Write an arrow-pushing mechanism for the base-catalyzed hydrolysis of 3-cyanopyridine to the carboxylate.

2. What extra steps would be necessary to isolate the product from this experiment if aqueous sodium hydroxide were used instead of the polymer-bound hydroxide?

3. Describe how you would prepare polymer-bound borohydride (BH_4^-) starting with the anion exchange resin IRA-400 in its chloride form.

4. What would be the advantage of using the polymer-bound borohydride compared with sodium borohydride in the reduction of a ketone? Assume the solvent is methanol for both reactions (Experiment 13).

Reactions of Salicylamide

salicylamide

Salicylamide, like salicylic acid, is an analgesic used for the treatment of mild pain and the common symptoms associated with a cold and the flu. It is most often prescribed in combination with other over-the-counter analgesics. For example, Frenadol is a brand name for the mixture of acetaminophen and salicylamide.

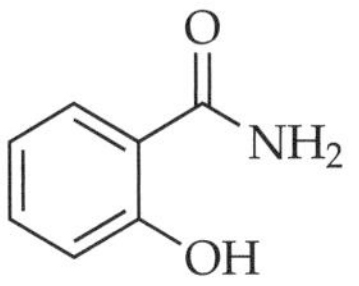

During the first laboratory period, you will determine the structure of the product of the reaction of salicylamide with the acylating-reagent mixture of acetic anhydride and pyridine. During the second laboratory period, you will work with two other students to explore the thermolysis of this product under three different reaction conditions.

Background Reading:

Before the laboratory period, read about ester and amide chemistry and acylation reactions in your lecture text.

Prelab Checklist:

Include the physical constants for all starting materials and solvents in your prelab write-up.

Laboratory Period 1: The Reaction of Salicylamide with Acetic Anhydride in Pyridine

Experimental Procedure

Don't forget to keep your safety glasses on during the entire laboratory period.

Put approximately 2 mmol of salicylamide in a reaction tube/small test tube and add 1 mL of diethyl ether. Add a magnetic stir bar and stir the solution. Use separate syringes to add 0.57 mL of acetic anhydride followed by 0.16 mL of pyridine.

CHEMICAL SAFETY NOTE Different syringes and syringe needles must be used to deliver the acetic anhydride and pyridine. In both cases, the syringes must be taken to the fume hood immediately after being used and washed before disposal. See the cleanup section for further details.

371

Continue stirring the solution for 10 minutes.

Filter the reaction mixture on a Hirsch funnel. Rinse the reaction tube with an additional 0.5 mL of diethyl ether and pass these rinses through the solid on the Hirsch funnel. Transfer the crude product from the Hirsch funnel to a watch glass and dry it to a constant weight. Record the weight and then recrystallize the solid from ethyl acetate.

Record the weight of the recrystallized product and measure the melting point. In addition, measure the melting point of a mixture of the product and salicylamide (Section 6.3.3).

Dissolve a small amount of the product in acetone and spot the solution on a TLC plate. On the same plate, spot a solution of salicylamide. Develop the plate in 70:30 ethyl acetate:hexane. Visualize the plate using a UV lamp before dipping the plate in a solution of ferric chloride ($FeCl_3$) in methanol.

Save On Solvents: Share your TLC solvent chamber with a neighbor.

Cleanup

Put the product in an appropriate container and label it with your name, the name of the experiment, and the date. The product is used as the starting material for the following thermolysis experiment.

The syringe used for delivering acetic anhydride should be rinsed with acetone and the syringe used for delivering the pyridine should be rinsed with 1M HCl followed by water. The rinsed syringes should be discarded in the container for contaminated laboratory debris. The needles must be discarded in the container for used syringe needles.

Pour the mother liquor, TLC chamber solvent, and syringe rinses into the appropriate liquid waste containers. Dispose of contaminated Pasteur pipettes and TLC and melting point capillaries in the container for contaminated laboratory debris.

Discussion

- Determine the structure of the product by analyzing the ^1H and $^{13}C[^1H]$ NMR and the EI mass spectrum of the product. Compare these with the spectra for the starting material and explain how you used all of this information to deduce the structure of the product. Assign the signals in the NMR spectra. You do not need to make any assignments on the EI mass spectra (Chapter 8). Your instructor will let you know how or where to obtain the spectra.

- Explain why the mixed melting point of the product with salicylamide was determined.

- There are two possible sites of acylation in salicylamide, not including Friedel-Crafts acylation of the aromatic ring (Experiment 18). Based only on the TLC results, can you conclude where acylation took place? Explain your answer.

- Is the product you obtained what you would have predicted? Explain.

- Calculate the percent yield of the product.

- Write an arrow-pushing mechanism leading to the product.

Laboratory Period 2: The Thermolysis of the Product from the Reaction of Salicylamide with Acetic Anhydride in Pyridine

Experimental Procedure

Don't forget to keep your safety glasses on during the entire laboratory period.

This experiment is done in groups of three students. Each student is responsible for performing one of the three procedures described below. Record all data from the other group members in your laboratory notebook.

Student 1: Heat a beaker of water on the sand bath to a final temperature of 95–100°C. Prepare a TLC plate to accommodate 4 lanes. In lane 1, spot a solution of the product from the reaction of salicylamide with acetic anhydride in pyridine. Label lanes 2–4 as 5, 10, and 15 minutes respectively.

Weigh 100 mg of the product from the acylation of salicylamide in a reaction tube/small test tube and add 0.8 mL of 1-propanol. Add a boiling stick to the mixture and place it in the hot-water bath after the desired temperature has been reached. Monitor the reaction every 5 minutes by TLC.

EXPERIMENTAL TIP If the TLC capillary clogs because of the crystallization of the starting material and/or product, dip it in a small volume of acetone. Because this solution is more dilute, it may require heavier spotting on the plate to achieve satisfactory results.

After 15 minutes, remove the reaction mixture from the water bath, spot the solution on the TLC plate in lane 4, and develop the plate in 70:30 ethyl acetate: hexane. Visualize it under a UV lamp before staining it with ferric chloride in methanol.

 **CHEMICAL SAFETY NOTE**   The ferric chloride dipping solution is very corrosive and all procedures involving this solution should be carried out in the fume hood. Do not touch the TLC plates dipped in this solution. Use tweezers to manipulate the plates and do not take the plates out of the fume hood.

Save On Solvents: Share your TLC solvent chamber with a neighbor.

- *If the reaction appears to be complete*, allow the mixture to cool slowly to room temperature. After crystallization begins, continue to cool the solution in an ice-water bath.

- *If it appears that the reaction is not complete after 15 minutes*, replace the boiling stick with a new one and continue heating the reaction mixture for an additional 15 minutes. Remove the reaction mixture from the water bath, spot the solution on a TLC plate, and develop the plate in 70:30 ethyl acetate:hexane. Visualize the plate with the UV lamp before staining it with the ferric chloride solution. Allow the mixture to cool slowly to room temperature. After crystallization begins, continue to cool the solution in an ice-water bath.

Isolate the crystals and wash them with ice-cold 1-propanol. Dry the crystals on a piece of filter paper. Determine the weight and the melting point of the product. Dissolve a small amount of the product in acetone and spot this solution on a TLC plate. Spot a solution of the product from the reaction of salicylamide with acetic anhydride in pyridine on the same plate. Develop the plate in 70:30 ethyl acetate:hexane and visualize the plate with the UV lamp before staining it with the ferric chloride solution.

Student 2: Heat the sand bath until the temperature remains constant at approximately 140°C.

Weigh 150 mg of the product from the reaction of salicylamide with acetic anhydride in pyridine in a clean, dry reaction tube/small test tube. Try to ensure that all of the solid is in the bottom of the reaction tube/small test tube. Place the tube in the 140°C sand bath for 30 minutes, then remove it. Allow the tube to cool to room temperature.

Spot an acetone solution of the product in lane 1 of a TLC plate that can accommodate three lanes. Spot a solution of the product from the acylation of salicylamide in lane 2.

Recrystallize the solid from ethyl acetate. Determine the weight and melting point of the product. Spot an acetone solution of the recrystallized product in lane 3 of the TLC plate. Develop the plate in 70:30 ethyl acetate:hexane and visualize it under a UV lamp before staining it with ferric chloride in methanol.

 **CHEMICAL SAFETY NOTE**   The ferric chloride dipping solution is very corrosive and all procedures involving this solution should be carried out in the fume hood. Do not touch the TLC plates dipped in this solution. Use tweezers to manipulate the plates and do not take the plates out of the fume hood.

Save On Solvents: Share your TLC solvent chamber with a neighbor.

Student 3: Heat a beaker of water on the sand bath to a final temperature of 95–100°C. Prepare a TLC plate to accommodate 4 lanes. In lane 1, spot a solution of the product from the reaction of salicylamide with acetic anhydride in pyridine. Label lanes 2–4 as 5, 10, and 15 minutes respectively.

Weigh 100 mg of the product from the acylation of salicylamide in a reaction tube/small test tube and add 0.5 mL of 1-propanol followed by 1 drop of 15% sulfuric acid. Add a boiling chip to the mixture and place it in the hot-water bath after the desired temperature has been reached.

EXPERIMENTAL NOTE   Do not use a boiling stick in this experiment. The acidic solution would leach colored compounds from the stick and contaminate the solution.

Monitor the reaction by spotting the TLC plate with some of the reaction mixture at 5 minute intervals. After 15 minutes, remove the reaction mixture from the water bath and allow it to cool to room temperature. Develop the TLC plate in 70:30 ethyl acetate:hexane. Visualize the plate under a UV lamp before staining it with ferric chloride in methanol.

CHEMICAL SAFETY NOTE   The ferric chloride dipping solution is very corrosive and all procedures involving this solution should be carried out in the fume hood. Do not touch the TLC plates dipped in this solution. Use tweezers to manipulate the plates and do not take the plates out of the fume hood.

Save On Solvents: Share your TLC solvent chamber with a neighbor.

Add 1.5 mL of distilled water to the reaction mixture and, after allowing it to sit an additional 5 minutes at room temperature, place it in an ice-water bath.

Once crystallization has ceased, isolate the crystals on a Hirsch funnel and wash them with a small amount of ice-cold 1-propanol followed by ice water. Remove the boiling chip and dry the crystals on a piece of filter paper. Determine the weight and melting point of the product.

Spot an acetone solution of the recrystallized product on a TLC plate along with a solution of the product from the reaction of salicylamide with acetic anhydride in pyridine. Develop the plate in 70:30 ethyl acetate:hexane and visualize it under a UV lamp before staining it with ferric chloride in methanol.

Cleanup

Put your product in an appropriate container. Put all used Pasteur pipettes, TLC and melting point capillaries, and boiling sticks and chips in the container for contaminated laboratory debris. Pour all solutions and washes into the appropriate liquid waste containers.

Discussion

- Use the ^{1}H and ^{13}C[^{1}H] NMR and the EI mass spectra to determine the structure of the products from these reactions. Assign the signals in the NMR spectra. You do not need to make any assignments for the EI mass spectra (Chapter 8). Your instructor will let you know how or where to obtain the spectra.

- Explain how you identified the product formed in each of the three thermolytic reactions performed in this experiment. For each experiment, was it necessary to have the NMR spectra of the product in order to identify the structure of the product? Explain your answers.

- Calculate the percent yield obtained in your reaction.

- Write rational arrow-pushing mechanisms leading to the products in each reaction. After you have written what you feel is the most likely mechanism for the reactions performed by Students 1 and 2, write an alternative mechanism for both of these experiments that is reasonable. Based on your hypotheses, describe further experiments that would potentially eliminate one or more of the mechanisms.

- Compare the results from each experiment and explain any differences in the outcome of each experiment based on the reaction conditions used.

Questions

1. Use an arrow to show the direction of the equilibrium for the following acid-base reaction (Section 4.1.1.b).

pK$_a$ = 4.7   K$_{equilibrium}$ pK$_a$ = 5.3

2. Explain why 2,4-pentandione is visualized on a TLC plate after staining with the ferric chloride stain reagent. Hint: Refer to information about β-dicarbonyl compounds in your lecture text.

3. Write an arrow-pushing mechanism for the following reaction. Discuss any similarities between this reaction and the one performed in this experiment.

4. Reaction of the product from the reaction of salicylamide with acetic anhydride in pyridine with potassium hydroxide in 1-propanol at room temperature gives a compound that is similar to, but not identical with the material isolated by Student 1. Determine the structure of this compound by writing an arrow-pushing mechanism.

5. Explain why the compound from the reaction discussed in Question 4 does not crystallize from 1-propanol.

The Hunsdiecker Reaction

The reaction of a metal carboxylate with a halogen, leading to an alkyl halide, carbon dioxide, and a metal halide, is generally referred to as the Hunsdiecker reaction. The husband and wife team of Heinz and Clare Hunsdiecker did not discover this reaction, but they popularized its use in organic synthesis[1]. The reaction was first reported in 1861 by Alexander Borodine, a Russian chemist, who prepared bromomethane from silver acetate and bromine[2] (Equation E33.1).

$$R\overset{O}{\underset{}{\overset{\|}{C}}}OAg \ + \ Br_2 \longrightarrow \ RBr \ + \ CO_2 \ + \ AgBr \qquad (E33.1)$$

The importance of the Hunsdiecker reaction results from its usefulness in synthesizing alkyl, alkenyl, and alkynyl halides by replacing a carboxy group with a halogen, typically bromine. The most significant drawback to using the reaction in modern organic synthesis is that it works best when silver salts of carboxylic acids are used. Silver is a precious metal and using it in stoichiometric amounts is considered wasteful. Fortunately, a less costly method for accomplishing the Hunsdiecker reaction with some classes of substrates was published by Chunxiang Kuang and coworkers[3]. This reaction uses lithium salts of carboxylic acids generated *in situ* and *N*-bromosuccinimide as the source of bromine.

Background Reading:

Before the laboratory period, read one of the articles by Kuang, et al.[3]

Prelab Checklist:

Include the structures and physical properties of the reactants and the solvents used in this experiment in your prelab write-up.

Experimental Procedure

Don't forget to keep your safety glasses on during the entire laboratory period.

Add 0.4 mmol of lithium acetate to a large test tube followed by 1.5 mL of a 92:8 mixture of acetonitrile:water. Add a stir bar to the solution and 1.9 mmol of *N*-bromosuccinimide (NBS). Using a magnetic stirrer, stir the solution until most of the NBS dissolves, then add 2 mmol of 4-methoxycinnamic

---

[1] Hunsdiecker, Clare and Hunsdiecker, Heinz *Ber.* **1942**, *76*, 291.
[2] Borodine, A. *Ann.* **1861**, *119*, 121.
[3] Kuang, Chunxiang; Yang, Qing; Senboku, Hisanori; Tokuda, Masao *Synlett* **2000**, 1439–1442 and *Synthesis* **2005** (8), 1319–1325.

acid. Quickly add another 0.5 mL of the 92:8 acetonitrile:water solution to the reaction mixture, using this solution to rinse down any solids that are on the sides of the test tube.

Within 1 minute of adding the 4-methoxycinnamic acid, stop stirring the reaction mixture and record the appearance of the reaction mixture in your laboratory notebook. Also, spot a sample of the reaction mixture in lane 1 of a TLC plate that can accommodate three lanes. Continue to stir the reaction for a total reaction time of 30 minutes. Spot the reaction mixture in lane 2 of the TLC plate and spot a solution of 4-methoxycinnamic acid in lane 3. Develop the plate in 60:40 ethyl acetate:hexane and visualize it using a UV lamp and iodine.

Save On Solvents: Share your TLC solvent chamber with a neighbor.

Prepare an ice-water mixture by adding approximately 4 g of ice and 1 mL of water to a 30–mL beaker. After the 30 minute reaction time, pour the reaction mixture into the ice water and stir the contents with a glass stirring rod. Continue stirring the mixture using the magnetic stir bar and the stir plate. Do not put the beaker directly on top of the stir plate because the top of the plate is warm and will melt the ice too quickly.

After stirring for a few minutes, filter the contents of the beaker on a Hirsch funnel using a vacuum. Add 1.5 mL of water to the beaker, stirring to suspend any remaining product, and pour this on the material in the funnel. Reapply the vacuum. Wash the solid on the filter with an additional 1.5 mL of water, stirring the solid with a spatula to make sure the water is in contact with all of the solid before reapplying the vacuum.

Pour the filtrate into a beaker and then reattach the filter flask to the vacuum source. Break up any large clumps of the solid on the Hirsch funnel with a spatula and continue drying it by pulling air through it using a vacuum. After a few minutes, transfer the solid from the Hirsch funnel to a piece of filter paper and blot it to remove more water. Transfer the solid to a tared watch glass, weigh it, and then place it in a 10–mL Erlenmeyer flask.

Add approximately 2.5 mL of *80%* ethanol and a boiling stick and heat the solution to a gentle boil. If an oil appears in the flask upon heating, add a few drops of warm *95%* ethanol, allowing the solution to resume boiling after each addition until all of the oil dissolves. Then, remove the solution from the heat and place the Erlenmeyer flask in a beaker packed with cotton to allow it to cool slowly. After crystallization has taken place at room temperature, cool the mixture in an ice-water bath.

Isolate the crystals by filtration and wash them with two or three 1–mL portions of ice-cold *80%* ethanol. Dry the crystals to a constant weight on a tared watch glass. Record the weight and measure the melting point.

If time permits, obtain a second crop of crystals by boiling the combined mother liquor and washes until it becomes slightly cloudy (Section 7.1.10). Let the solution cool to room temperature, cool the mixture in an ice-water bath, and then follow the procedure used for crop one. Determine the weight and melting point and a mixed melting point with the first crop. Do not combine this with the first crop of crystals unless the mixed melting point of the two crops indicates that they are the same material.

Cleanup

Put the product in a suitable container. If the mixed melting point indicates that the first and second crops are not the same, put them in separate containers. Put all used Pasteur pipettes, weigh paper, TLC and melting point capillaries, and TLC plates in the container for contaminated laboratory debris. Pour all solutions and washes into the appropriate liquid waste containers.

Discussion

- Make reasonable assignments for the ^1H and $^{13}C[^1H]$ NMR spectra. Calculate any useful coupling constants from the ^1H NMR spectrum of the product and explain the information these provide about the structure of the product. In the EI mass spectrum of the product, assign the ions at m/z 197, 199, 212, 213, 214, and 215 (Chapter 8). Your instructor will let you know how or where to obtain the spectra.

- Discuss how you used all of the spectral data to determine the structure of the product.

- Calculate the percent yield of the first crop or of the combination of both crops if the mixed melting point confirms that the two crops are the same compound.

- Write a rational arrow-pushing mechanism leading to the product obtained from this reaction.

- Explain any observation(s) that you made during the experiment that are consistent with the mechanism you wrote.

- Based on the actual amounts of reactants used in this experiment, calculate the theoretical yield of succinimide.

- Explain what happened to the succinimide. **Hint:** Look up the solubility of succinimide in the solvent used during the workup of this reaction.

Questions

1. What is the expected product(s) from the following reaction?

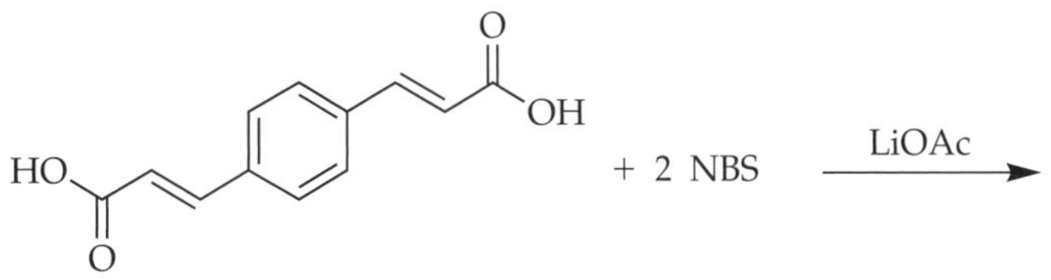

2. Write the arrow-pushing mechanism for one other known organic reaction that involves the loss of carbon dioxide (a decarboxylation).

3. Write a balanced equation for another type of organic reaction that uses *N*-bromosuccinimide as a reactant.

4. The instructions for the recrystallization of the product from this experiment state the following, "Add approximately 2.5 mL of 80% ethanol and a boiling stick and heat the solution to a gentle boil. If an oil appears in the flask upon heating, add a few drops of warm 95% ethanol, allowing

the solution to resume boiling after each addition until all of the oil dissolves." Explain what the oil is and why it formed (Section 7.1.2).

5. Propose a synthesis of the compound shown below starting with 4-methoxycinnamic acid. Use the Hunsdiecker reaction as one of the steps in your synthesis.

The Investigation of the Mechanism of a Reaction[1]

The arrow-pushing mechanisms found in textbooks and in the literature show the detailed steps by which reactions are presently thought to occur. They detail the movement of the electrons in the breaking and formation of bonds during a reaction and also the rate order, the relative rate, and the energy requirements of each step. These reaction pathways have been tested by carefully designed experiments and are consistent with the results of these experiments, but they are modified when new facts about the reaction are discovered. Researchers in the field of physical organic chemistry look for alternate possible mechanisms and design experiments to distinguish and test the possible alternatives that have been proposed. They attempt to explore the exact details of each of the steps and the precise position and function of all of the reactants and the solvent to clarify the mechanism.

There are several requirements that must be met by any mechanism that is proposed, including:

- It must account for the necessary reactants and reaction conditions.
- It must predict all resulting products and side-products.
- It must account for the rate law observed for the reaction.
- If the mechanism invokes an intermediate, that compound can sometimes be confirmed as being involved in the reaction pathway by isolating it from the reaction or by observing it using spectroscopic methods. If the intermediate is a stable compound, it can be subjected to identical reaction conditions to determine if the same products and by-products are obtained in order to test if it is a possible intermediate.
- It must account for any stereochemistry of the products.
- It must account for the effect on the rate of the reaction caused by the inductive and resonance effects of substituents on the reactants.
- If isotopic labeling is used, the results must be consistent with the products and/or the rate of the reaction predicted by the mechanism.

Mechanisms can never be proven, but are proposed to account for all known experimental facts. In fact, more than one reaction pathway is often consistent with all of the available data.

In many of the experiments you have done, you have been asked to write arrow-pushing mechanisms to account for the product(s) formed. Some of these

---

[1] Ritter, John J. and Minieri, P. Paul *J. Am. Chem. Soc.* **1948**, *70*, 4045.

mechanisms are well understood and based on all of the available experimental data. However, some may not have been verified by experiment and would need to be tested to determine if they are feasible pathways for the reaction.

In this experiment, you will investigate the possible presence of an intermediate in the reaction of a nitrile and a tertiary alcohol in aqueous sulfuric acid. When benzonitrile is reacted with *tert*-butyl alcohol in aqueous sulfuric acid, the product is *N*-*tert*-butylbenzamide (Equation 34.1).

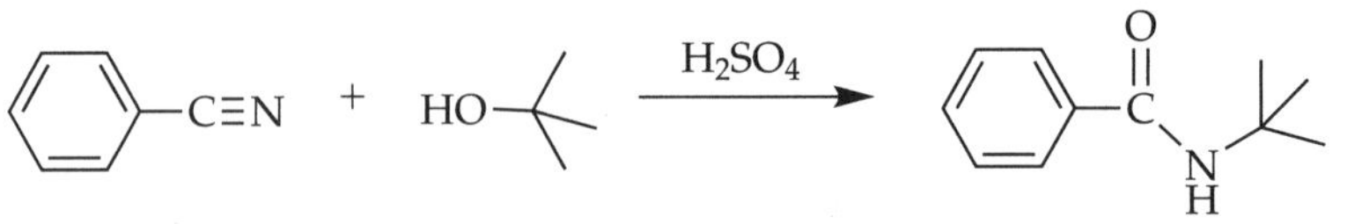

$$\text{benzonitrile} \quad + \quad HO-\!\!\!<\!\!\!\text{\quad} \xrightarrow{H_2SO_4} \quad \text{\quad} \qquad (E34.1)$$

benzonitrile   ***N-tert*-butylbenzamide**

In one possible mechanism, benzamide is formed as an intermediate by acid hydrolysis of benzonitrile. The benzamide then reacts with the *tert*-butyl carbocation formed from *tert*-butyl alcohol and sulfuric acid. In an alternative mechanism, the *tert*-butyl carbocation adds to the benzonitrile to form a carbon-nitrogen bond, resulting in a nitrilium ion. Water then adds to this intermediate to give the final product.

Background Reading:

Review the acid-catalyzed hydrolysis of nitriles in your lecture text.

Prelab Checklist:

Include the structures and physical properties of the reactants and the solvents used in this experiment in your prelab write-up. Write at least two reasonable arrow-pushing mechanisms for the reaction.

Experimental Procedure

Wear your safety glasses at all times in the lab.

In each of two reaction tubes/small test tubes, measure 0.7 mL glacial acetic acid and 0.2 mL 2-methyl-2-propanol (*tert*-butyl alcohol) and mix them well by stirring.

TECHNIQUE TIP   The glass stirring rod displaces too much volume so it is preferable to use the stainless steel spatula for stirring these solutions.

SAFETY NOTE   The stainless steel spatula has acid on it. Have a test tube of 10% aqueous $NaHCO_3$ available and place the spatula in this solution immediately after removing it from the reaction mixture. Care should also be taken with micropipettes used to sample the reaction solution for thin layer chromatography because these are also contaminated with the acid solution.

CHEMICAL SAFETY NOTE   Sulfuric and glacial acetic acid are corrosive materials and should be dispensed in the fume hood. If either acid contacts your skin, you should immediately flush the affected area with water for five minutes and have another student notify the instructor.

Add 0.1 mL benzonitrile to one of the tubes and 120 mg benzamide to the other. Stir the mixtures well using a clean spatula for each of the solutions.

CHEMICAL SAFETY NOTE   Benzonitrile has a disagreeable odor to some people. Rinse the pipette or syringe used to deliver this material with some acetone in the fume hood and place it in the appropriate waste container.

Prepare two TLC plates to accommodate 4 lanes on each plate. Spot a small amount of the benzonitrile and the benzamide solutions in lanes 1 and 2 respectively on both TLC plates.

Add 0.1 mL concentrated sulfuric acid (about 6 drops) dropwise with stirring to each of the tubes. Mix the contents by stirring with a clean spatula. After about 10 minutes, spot a small amount of one of the reaction mixtures in lane 3 and the other reaction mixture in lane 4 of one of the TLC plates.

After 30 minutes, spot the solutions in lanes 3 and 4 of the second TLC plate.

Pour each solution into 3.0 mL of ice-cold water. Isolate any solid product(s) that form(s) by filtering on a Hirsch funnel and wash the product(s) with ice-cold water. Dry the solid(s) on a piece of filter paper.

Spot acetone solutions of benzonitrile, benzamide, and the product(s) obtained on a third TLC plate. Develop the plates in a mixture of 5:1 hexane:ethyl acetate and visualize them using UV light and iodine.

Save On Solvents: Share your TLC solvent chamber with a neighbor.

Determine the weight and melting point of the product(s). Determine a mixed melting point of the product(s) with benzamide.

Cleanup

The aqueous filtrate from the isolation of the crude product is acidic. Pour the filtrate into a beaker of water and neutralize it by slowly adding saturated sodium bicarbonate while stirring. Pour the neutralized solution and any remaining organic solvents, including TLC chamber solvents, into the appropriate liquid waste containers. Dispose of the contaminated Pasteur pipettes, used weigh paper, and TLC plates in the container for contaminated laboratory debris. Clogged syringe needles must be placed in the container for used syringe needles. Put the product(s) in an appropriate container(s).

Discussion

- Assign the signals in the ¹H and ¹³C[¹H] NMR spectra of the product(s) and benzamide (Chapter 8). Your instructor will let you know how or where to obtain the spectra.
- Calculate the percent yield of the product(s).
- Discuss the mechanistic details of this reaction in depth, as inferred from the results of the two experiments you performed.

Questions

1. Draw all reasonable resonance structures for the nitrilium ion shown in the mechanism you have drawn.
2. Which of the resonance structures drawn in question 1 explains the position of the addition of water in the hydrolysis of the nitrilium ion?
3. Based on the resonance structures you drew in question 1, is it possible that water could add to the aromatic ring of the nitrilium ion? Explain why this product is not formed in these reactions.
4. Starting with 1-bromopropane, propose a synthesis of *N-tert*-butylbutanamide.

5. Based on the mechanistic details obtained from this experiment, write a possible mechanism for the reaction below.

Monoterpenes and the Ritter Reaction

In 1948, John Ritter reported a synthesis of amides from the reaction of alkenes and nitriles in the presence of H_2SO_4[1] (Equation E35.1). The reaction works with various alcohols under the same conditions[2] (Equation E35.2 and Experiment 34). The Ritter reaction is useful for the synthesis of amides and also for amines, which are obtained from the hydrolysis of the amides.

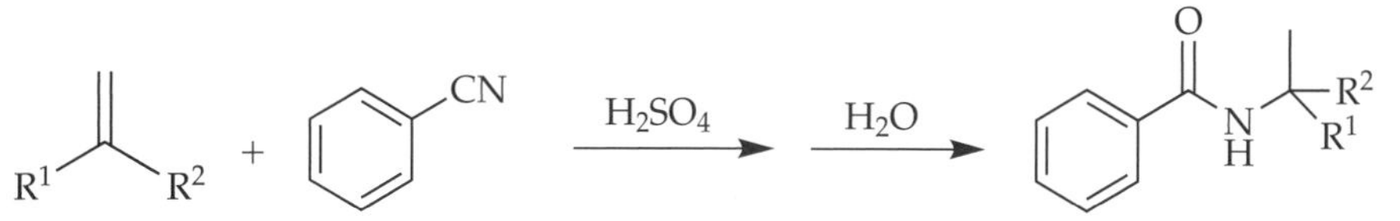

(E35.1)

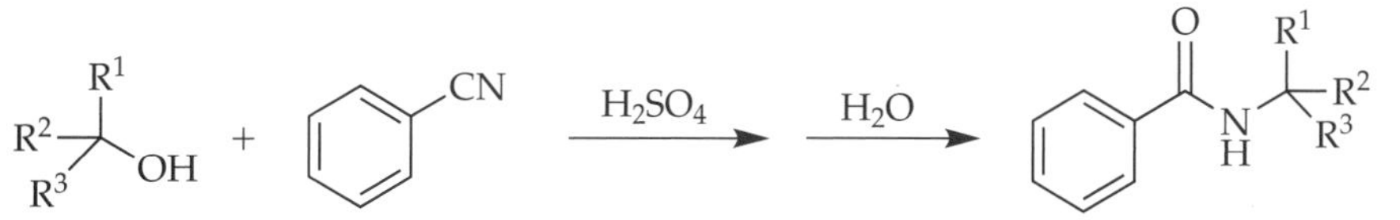

(E35.2)

In this experiment, you will work in pairs to investigate the reaction of benzonitrile with two different monoterpenes, camphene and isoborneol (Figure E35.1). At the end of the reaction, you will compare your product with that obtained by your partner.

Figure E35.1   The Structures of Camphene and Isoborneol

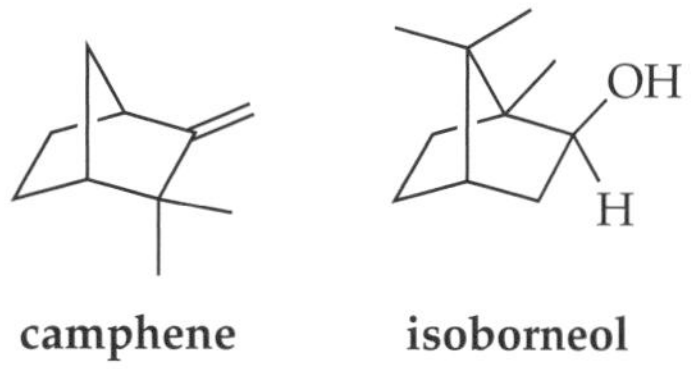

camphene   isoborneol

Background Reading:

Before the laboratory period, review Experiment 34. Read about carbocation rearrangements in your lecture text.

Prelab Checklist:

Provide all physical constants for the starting materials and solvents in your prelab write-up. Include the structures of the compounds you expect to get from the reaction of benzonitrile with camphene and with isoborneol in the presence of sulfuric acid followed by an aqueous work-up.

[1] Ritter, John J. and Minieri, P. Paul *J. Am. Chem. Soc.* **1948**, *70*, 4045.
[2] *Org. React.* **1969**, *17*, 213.

Experimental Procedure

Don't forget to keep your safety glasses on during the entire laboratory period.

This experiment is to be done in pairs with one student using camphene and the other student using isoborneol. Before weighing any materials, put 15 mL of water in a 30–mL beaker and start heating it on a sand bath.

Add 1 mmol of camphene or isoborneol to a reaction tube/small test tube followed by 0.7 mL of glacial acetic acid and 1.5 mmol of benzonitrile. Add 7 drops of concentrated sulfuric acid to this mixture. Mix the solution with a stainless steel spatula and add a boiling chip.

EXPERIMENTAL NOTE   Do not use a boiling stick in this experiment. The acidic solution would leach colored compounds from the stick and contaminate the solution.

SAFETY NOTE   The spatula has strong acid on it. Have a beaker of 10% aqueous $NaHCO_3$ available and place the spatula in this solution immediately after removing it from the reaction mixture. Care should also be taken with micropipettes used to sample the reaction solution for thin layer chromatography because these are contaminated with the sulfuric acid solution.

CHEMICAL SAFETY NOTE   Sulfuric and glacial acetic acid are corrosive materials. Both of these acids should be dispensed in the fume hood. If either acid contacts your skin, you should immediately flush the affected area with water for five minutes and have another student notify the instructor.

CHEMICAL SAFETY NOTE   Benzonitrile has a disagreeable odor to some people. Rinse the pipette or syringe used to deliver this material with some acetone in the fume hood and place it in the appropriate waste container.

Spot a solution of benzonitrile in lane 1 of a TLC plate that can accommodate four lanes. Spot some of the reaction mixture in lane 2 and label it $t = 0$.

Fit an air condenser with a septum and needle to provide pressure relief. Attach the air condenser to the reaction tube/small test tube. When the temperature of the hot water bath is 95–100°C, place the tube in the water bath.

After 10 minutes, remove the tube from the hot water bath and remove the air condenser. Spot the solution on the TLC plate in lane 3.

Replace the air condenser and return the tube to the hot water bath. After heating for an additional 20 minutes, remove the reaction mixture from the hot water bath and spot some of the solution on the TLC plate in lane 4.

Develop the plate in 5:1 hexane:ethyl acetate and visualize it using UV light and iodine.

Save On Solvents: Share your TLC solvent chamber with a neighbor.

Pour the solution into a 10–mL Erlenmeyer flask containing 3 mL of ice-cold water. Put the Erlenmeyer in an ice-water bath and stir/scratch the inside of the flask with a glass stirring rod until solid forms. Once a solid appears, leave the flask in the ice-water bath for an additional five minutes.

Isolate the solid by filtering it on the Hirsch funnel and wash it with cold water. Dry the solid by placing it on a piece of filter paper.

Dissolve the dry product in a minimum amount of hot hexane (2–8 mL). When all of the solid has dissolved, evaporate the solution to a volume of 1.5–2 mL. Cool the solution to room temperature and then place it in an ice-water bath. After crystallization has ceased, isolate the product by filtering it on the Hirsch funnel, wash it with a small amount of cold hexane, and dry it. Weigh the product and determine the melting point.

Compare your recrystallized product with that of your partner by TLC. Spot your product, your partner's product, and cospot both products on a TLC plate (Section 7.8.3.f). Develop the plate in 5:1 hexane:ethyl acetate and visualize it using UV light and iodine. Also, take a mixed melting point of the two products. Record your results and those of your partner.

Cleanup

The aqueous filtrate from the isolation of the crude product is acidic. Place the filter flask containing the solution in an ice-water bath in the fume hood and neutralize it by slowly adding saturated sodium bicarbonate while stirring. Pour the neutralized material and any remaining organic solvents, including TLC chamber solvents, into the appropriate liquid waste containers. Dispose of the contaminated Pasteur pipettes, used weigh paper, and TLC plates in the container for contaminated laboratory debris. Clogged syringe needles must be placed in the container for used syringe needles. Put the recrystallized product in an appropriate container.

Discussion

- Analyze the ^{1}H and ^{13}C[^{1}H] NMR spectra of camphene, isoborneol, and the product you obtained. In the EI mass spectrum of the product, assign the ions at m/z 257, 105, and 77 (Chapter 8). Your instructor will let you know how or where to obtain the spectra.

- Calculate the percent yield of the product you obtained.

- Write detailed arrow-pushing mechanisms for the formation of the product(s) you and your partner obtained. That is, one mechanism should start with camphene and one should begin with isoborneol.

- Based on what you know about the relative stabilities of carbocations, are your results surprising? Were there any differences between what you predicted and what you determined to be the final product? Explain.

Questions

1. 1-Aminoadamantane, known as amantadine, is used to reduce the symptoms associated with flu in patients that are at risk for a lengthy bout of influenza. This pharmaceutical is prepared from 1-bromoadamantane and acetonitrile (Experiment 10 and the Did you know?). Provide the conditions and other reagents that are necessary to complete this synthesis.

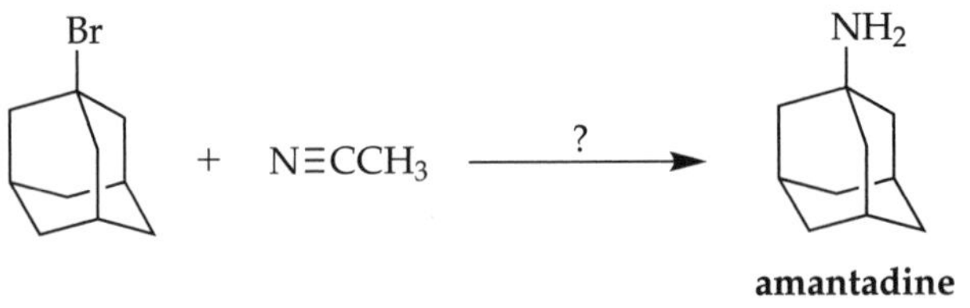

amantadine

2. Propose a synthesis of isobornylamine from camphor.

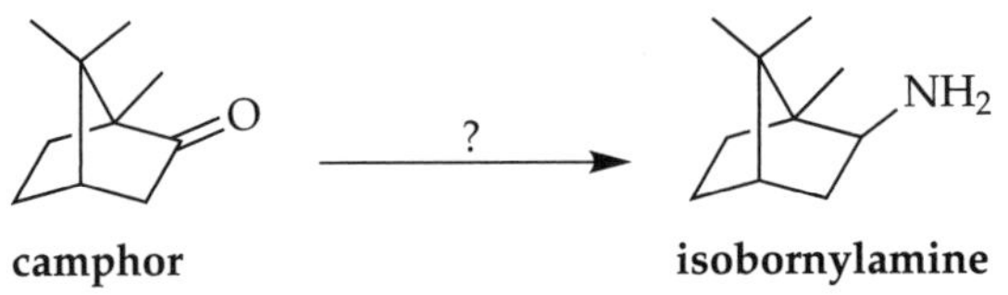

camphor   **isobornylamine**

3. Borneol is the endo isomer of isoborneol. Would borneol give the the same product as was obtained from isoborneol in this experiment? Explain.

4. Write a rational arrow-pushing mechanism for the reaction shown below[3].

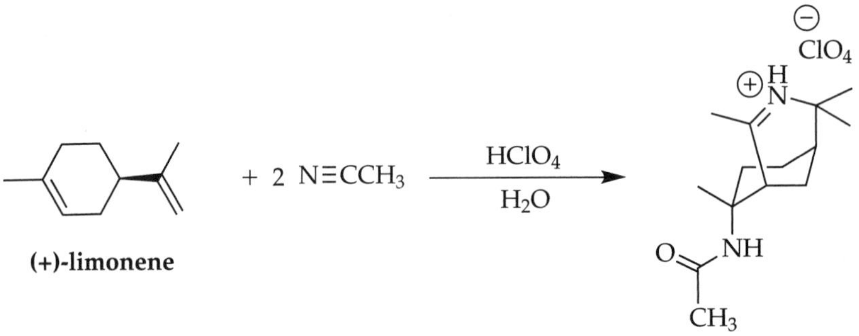

(+)-limonene

[3] Caram, J.; Martins, M. E.; Marschoff, C. M.; Cafferata, L. F. R.; Gros, E. G.; Z. Naturforsch. B **1984**, *37B*, (7), 972.

Chemoselectivity in Transfer-Hydrogenation Reactions

When chemists seek to develop new methods for synthesizing organic compounds, it is often important to find reagents that react with only one functional group in a polyfunctional molecule. This type of reagent is said to be **chemoselective.**

In the introduction for Experiment 13, the chemoselectivity of sodium borohydride was compared with that of lithium aluminum hydride. For example, sodium borohydride reduces aldehydes, ketones, and acid chlorides at 25°C. Unlike lithium aluminum hydride, it has no effect on carboxylic acids, epoxides, nitro groups, nitriles, amides, or alkyl halides (Equations E36.1 and E36.2).

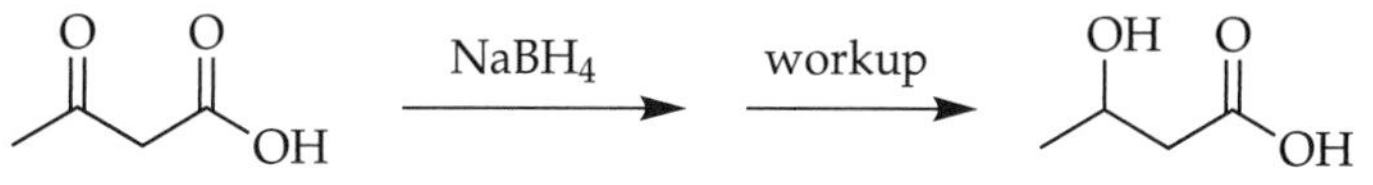

$$\text{(E36.1)}$$

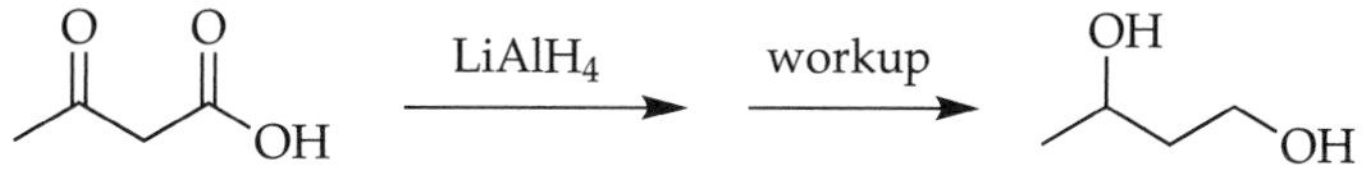

$$\text{(E36.2)}$$

Both of these reagents reduce polarized double bonds, but have no effect on carbon-carbon double bonds (Equations E36.3 and E36.4).

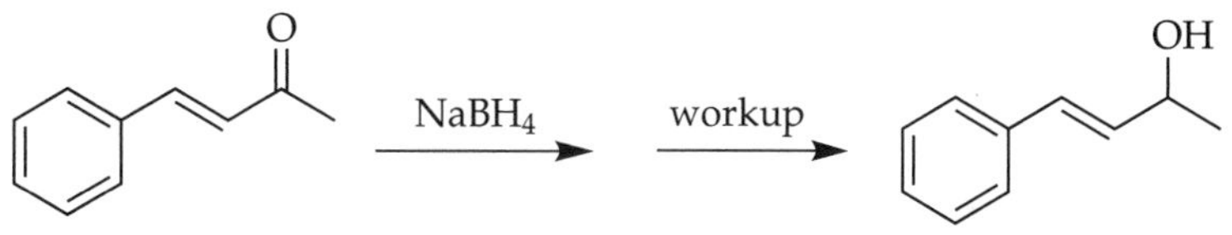

$$\text{(E36.3)}$$

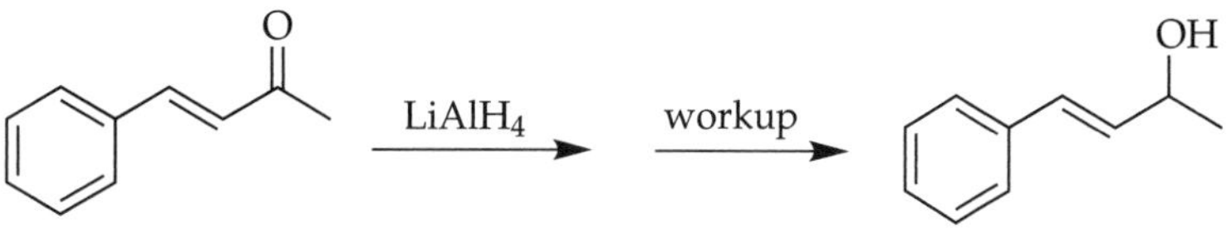

$$\text{(E36.4)}$$

Hydrogen can be added across a carbon-carbon double bond in the presence of a suitable catalyst to give the corresponding alkane (Equation E36.5).

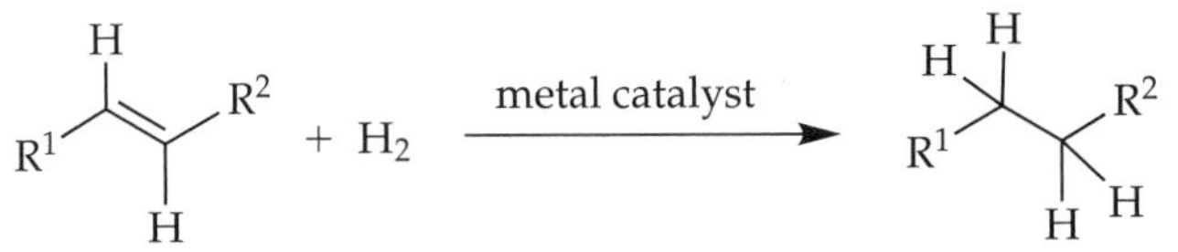

$$(E36.5)$$

As discussed in the introduction to Experiment 17, the use of hydrogen gas and a metal catalyst presents safety hazards because of the flammability of hydrogen gas. In addition, specialized equipment is sometimes required because of the high temperatures and pressures needed. Ammonium formate can be used as a hydrogen donor to hydrogenate a carbon-carbon double bond in the presence of a Pd/C catalyst (introduction to Experiment 17). Under these conditions, the carbon-carbon double bond of 3-phenylpropenoic acid is hydrogenated to leave the carboxylate group intact (Equation E36.6).

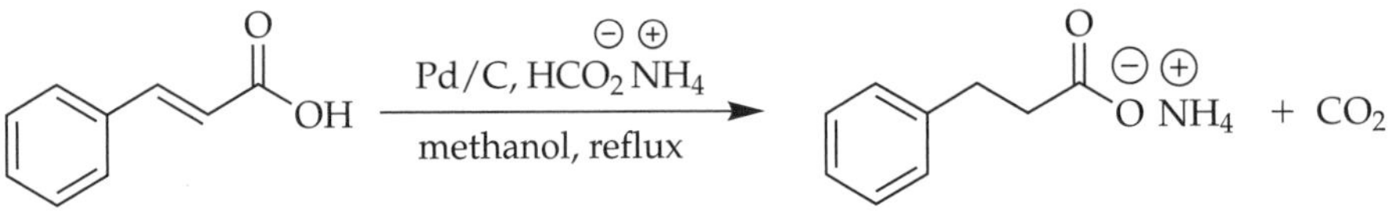

$$(E36.6)$$

In this experiment, the chemoselectivity of transfer-hydrogenation reactions using ammonium formate and palladium on carbon as a catalyst are examined using derivatives of 3-phenylpropenoic acid (cinnamic acid) and the structurally related ketone, *trans*-1,3-diphenyl-2-propen-1-one (chalcone) as substrates (Figure E36.1).

Did you know?

Finely divided palladium is able to adsorb up to 900 times its own volume of hydrogen. In the palladium on carbon catalyst, the palladium is distributed over finely divided carbon to increase the surface area and increase the activity of the catalyst. Because hydrogen diffuses through the catalyst when it is heated, palladium on carbon can be used to purify hydrogen.

Figure E36.1   Substrates for the Transfer-Hydrogenation Reactions

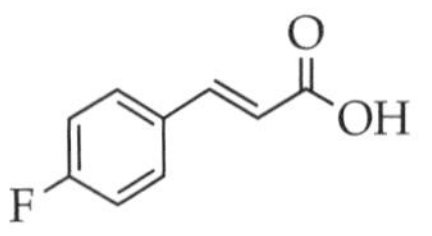

4-fluorocinnamic acid 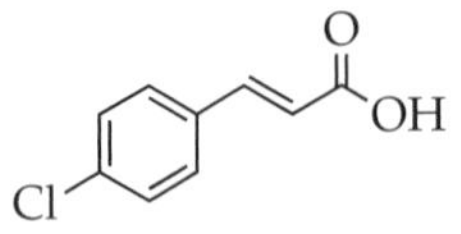   4-chlorocinnamic acid 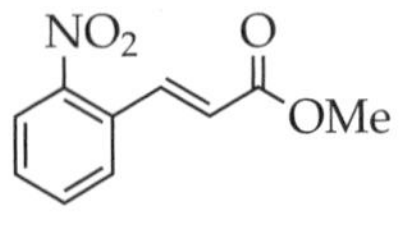   methyl 2-nitrocinnamate

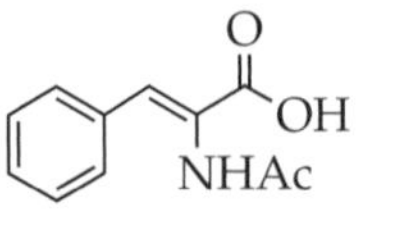

α-acetamidocinnamic acid 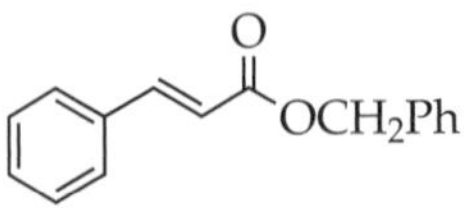   benzyl cinnamate 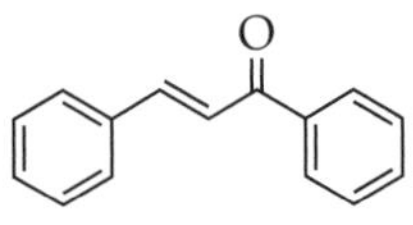   1,3-diphenyl-2-propen-1-one

Background Reading:

Before the laboratory period, review the introductions to Experiments 13 and 17. Read about the catalytic hydrogenation of carbon-carbon double bonds in your lecture text.

Prelab Checklist:

Your instructor will assign one of the substrates the week before the experiment is scheduled. Each student must run his or her own experiment, but more than one student may be using the same substrate. The prelab write-up should include all relevant physical data for the substrate you have been assigned and for all of the starting materials and solvents used in this experiment.

Laboratory Period 1: The Hydrogenation Reaction

Experimental Procedure

Wear your safety glasses at all times in the lab.

Spot a solution of your substrate in lane 1 of a TLC plate that can accommodate three lanes.

Set up a gas trap as described in Sections 7.4.1 and 7.4.1.a and shown in Figure 7.16. Insert the tubing into the septum and then attach the septum to the side arm of an inverted distillation head. Place a second septum on the top of the distillation head as shown in Figure 7.16.

> **SAFETY NOTE** Make sure that the piece of tubing is not clogged and that the damp cotton is packed loosely in the inverted test tube to prevent the buildup of pressure.

Weigh 150 mg of the assigned substrate and place it in a 5–mL long-neck round-bottom flask along with a stir bar. Add 30 mg of 10% palladium on carbon and 1 mL of methanol. Calculate the number of mmol of the substrate weighed and add 7.5 equivalents of ammonium formate followed by an additional 1 mL of methanol.

> **CHEMICAL SAFETY NOTE** All of the substrates and the reaction products are mild irritants and dermal contact should be minimized. If you spill a solution of any of these materials on your skin, wash the affected area with soap and water and inform your instructor.

> **CHEMICAL SAFETY NOTE** Do not weigh the Pd/C and the ammonium formate on the same weighing paper. Hydrogen gas is generated when the two come in contact and a fire can result. Hydrogen is also generated during these reactions, so make sure that there are no flames in the lab during this experiment. Palladium on carbon can be pyrophoric when exposed to air, especially when it is dry. Immediately after use, the weigh paper used for the Pd/C should be submerged in the container of water for Pd/C waste in the hood. Make sure you put the top back on the container of Pd/C.

Quickly spot the reaction mixture in lane 2 of the TLC plate and label it *t* = 0.

CHEMICAL SAFETY NOTE   Do not place your face over the reaction flask when you are sampling for TLC. The ammonia generated is an irritant.

TECHNIQUE TIP   In order to reach the reaction solution from the top of the inverted distillation head, attach a micropipette to a boiling stick using a piece of tape. You can reuse this pipette by dipping it in some methanol two or three times, blotting it on a paper towel each time.

EXPERIMENTAL NOTE   Some of the hydrogenation reactions proceed at room temperature. It is important to spot the reaction mixture as soon as possible after the addition of the methanol.

Immediately attach the inverted distillation head that is connected to the gas trap. Heat the reaction on the sand bath to maintain a gentle reflux while stirring the reaction mixture.

TECHNIQUE TIP   For efficient stirring, the stir plate must be centered underneath the sand bath. Slowly adjust the motor to achieve a rate of stirring that generates a slight vortex in the reaction flask.

SAFETY NOTE   It is important that the sand bath temperature remain below 80°C to prevent the ammonium formate from subliming and clogging the gas trap tubing which would result in a closed system. A closed system will cause pressure to buildup and the septum to blow off.

EXPERIMENTAL NOTE   The reaction will start bubbling before it begins to reflux because of the decomposition of the ammonium formate (Experiment 17, Equation E17.2).

Monitor the reaction every 10 minutes by TLC using lane 3 of the first TLC plate and additional plates as necessary.

All TLC plates of cinnamic acid derivatives should be developed in 98:2 ethyl acetate:acetic acid and all plates of the ketone should be developed in 80:20 hexane:ethyl acetate. The plates should be visualized using UV light and iodine.

Save On Solvents: Share your TLC solvent chamber with a neighbor.

When the TLC indicates that the reaction is complete or after 50 minutes, whichever comes first, remove the reaction mixture from the sand bath and allow it cool to room temperature. Filter the reaction mixture into a large test tube using the glass-fiber Pasteur pipette method (Section 7.2.2). Rinse the reaction flask with an additional 0.5 mL of methanol and pass this through the filter. Cool the filtrate in an ice-water bath for 10 minutes.

SAFETY NOTE   Palladium on carbon can be pyrophoric when exposed to air, especially when it is dry. Immediately after use, the Pasteur pipette containing the Pd/C should be submerged in the specially labeled container of water in the fume hood.

- *If crystals are apparent after ten minutes,* allow the system to stand undisturbed until no more crystals appear to be forming (approximately 10 minutes). Isolate the crystals by filtering on a Hirsch funnel and wash them with cold 2-propanol. Dry the crystals and obtain a weight and a melting point. Dissolve some of the crystals in an appropriate solvent and spot the solution on a TLC plate. Compare the result obtained with the first TLC plate that contained the starting material and the reaction mixture.

- *If no crystals are apparent after ten minutes,* transfer the filtrate to a filter flask and remove the methanol by evaporation using a vacuum source that is equipped with a cold trap and heating gently. Do not overheat the solution. The product may be an oil that should not be mistaken for residual methanol.

    Add 4.5 mL of 1M HCl to the flask and 3 mL of ether. Swirl the contents of the flask until everything has dissolved. Transfer this mixture to a centrifuge tube, cover it, and shake to mix the contents well. After the layers separate, transfer the organic layer to a large test tube.

    Extract the aqueous layer with an additional 1.5 mL of ether and add the ether layer to the organic layer in the large test tube. Dry the combined organic layers with sodium sulfate (Section 7.7). Spot some of this solution on a TLC plate.

    Transfer the dried ether solution to a tared 10–mL beaker and evaporate the ether in the fume hood using a stream of air.

If you obtain a solid product after evaporation of the ether, take a melting point and dissolve some of the crystals in an appropriate solvent. Spot the solution on a TLC plate. Compare the result obtained with the first TLC plate that contained the starting material and the reaction mixture.

If you obtain an oil, compare the TLC result obtained from the dry ether solution with the first TLC plate that contained the starting material and the reaction mixture.

Cleanup

Put the product in an appropriate container. Pour all organic solvents, including TLC solvents, and aqueous waste into the appropriate liquid waste containers. Put the pipettes used for filtering the palladium on carbon and the weigh paper used for the Pd/C in the specially labeled containers of water found in the fume hood. Dispose of all contaminated Pasteur pipettes, used TLC plates, and melting point and TLC capillaries in the container for contaminated laboratory debris.

Discussion for Laboratory Period 1:

- Analyze the ^{1}H and ^{13}C[^{1}H] NMR spectra of the substrate and the product. Assign the molecular ions in the EI mass spectra of the starting material and the product (Chapter 8). Your instructor will let you know how or where to obtain the spectra.

- Discuss how you used the spectral data to determine the structure of your product.
- Write a balanced equation for the reaction.
- Calculate the percent yield of the product.

Laboratory Period 2: Analysis of the Data from the Chemoselectivity in Transfer-Hydrogenation Reactions Experiment

Your instructor will assign Alternative A or Alternative B.

Alternative A

You will collaborate with the other students who hydrogenated the same substrate. Each group of students who hydrogenated the same substrate should provide an oral presentation to the laboratory section that summarizes the results from the transfer-hydrogenation reaction of the assigned substrate. The presentation should include a balanced equation for the reaction and a discussion of the spectral data that supports your conclusion. For supporting evidence, it is sometimes important to consider the spectra of the starting material compared with the product. Each member of the group should present some aspect of the information. Presentations should include responses to questions from the audience.

You should take notes on the results from the remaining five substrates during the presentations in order to include them in your laboratory report.

Alternative B

During the laboratory period, your instructor will lead a discussion on the results obtained from the six substrates. Be prepared to answer questions about your experimental data and defend your conclusions. During the discussion, take notes about the data obtained for the other five substrates in order to include all of the results in your laboratory report.

Discussion for Laboratory Period 2:

- The purpose of this experiment is to evaluate the chemoselectivity of transfer-hydrogenations using ammonium formate as the hydrogen donor. In the discussion, compare the functional groups present in all six substrates and their relative reactivities. Include balanced equations for the reactions for all substrates. In addition, discuss the results of the reactions shown in Equations E36.7 and E36.8.

$$\text{(E36.7)}$$

$$\text{(E36.8)}$$